Epigenetics and Human Health

Edited by
Alexander G. Haslberger

Co-edited by
Sabine Gressler

Related Titles

Knasmüller, S., DeMarini, D. M., Johnson, I., Gerhäuser, C. (Eds.)

Chemoprevention of Cancer and DNA Damage by Dietary Factors

2009
ISBN: 978-3-527-32058-5

Allgayer, H., Rehder, H., Fulda, S. (Eds.)

Hereditary Tumors

From Genes to Clinical Consequences

2009
ISBN: 978-3-527-32028-8

Kahl, G.

The Dictionary of Genomics, Transcriptomics and Proteomics

2009
ISBN: 978-3-527-32073-8

Ziegler, A., Koenig, I. R.

A Statistical Approach to Genetic Epidemiology

Concepts and Applications

Second Edition
2010
ISBN: 978-3-527-32389-0

Janitz, M. (Ed.)

Next Generation Genome Sequencing

Towards Personalized Medicine

2008
ISBN: 978-3-527-32090-5

Kaput, J., Rodriguez, R. L. (Eds.)

Nutritional Genomics

Discovering the Path to Personalized Nutrition

2005
ISBN: 978-0-471-68319-3

Epigenetics and Human Health

Linking Hereditary, Environmental and Nutritional Aspects

Edited by
Alexander G. Haslberger

Co-edited by
Sabine Gressler

WILEY-VCH Verlag GmbH & Co. KGaA

The Editor

Dr. Alexander G. Haslberger
University of Vienna
Department of Nutritional Sciences
Althanstrasse 14
1090 Vienna
Austria

The Co-editor

Mag. Sabine Gressler
Barichgasse 30/4/54
1030 Vienna
Austria

All books published by **Wiley-VCH** are carefully produced. Nevertheless, authors, editors, and publisher do not warrant the information contained in these books, including this book, to be free of errors. Readers are advised to keep in mind that statements, data, illustrations, procedural details or other items may inadvertently be inaccurate.

Library of Congress Card No.: applied for

British Library Cataloguing-in-Publication Data
A catalogue record for this book is available from the British Library.

Bibliographic information published by the Deutsche Nationalbibliothek
The Deutsche Nationalbibliothek lists this publication in the Deutsche Nationalbibliografie; detailed bibliographic data are available on the Internet at <http://dnb.d-nb.de>.

© 2010 WILEY-VCH Verlag GmbH & Co. KGaA, Weinheim

All rights reserved (including those of translation into other languages). No part of this book may be reproduced in any form – by photoprinting, microfilm, or any other means – nor transmitted or translated into a machine language without written permission from the publishers. Registered names, trademarks, etc. used in this book, even when not specifically marked as such, are not to be considered unprotected by law.

Cover Design: Formgeber, Eppelheim
Composition   SNP Best-set Typesetters Ltd., Hong Kong
Printing   betz-druck GmbH, Darmstadt
Bookbinding   Litges & Dopf GmbH, Heppenheim

Printed in the Federal Republic of Germany
Printed on acid-free paper

ISBN: 978-3-527-32427-9

For Conny

Contents

Preface *XV*
List of Contributors *XVII*

Part I **General Introduction**

1 **The Research Program in Epigenetics: The Birth of a New Paradigm** *3*
Paolo Vineis
References *5*

2 **Interactions Between Nutrition and Health** *7*
Ibrahim Elmadfa
2.1 Introduction *7*
2.2 Epigenetic Effects of the Diet *8*
2.3 Current Nutrition Related Health Problems *8*
References *9*

3 **Epigenetics: Comments from an Ecologist** *11*
Fritz Schiemer
References *12*

4 **Interaction of Hereditary and Epigenetic Mechanisms in the Regulation of Gene Expression** *13*
Thaler Roman, Eva Aumüller, Carolin Berner, and Alexander G. Haslberger
4.1 Hereditary Dispositions *13*
4.2 The Epigenome *14*
4.3 Epigenetic Mechanisms *15*
4.3.1 Methylation *15*
4.3.2 Histone Modifications *18*
4.3.3 Micro RNAs *20*
4.4 Environmental Influences *20*
4.4.1 Nutritional and Environmental Effects in Early Life Conditions *20*
4.4.2 Environmental Pollution and Toxins *22*
4.5 Dietary Effects *22*

Epigenetics and Human Health
Edited by Alexander G. Haslberger, Co-edited by Sabine Gressler
Copyright © 2010 WILEY-VCH Verlag GmbH & Co. KGaA, Weinheim
ISBN: 978-3-527-32427-9

4.5.1	Nutrition and the Immune System 26
4.5.2	Nutrition and Aging 26
4.6	Inheritance and Evolutionary Aspects 28
4.7	Conclusion 29
	References 30

Part II Hereditary Aspects

5 Methylenetetrahydrofolate Reductase C677T and A1298C Polymorphisms and Cancer Risk: A Review of the Published Meta-Analyses 37
Stefania Boccia

5.1	Key Concepts of Population-Based Genetic Association Studies 37
5.1.1	Definition and Goals of Genetic Epidemiology 37
5.1.2	Study Designs in Genetic Epidemiology 38
5.1.3	The Human Genome 38
5.1.4	Meta-Analysis in Genetic Epidemiology 39
5.1.5	Human Genome Epidemiology Network 40
5.1.6	"Mendelian Randomization" 41
5.2	Methylenetetrahydrofolate Reductase Gene Polymorphisms (C677T and A1298C) and Its Association with Cancer Risk 41
5.2.1	Gene and Function 41
5.2.2	C677T and A1298C Gene Variants 43
5.2.3	Gene–Environment Interaction 43
5.3	Meta-Analyses of Methylenetetrahydrofolate Reductase C677T and A1298C Polymorphisms and Cancer 44
	References 47

6 The Role of Biobanks for the Understanding of Gene–Environment Interactions 51
Christian Viertler, Michaela Mayrhofer, and Kurt Zatloukal

6.1	Background 51
6.1.1	What Purpose Do Different Biobank Formats Serve? 52
6.1.2	Why Do We Need Networks of Biobanks? 53
6.2	The Investigation of Gene–Environment Interactions as a Challenge for Biobanks 55
6.2.1	How to Evaluate Risk Factors for Metabolic Syndrome and Steatohepatitis? 58
6.2.2	Why Are Biobanks Needed in This Context and What Challenges Do They Have to Face? 58
	References 60

7 Case Studies on Epigenetic Inheritance 63
Gunnar Kaati

7.1	Introduction 64
7.2	Methodology 65

7.2.1	On the Study of Epigenetic Inheritance	65
7.2.2	The Ideal Study Design	66
7.2.3	The Överkalix Cohorts of 1890, 1905 and 1920	67
7.2.4	The ALSPAC Data Set	68
7.2.5	The Proband's Childhood	68
7.2.6	Food Availability	69
7.2.7	Ancestors' Experience of Crises	69
7.2.8	Growth Velocity	69
7.3	Patterns of Transgenerational Responses	70
7.3.1	The Social Context	70
7.3.2	The Ancestors' Nutrition	71
7.3.3	Longevity and Paternal Ancestors' Nutrition	71
7.3.4	The Influence of Nutrition During the Slow Growth Period on Cardiovascular and Diabetes Mortality	72
7.3.5	Is Human Epigenetic Inheritance Mediated by the Sex Chromosomes?	73
7.3.5.1	Paternal Initiation of Smoking and Pregnancy Outcome	75
7.3.6	Epigenetic Inheritance, Early Life Circumstances and Longevity	75
7.3.7	How to Explain the Effects of Food Availability During SGP on Human Health?	76
7.3.7.1	Genetic Selection Through Differential Survival or Fertility?	76
7.3.7.2	Chromosomal Transmission of Nutritionally Induced Epigenetic Modifications	76
7.4	Epigenetic Inheritance	77
7.4.1	Fetal Programming and Epigenetic Inheritance	78
7.5	Future Directions	80
7.6	Conclusions	81
	References	83

Part III   Environmental and Toxicological Aspects

8	**Genotoxic, Non-Genotoxic and Epigenetic Mechanisms in Chemical Hepatocarcinogenesis: Implications for Safety Evaluation** *89*	
	Wilfried Bursch	
8.1	Introduction	90
8.2	Genotoxic and Non-Genotoxic Chemicals in Relation to the Multistage Model of Cancer Development	91
8.2.1	Tumor Initiation	91
8.2.2	Tumor Promotion	92
8.2.3	Tumor Progression	93
8.2.4	Cellular and Molecular Mechanisms of Tumor Initiation and Promotion	93
8.2.5	Epigenetic Effects of Genotoxic and Non-Genotoxic Hepatocarcinogens	96

8.2.6	How Carcinogens Alter the Microenvironment – Crucial Roles of Inflammation 97
8.3	Concluding Remarks 97
	References 99

9 Carcinogens in Foods: Occurrence, Modes of Action and Modulation of Human Risks by Genetic Factors and Dietary Constituents 105
M. Mišík, A. Nersesyan, W. Parzefall, and S. Knasmüller

9.1	Introduction 105
9.2	Genotoxic Carcinogens in Human Foods 106
9.2.1	Polycyclic Aromatic Hydrocarbons 107
9.2.2	Nitrosamines 107
9.2.3	Heterocyclic Aromatic Amines (HAAs) and Other Thermal Degradation Products 108
9.2.4	Mycotoxins 109
9.2.5	Food Additives and Carcinogens in Plant-Derived Foods 109
9.2.6	Alcohol 111
9.3	Contribution of Genotoxic Dietary Carcinogens to Human Cancer Risks 111
9.4	Protective Effects of Dietary Components Towards DNA-Reactive Carcinogens 112
9.5	Gene Polymorphisms Affecting the Metabolism of Genotoxic Carcinogens 114
9.6	Concluding Remarks, Epigenetics and Outlook 118
	References 118

Part IV Nutritional Aspects

10 From Molecular Nutrition to Nutritional Systems Biology 127
Guy Vergères

10.1	Impact of Life Sciences on Molecular Nutrition Research 127
10.2	Nutrigenomics 129
10.2.1	Genomics and Nutrition Research 129
10.2.2	Transcriptomics and Nutrition Research 130
10.2.3	Proteomics and Nutrition Research 131
10.2.4	Metabolomics and Nutrition Research 132
10.3	Nutrigenetics 133
10.4	Nutri-Epigenetics 135
10.5	Nutritional Systems Biology 137
10.6	Ethics and Socio-Economics of Modern Nutrition Research 137
	References 139

11	**Effects of Dietary Natural Compounds on DNA Methylation Related to Cancer Chemoprevention and Anticancer Epigenetic Therapy** *141*	
	Barbara Maria Stefanska and Krystyna Fabianowska-Majewska	
11.1	Introduction *141*	
11.2	DNA Methylation Reaction *142*	
11.3	Implication of the Selected Natural Compounds in DNA Methylation Regulation *144*	
11.3.1	ATRA, Vitamin D_3, Resveratrol, and Genistein *144*	
11.3.1.1	Involvement of $p21^{WAF1/CIP1}$ and Rb/E2F Pathway in Regulation of DNMT1 *147*	
11.3.1.2	Involvement of the AP-1 Transcriptional Complex in Regulation of DNMT1 *148*	
11.3.2	Polyphenols with a Catechol Group *149*	
11.4	Conclusions and Future Perspectives *151*	
	References *152*	
12	**Health Determinants Throughout the Life Cycle** *157*	
	Petra Rust	
12.1	Introduction *157*	
12.2	Pre- and Postnatal Determinants *159*	
12.3	Determinants During Infancy and Adulthood *160*	
12.4	Determinants in Adults and Older People *160*	
12.5	Interactions Throughout the Lifecycle *161*	
12.6	Intergenerational Effects *161*	
	References *162*	
Part V	**Case Studies**	
13	**Viral Infections and Epigenetic Control Mechanisms** *167*	
	Klaus R. Huber	
13.1	The Evolutionary Need for Control Mechanisms *167*	
13.2	Control by RNA Silencing *168*	
13.3	Viral Infections and Epigenetic Control Mechanisms *169*	
13.3.1	RNA Silencing in Plants *169*	
13.3.2	RNA Silencing in Fungi *170*	
13.3.3	RNA Silencing in Mammals *170*	
13.4	Epigenetics and Adaptive Immune Responses *171*	
	References *171*	
14	**Epigenetics Aspects in Gyneacology and Reproductive Medicine** *173*	
	Alexander Just and Johannes Huber	
	References *178*	

15 Epigenetics and Tumorigenesis 179
Heidrun Karlic and Franz Varga

15.1 Introduction 179
15.2 Role of Metabolism Within the Epigenetic Network 181
15.3 Epigenetic Modification by DNA Methylation During Lifetime 183
15.4 Interaction of Genetic and Epigenetic Mechanisms in Cancer 184
15.5 DNA Methylation in Normal and Cancer Cells 185
15.6 Promoter Hypermethylation in Hematopoietic Malignancies 186
15.7 Hypermethylated Gene Promoters in Solid Cancers 187
15.8 Interaction DNA Methylation and Chromatin 188
References 190

16 Epigenetic Approaches in Oncology 195
Sabine Zöchbauer-Müller and Robert M. Mader

16.1 Introduction 195
16.2 DNA Methylation, Chromatin and Transcription 196
16.3 Methods for Detecting Methylation 197
16.4 The Paradigm of Lung Cancer 198
16.4.1 Frequently Methylated Tumor Suppressor Genes and Other Cancer-Related Genes in Lung Carcinomas 199
16.4.2 Monitoring of DNA Methylation in Blood Samples 200
16.5 Epigenetics and Therapy 200
16.6 Epigenetic Alterations Under Cytotoxic Stress 201
16.7 Therapeutic Applications of Inhibitors of DNA Methylation 202
16.8 How May Methylation Become Relevant to Clinical Applications? 203
16.9 Conclusions 204
References 205

17 Epigenetic Dysregulation in Aging and Cancer 209
Despina Komninou and John P. Richie

17.1 Introduction 210
17.2 The Cancer-Prone Metabolic Phenotype of Aging 210
17.3 Age-Related Epigenetic Silencing Via DNA Methylation 212
17.4 Inflammatory Control of Age-Related Epigenetic Regulators 214
17.5 Lessons from Anti-Aging Modalities 215
17.6 Conclusions 217
References 218

18 The Impact of Genetic and Environmental Factors in Neurodegeneration: Emerging Role of Epigenetics 225
Lucia Migliore and Fabio Coppedè

18.1 Neurodegenerative Diseases 225
18.2 The Role of Causative and Susceptibility Genes in Neurodegenerative Diseases 226

18.3	The Contribution of Environmental Factors to Neurodegenerative Diseases *231*
18.4	Epigenetics, Environment and Susceptibility to Human Diseases *233*
18.5	Epigenetics and Neurodegenerative Diseases *234*
18.6	The Epigenetic Role of the Diet in Neurodegenerative Diseases *237*
18.7	Concluding Remarks *238*
	References *239*

19   Epigenetic Biomarkers in Neurodegenerative Disorders *245*
Borut Peterlin

19.1	Introduction *245*
19.2	Epigenetic Marks in Inherited Neurological and Neurodegenerative Disorders *246*
19.3	Epigenetic Dysregulation in Neurodegenerative Disorders *247*
19.4	Gene Candidates for Epigenetic Biomarkers *248*
19.5	Conclusions *249*
	References *250*

20   Epigenetic Mechanisms in Asthma *253*
Rachel L. Miller and Julie Herbstman

20.1	Introduction *253*
20.2	Epigenetic Mechanisms *255*
20.3	Fetal Basis of Adult Disease *255*
20.4	Fetal Basis of Asthma *256*
20.5	Experimental Evidence *257*
20.6	Epigenetic Mechanisms in Asthma *257*
20.7	Cell-Specific Responses *259*
20.8	Conclusion *259*
	References *260*

Part VI   Ways to Translate the Concept

21   Public Health Genomics – Integrating Genomics and Epigenetics into National and European Health Strategies and Policies *267*
Tobias Schulte in den Bäumen and Angela Brand

21.1	Public Health and Genomics *267*
21.2	The Bellagio Model of Public Health Genomics *268*
21.3	The Public Health Genomics European Network *271*
21.4	From Public Health Genomics to Public Health and Epigenetics/Epigenomics *272*
21.5	Health in All Policies – Translating Epigenetics/Epigenomics into Policies and Practice *272*
21.6	Health in All Policies as a Guiding Concept for European Policies *273*
21.7	Relative Risk and Risk Regulation – A Model for the Regulation of Epigenetic Risks? *274*

21.8	Attributable Risks and Risk Regulation	*275*
21.9	Translating Attributable Risks into Policies	*275*
21.10	Limits to the Concept of Health in All Policies in Genomics and Epigenetics	*277*
21.11	Conclusion	*278*
	References	*278*

22 Taking a First Step: Epigenetic Health and Responsibility *281*
Astrid H. Gesche

22.1	Introduction	*281*
22.2	Responding to Epigenetic Challenges	*282*
22.3	Responsibility and Public Health Care Policy	*283*
22.4	Conclusion	*284*
	References	*285*

Index *287*

Preface

We all know only too well that our way of life, the food we eat, smoking, stress or environmental toxins influence our health. But we have just started to learn how these environmental factors cooperate with our hereditary genetic dispositions to determine health or the development of diseases.

Moreover, we did not know until recently that all these factors may also influence the health of our children and grandchildren to whom we may transmit functional changes of our genes. Are we really responsible for the well-being of our unborn descendants?

Does nutrition or stress in our early childhood and in our daily life determine functions of genes and tissues by epigenetic mechanisms? And how does this influence change during life affecting ageing and longevity? To what extent is there an inheritance of environmentally acquired characteristics? These are main questions in epigenetics, a new and exciting hot topic in natural sciences linking multiple hereditary and environmental impacts on our health (http://www.integratedhealthcare.eu).

It has been noted in an article "Epigenetics: The Science of Change" of the *Environmental Health Perspectives*, that interest in epigenetics is increasing "as it has become clear that understanding epigenetics and epigenomics – the genome-wide distribution of epigenetic changes – will be essential in work related to many other topics requiring a thorough understanding of all aspects of genetics, such as stem cells, cloning, aging, synthetic biology, species conservation, evolution, and agriculture" (http://www.ehponline.org/members/2006/114-3/focus.html). Because of this interaction of epigenetics with so many scientific and technological fields, epigenetics will be at the center of public, governmental and scientific interest.

There are now great books available which thoroughly describe mechanisms of epigenetics. The idea for this book was born at a meeting at the University of Vienna where participants from different areas of nutrition, environmental and molecular biology stressed the need of more discussion on concepts towards environmental health interactions between scientists of these disciplines.

Clearly the more optimistic aspect of the possibility to prevent, interfere or even correct epigenetic marks which could result in hazards for diseases, e.g. by dietary concepts or changes of our personal environment and lifestyle, encourages the work in epigenetics. In contrast many results from the analysis of hereditary genetic dispositions can only be respected.

This book picks a transdisciplinary approach focusing on the new understanding of epigenetic and gene-environment interactions for scientific, biomedical, toxicological, environmental, nutritional, evolutionary and regulatory aspects. It focuses on the views of the many exceptional scientists working in these different areas towards epigenetics and health environmental interactions.

The articles in Part I of the book emphasize the interactions between concepts of genetic diversity, epigenetics, environmental health, molecular epidemiology, nutrition and evolution theory. Part II focuses on hereditary aspects, Part III on environmental and toxicological aspects. Part IV extends on nutritional aspects. In Part V the new understanding of epigenetics and environmental health interactions is detailed in case studies of fields such as gynecology, oncology, infectious diseases, asthma, or neurodegenerative diseases. The last Part VI explores concepts to translate the new understanding into public health policies and strategies including principal ethical aspects.

The book is targeted at scientists, environmental, nutritional and health experts, geneticists, experts in science communication, policy makers, experts from standard setting authorities, teachers as well as scientifically experienced consumers interested in interdisciplinary aspects in this area.

The major objective of the book is to strengthen the understanding of interactions between hereditary, genetic and environmental interactions and to bridge gaps which often have evolved between scientific disciplines of molecular, genetic and biotechnological areas on one side and environmental oriented sciences, conservation biology and environmental health on the other.

We thank all the brilliant authors who have contributed, as the summary of their distinguished views on this complex area is the essence of this book.

We do hope that you all enjoy the rather rough ride through the newly emerging and exciting fields of epigenetics!

Vienna, September 2009 *Alexander G. Haslberger*

Acknowledgment

We thank the Austrian Federal Ministry of Science and Research and the Forum of Austrian Scientists for Environmental Protection (http://www.fwu.at/english.htm) for support and many dedicated scientists and colleagues, especially MR Dr. Christian Smoliner for stimulating discussions.

List of Contributors

Eva Aumüller
University of Vienna
Department of Nutritional
Sciences
Althanstrasse 14
1090 Vienna
Austria

Tobias Schulte in den Bäumen
Maastricht University
European Centre for Public
Health Genomics
Faculty of Health
Medicine and Life Sciences
Universiteitssingel 40 West
6229 ER Maastricht-Randwijck
The Netherlands

Carolin Berner
University of Vienna
Department of Nutritional
Sciences
Althanstrasse 14
1090 Vienna
Austria

Stefania Boccia
Università Cattolica del Sacro Cuore
Institute of Hygiene
Genetic Epidemiology and
Molecular Biology Unit
Largo Francesco Vito, 1
00168 Rome
Italy

Angela Brand
Maastricht University
European Centre for Public
Health Genomics
Faculty of Health
Medicine and Life Sciences
Universiteitssingel 40 West
6229 ER Maastricht-Randwijck
The Netherlands

Wilfried Bursch
Medical University of Vienna
Department of Medicine
Institute of Cancer Research
Chemical Safety and Cancer
Prevention
Borschkegasse 8a
1090 Vienna
Austria

List of Contributors

Fabio Coppedè
University of Pisa
Department of Neuroscience
Via Roma 67
56126 Pisa
Italy

Ibrahim Elmadfa
University of Vienna
Department of Nutritional
Sciences
Althanstrasse 14
1090 Vienna
Austria

Krystyna Fabianowska-Majewska
Medical University of Lodz
Department of Biomedical
Chemistry
Lindleya 6
90–131 Lodz
Poland

Astrid H. Gesche
University of New England
Faculty of Arts and Sciences
Armidale, NSW 2351
Australia

Alexander G. Haslberger
University of Vienna
Department of Nutritional
Sciences
Althanstrasse 14
1090 Vienna
Austria

Julie Herbstman
Columbia University
Mailman School of Public Health
Department of Environmental
Health Sciences
100 Haven Ave #25F
New York, NY 10032
USA

Johannes Huber
University Hospital Vienna
Department of Gynecological
Endocrinology and Reproductive
Medicine
Währinger Gürtel 18–20
1090 Vienna
Austria

Klaus R. Huber
Danube Hospital Vienna
Institute of Laboratory Medicine
Langobardenstrasse 122
1120 Vienna
Austria

Alexander Just
University Hospital Vienna
Department of Gynecological
Endocrinology and Reproductive
Medicine
Währinger Gürtel 18–20
1090 Vienna
Austria

Gunnar Kaati
University of Umeå
Department of Public Health and
Clinical Medicine
Building 1A
90185 Umea
Sweden

Heidrun Karlic
Hanusch Hospital
Ludwig Boltzmann Institute
for Leukemia Research
Heinrich Collin Strasse 30
1140 Vienna
Austria

Siegfried Knasmüller
Medical University of Vienna
Department of Medicine
Institute of Cancer Research
Chemical Safety and Cancer
Prevention
Borschkegasse 8A
1090 Vienna
Austria

Despina Komninou
Nutrition and Disease
Prevention Center
31-11, 31st Avenue
Long Island City, NY 11106
USA

Robert M. Mader
Medical University of Vienna
Department of Medicine
Division of Oncology
Währinger Gürtel 18–20
1090 Vienna
Austria

Michaela Theresia Mayrhofer
Medical University of Graz
Institute of Pathology
Auenbrugger Platz 25
8036 Graz
Austria

Lucia Migliore
University of Pisa
Department of Human and
Environmental Science
Via San Giuseppe 22
56126 Pisa
Italy

Rachel L. Miller
Columbia University
College of Physicians and Surgeons
Department of Medicine
Division of Pulmonary Allergy
and Critical Care Medicine
630 West 168th Street
New York, NY 10032
USA

Miroslav Mišík
Medical University of Vienna
Department of Medicine
Institute of Cancer Research
Chemical Safety and Cancer
Prevention
Borschkegasse 8A
1090 Vienna
Austria

Armen Nersesyan
Medical University of Vienna
Department of Medicine
Institute of Cancer Research
Chemical Safety and Cancer
Prevention
Borschkegasse 8A
1090 Vienna
Austria

Wolfram Parzefall
Medical University of Vienna
Department of Medicine
Institute of Cancer Research
Chemical Safety and Cancer Prevention
Borschkegasse 8A
1090 Vienna
Austria

Borut Peterlin
University Medical Center
Ljubljana
Institute of Medical Genetics
Department of Obstetrics and
Gynecology
Slajmerjeva 3
1000 Ljubljana
Slovenia

John P. Richie
Pennsylvauia State University
College of Medicine
Department of Public Health
Sciences
500 University Drive
Hershey, PA 17033
USA

Thaler Roman
University of Vienna
Department of Nutritional
Sciences
Althanstrasse 14
1090 Vienna
Austria

Petra Rust
University of Vienna
Department of Nutritional
Sciences
Althanstrasse 14
1090 Vienna
Austria

Fritz Schiemer
University of Vienna
Center for Ecology
Althanstrasse 14
1090 Vienna
Austria

Barbara Maria Stefanska
Medical University of Lodz
Department of Biomedical Chemistry
Lindleya 6
90–131 Lodz
Poland

Franz Varga
Hanusch Hospital
Ludwig Boltzmann Institute of
Osteology
Heinrich Collin Strasse 30
1140 Vienna
Austria

Guy Vergères
Federal Department of Economics
Affairs
Agroscope Liebefeld-Posieux
Research Station
Schwarzenburgstrasse 161
3003 Berne
Switzerland

Christian Viertler
Medical University of Graz
Institute of Pathology
Auenbrugger Platz 25
8036 Graz
Austria

Paolo Vineis
Imperial College London
MRC Centre for Environment
and Health
St Mary's Campus
Norfolk Place
London W2 1PG
UK

Kurt Zatloukal
Medical University of Graz
Institute of Pathology
Auenbrugger Platz 25
8036 Graz
Austria

Sabine Zöchbauer-Müller
Medical University of Vienna
Department of Medicine
Division of Oncology
Währinger Gürtel 18–20
1090 Vienna
Austria

Part I
General Introduction

1
The Research Program in Epigenetics: The Birth of a New Paradigm
Paolo Vineis

Abstract

This introductory chapter sketches a short history of the concept of epigenetics, from Waddington to today. The chapter outlines the promises associated with the development of epigenetic research, particularly in the field of cancer, and the still unmet challenges, with several examples.

The recent discovery that humans and chimpanzees have essentially the same DNA sequence is simply revolutionary. The obvious question is "why then do they differ so widely"? Obviously, there is something else other than the DNA sequence that explains differences among species. An even more revolutionary advancement could then be the discovery that what makes the difference is a certain pattern of methylation of CpG islands in key genes, for example for the olfactory receptors in chimpanzees (unmethylated) and for brain development in humans. Though this is still speculation, there are great expectations from epigenetics/omics to fill the gaps left by genetics/omics.

If we consider Thomas Kuhn's description of the advancement of science through a sequence of revolutions (leading to paradigmatic leaps), we can probably conclude that epigenetics is definitely a new paradigm. According to Kuhn there are several ways in which a new paradigm arises. Usually this implies a more or less profound crisis of the existing theory, the development of alternative theories – without sound observations yet – and possibly a technological leap forward. These three conditions hold for the shift from genetics to epigenetics, though not necessarily in the order I have suggested.

In a way, a theoretical model for epigenetics (the one by Waddington, who coined the term) came first historically, when genetics was still flourishing. Then several signs of crisis emerged, and now the technological developments allow one to study epigenetic changes properly. To be clear, when I say that the genetic paradigm is in a crisis, this may seem at odds with the successes of genome-wide association studies (GWAS) in 2007–2008. In fact, by crisis I mean (i) the obvious

Epigenetics and Human Health
Edited by Alexander G. Haslberger, Co-edited by Sabine Gressler
Copyright © 2010 WILEY-VCH Verlag GmbH & Co. KGaA, Weinheim
ISBN: 978-3-527-32427-9

gap – referred to above – between DNA sequencing and the ability to explain, for example, differences between species; and (ii) the emerging failures of the paradigm that until very recently strictly separated genes from the environment, according to the neo-Darwinian view. On the one hand we had the environmental exposures, that could cause somatic mutations, or cause chronic diseases by several mechanisms not involving DNA. On the other hand, we had inherited variation, but the link between the two was not straightforward. Recently, to fill the gap the theory of gene–environment interactions (GEI) was coined, with not much success, or at least not the kind of success that was expected. Not many good examples of *bona fide* GEI are available today. Ten years ago, for example, people expected that variants in DNA repair could explain much of cancer variation, in particular in relation to exposure to carcinogens, but a recent synopsis on DNA repair variants in cancer done by us [1] showed surprisingly few associations. Also GWAS led to the discovery of not many variants strongly associated with cancer (with relative risks usually lower than 1.5). In addition, the patterns of association were rather unusual with some regions or SNP associated with several cancers or several diseases, like in the case of 5p15 [2]. Ironically, for 8q24 not only have multiple associations been found, but also the implicated regions are non-coding regions, shedding light probably on some regulatory mechanisms involved, that is, exactly epigenetics.

Well before the gene–environment divide fell into a crisis, Waddington coined his theory of phenotypic plasticity and epigenetics. Waddington referred to epigenetics as an amalgam between genetics and epigenesis, where the latter is the progressive development of new structures. Waddington related epigenetics very much to embryonic development, and put forward the idea that the latter is not entirely due to the "program" encoded in DNA, but depends on environmental influences [3]. His definition of epigenetics is extremely modern: "the causal interactions between genes and their products, which bring the phenotype into being", that echoes a contemporary definition: "the inheritance of DNA activity that does not depend on the naked DNA sequence" [4].

Coming to the present time, the study of epigenetics has definitely been enabled by recent technological advancements, that allow us to investigate DNA methylation, histone acetylation, RNA interference, chromatine formation and other signs of epigenetic events.

What is new in this paradigm? First, it refers not to structural but to functional changes in DNA (gene regulation). Second, we are observing continuous quantitative changes, that is, nature seems to work in degrees, not according to leaps like mutations: the ratio between hypo- and hyper-methylation, for example, seems to be very relevant to cancer. Third, epigenetic changes are reversible: as some chapters in this book show, nice animal experiments have been conducted with dietary supplements that were able to reverse methylation patterns. Fourth, epigenetic patterns seem to be heritable (though this may be the weakest part, since the evidence is not entirely persuasive). Fifth, epigenetic changes fill the gap between genes and the environment: the mysterious relationships between (spontaneous) heritable mutations and selection in neo-Darwinian theory may be overcome by a more sophisticated paradigm that resembles Lamarck's research program – but of course we have to be cautious. Sixth, a successful new theory according to Popper, Lakatos and Kuhn is one that explains unexplained findings in the previous theory and is able to predict new findings.

Are we already in the position to say that the epigenetic theory is able to overcome the old divide between genetics and the environment? I am not aware of any prediction made by epigenetics on theoretical grounds that was subsequently verified, but we can wait. One good candidate is what I said at the start about humans and chimpanzees.

To be sure, some recent research involving epigenetics is extremely promising [5]. In addition to the studies mentioned above, it is worth mentioning the fact that Inuit populations exposed to persistent organic pollutants (POPs) also had detectable hypomethylation of their DNA [6]; this kind of investigation can prove very effective in finding a link between low-level environmental exposures and the risk of disease, through the investigation of sensitive intermediate markers. Exposures that have been found to interact with "metastable epialleles" are, for example, genistein, a component of diet that seems to protect from epigenetic damage, the drug valproic acid, arsenic, and of course vinclozoline (see the current book). But the research is just in its infancy, and many more examples are likely to follow.

In addition to clarifying the relationships between genes and the environment, there is a further dimension in epigenetics, that is the fact that it may explain a feature of evolution that has been slightly neglected, except in developmental studies: self-organization of the living being. In fact a modern theory of evolution should encompass two big chapters, both the selection–adaptation component, and the self-organization component (the latter very often overlooked). This is in fact a promising component of the new revolutionary paradigm of epigenetics; for example, one might speculate that cancer is explained by a Darwinian paradigm (since it is due to selective advantage of mutated/epimutated cells) [7] but without the self-organization element that has characterized the evolution of organisms and species.

The next years will probably show the ability of the new paradigm to explain unexplained findings, and to make correct predictions.

References

1 Vineis, P., Manuguerra, M., Kavvoura, F.K., Guarrera, S., Allione, A., Rosa, F., Di Gregorio, A., Polidoro, S., Saletta, F., Ioannidis, J.P., and Matullo, G. (2009) A field synopsis on low-penetrance variants in DNA repair genes and cancer susceptibility. *J. Natl. Cancer Inst.*, **101** (1), 24–36.

2 Rafnar, T., Sulem, P., Stacey, S.N., Geller, F., Gudmundsson, J., Sigurdsson, A., Jakobsdottir, M., Helgadottir, H., Thorlacius, S., Aben, K.K., Blöndal, T., Thorgeirsson, T.E., Thorleifsson, G., Kristjansson, K., Thorisdottir, K., Ragnarsson, R., Sigurgeirsson, B., Skuladottir, H.,

Gudbjartsson, T., Isaksson, H.J., Einarsson, G.V., Benediktsdottir, K.R., Agnarsson, B.A., Olafsson, K., Salvarsdottir, A., Bjarnason, H., Asgeirsdottir, M., Kristinsson, K.T., Matthiasdottir, S., Sveinsdottir, S.G., Polidoro, S., Höiom, V., Botella-Estrada, R., Hemminki, K., Rudnai, P., Bishop, D.T., Campagna, M., Kellen, E., Zeegers, M.P., de Verdier, P., Ferrer, A., Isla, D., Vidal, M.J., Andres, R., Saez, B., Juberias, P., Banzo, J., Navarrete, S., Tres, A., Kan, D., Lindblom, A., Gurzau, E., Koppova, K., de Vegt, F., Schalken, J.A., van der Heijden, H.F., Smit, H.J., Termeer, R.A., Oosterwijk, E., van Hooij, O., Nagore, E., Porru, S., Steineck, G., Hansson, J., Buntinx, F., Catalona, W.J., Matullo, G., Vineis, P., Kiltie, A.E., Mayordomo, J.I., Kumar, R., Kiemeney, L.A., Frigge, M.L., Jonsson, T., Saemundsson, H., Barkardottir, R.B., Jonsson, E., Jonsson, S., Olafsson, J.H., Gulcher, J.R., Masson, G., Gudbjartsson, D.F., Kong, A., Thorsteinsdottir, U., and Stefansson, K. (2009) Sequence variants at the TERT-CLPTM1L locus associated with many cancer types. *Nat. Genet.*, **41** (2), 221–227.

3 Feinberg, A.P. (2007) Phenotypic plasticity and the epigenetics of human diseases. *Nature*, **447**, 433–440.

4 Esteller, M. (2008) Epigenetics in evolution and disease. *Lancet*, **372**, S90–S96.

5 Jirtle, R.L., and Skinner, M.K. (2007) Environmental epigenomics and disease susceptibility. *Nat. Rev. Genet.*, **8**, 253–262.

6 Rusiecki, J.A., Baccarelli, A., Bollati, V., Tarantini, L., Moore, L., and Bonefeld-Jorgensen, E.C. (2008) Global DNA hypomethylation is associated with high serum-persistent organic pollutants in Greenlandic Inuits. *Environ. Health Perspect.*, **116**, 1547–1552.

7 Vineis, P., and Berwick, M. (2006) The population dynamics of cancer: a Darwinian perspective. *Int. J. Epidemiol.*, **35** (5), 1151–1159.

2
Interactions Between Nutrition and Health

Ibrahim Elmadfa

Abstract

Nutrition is a major contributor to health providing the organism with the energy, essential nutrients and biologically active plant cell components necessary for its maintenance and proper functioning. More recently, food components have also been discovered as regulators of a number of physiological pathways often involving their own metabolism. This regulation is to a large extent mediated via gene expression in which epigenetic effects play an important part. Methylation of DNA is a major regulatory mechanism in the transcription of genes and is influenced by food components providing methyl groups. Due to the universality of this mechanism and depending on the genes and tissues involved, alterations of DNA methylation can have a number of consequences. There is evidence that they play a role in the development of certain cancer types that are related to exposure to carcinogens. Epigenetic alterations of gene expression were also shown to be involved in some animal models of obesity. As many of these changes are inheritable, the diet of the parents could have a far-reaching influence on their offspring and possibly contribute to the recent rise in the prevalence of overweight and related metabolic diseases.

In light of the impact of nutrition on gene regulation, molecular approaches will contribute to our understanding of the relationship between nutrition and health.

2.1
Introduction

The close relationship between nutrition and health is not a recent discovery. In fact, the deep impact of food on health has been known for centuries and even millennia. However, knowledge about the effects of health status on the metabolism of food is more recent. Insights in genetic make-up and regulations show that alterations in nutrient processing are not necessarily restricted to certain diseases but can also occur in healthy subjects. Epigenetic modifications are important determinants of such variances and can be influenced by food and nutrient intake beside other environmental factors.

2.2
Epigenetic Effects of the Diet

The pathways of nutrient metabolism are encoded in the genes. Hence, mutations can lead to disturbances in the breakdown of a certain compound, as is the case in galactose or fructose intolerance. However, the regulation of gene expression is as important and is directly influenced by dietary components. A well-known example for the epigenetic effects of a nutrient is the methylation of DNA, a major regulatory mechanism, by methyl group donors like folic acid, vitamin B_{12}, betain and choline. It was shown that, in mice, supplementation of these nutrients to pregnant dams had an influence on the offspring, manifesting in alterations of the coat color [1].

2.3
Current Nutrition Related Health Problems

In wealthy societies, the major health problems arising from nutrition are overweight and obesity. Both have been increasing at an alarming rate for the past 50 years. While unlimited access to food provides the residents of industrialized nations with the necessary energy sources, this wide choice is not the only cause of increased body weight. Lack of physical activity is another important contributor. However, although both account for the majority of cases of overweight, additional factors play a role. As the increase in obesity has occurred very rapidly, changes in the genome itself are unlikely. Therefore, epigenetic modifications might be involved. Maternal obesity and nutrition may lead to epigenetic modifications that establish overweight in the infant as well [2]. For example, hypomethylation of the agouti gene in mice causes an over-expression of the agouti protein that, by binding antagonistically to the melanocortin receptor (MCR) 4, induces hyperphagia [3]. Differences in gene expression were also observed between low and high weight gainers in a diet-induced obesity study in mice [4, 5].

There is evidence that diseases associated with obesity, like cardiovascular diseases and diabetes mellitus type II, also have epigenetic backgrounds [6, 7]. Thus, a subject's exposure to food scarcity correlated with a lower risk for cardiovascular death and diabetes mellitus in his grandchildren. Interestingly, this legacy was transmitted through the male line [8].

The role of epigenetic modifications in cancer development is well established. Altered methylation patterns are observed in many tumors with hypo- and hypermethylation occurring at the same time. This methylation is partly influenced by nutritional factors. Notably, hypermethylation is particularly frequent in gastrointestinal tumors and this may be related to exposure to carcinogens [9, 10].

Nutrition has an important influence on health and disease. While this knowledge is not new, novel technologies allow insights into the mechanisms behind this relationship revealing nutrients not only as building material for body tissues and co-factors of enzymes but as modulators of gene expression. The involvement of epigenetic events is supported by the apparent heredity of certain diet-related diseases.

Understanding the influence of an individual's genetic make-up on the metabolism of nutrients and of nutrients on gene regulation presents a great challenge to modern nutritional scienctists. Molecular genetic approaches have found their way into research in nutritional sciences adding to its interdisciplinarity. Applied nutrition and dietetics will be increasingly shaped by the emerging field of nutrigenetics and nutrigenomics. In this sense, the following chapters are meant to give an overview of the plethora of health conditions that are influenced by the interplay of nutrition and the genome.

References

1 Cropley, J.E., Suter, C.M., Beckman, K.B., and Martin, D.I. (2006) Germ-line epigenetic modification of the murine A vy allele by nutritional supplementation. *Proc. Natl. Acad. Sci. U S A*, **103**, 17308–17312.

2 Waterland, R.A., and Michels, K.B. (2007) Epigenetic epidemiology of the developmental origins hypothesis. *Annu. Rev. Nutr.*, **27**, 363–388.

3 Wolff, G.L., Roberts, D.W., and Mountjoy, K.G. (1999) Physiological consequences of ectopic agouti gene expression: the yellow obese mouse syndrome. *Physiol. Genomics*, **1**, 151–163.

4 Samama, P., Rumennik, L., and Grippo, J.F. (2003) The melanocortin receptor MCR4 controls fat consumption. *Regul. Pept.*, **113**, 85–88.

5 Koza, R.A., Nikonova, L., Hogan, J., Rim, J.S., Mendoza, T., Faulk, C., Skaf, J. and Kozak, L.P. (2006) Changes in gene expression foreshadow diet-induced obesity in genetically identical mice. *PLoS Genet.* **2**, e81.

6 Lund, G., Andersson, L., Lauria, M., Lindholm, M., Fraga, M.F., Villar-Garea, A., Ballestar, E., Esteller, M., and Zaina, S. (2004) DNA methylation polymorphisms precede any histological sign of atherosclerosis in mice lacking apolipoprotein E. *J. Biol. Chem.*, **279**, 29147–29154.

7 Wren, J.D., and Garner, H.R. (2005) Data-mining analysis suggests an epigenetic pathogenesis for type 2 diabetes. *J. Biomed. Biotechnol.*, **2**, 104–112.

8 Kaati, G., Bygren, L.O., and Edvinsson, S. (2002) Cardiovascular and diabetes mortality determined by nutrition during parents' and grandparents' slow growth period. *Eur. J. Hum. Genet.*, **2**, 682–688.

9 Lopez, J., Percharde, M., Coley, H.M., Webb, A., and Crook, T. (2009) The context and potential of epigenetics in oncology. *Br. J. Cancer*, **100**, 571–577.

10 Nyström, M., and Mutanen, M. (2009) Diet and epigenetics in colon cancer. *World J. Gastroenterol.*, **15**, 257–263.

3
Epigenetics: Comments from an Ecologist
Fritz Schiemer

Abstract

The "Comments from an Ecologist" are based on the results of a workshop initiated by the "Forum of Austrian Scientists for Environmental Protection". It emphasizes epigenetics as a main research priority for an improved understanding of the interactions between human societies and their environment.

In 2004 Leslie Pray summarized new scientific findings in the area of epigenetics [1] saying that the environmental lability of epigenetic inheritance may not necessarily bring to mind Lamarckian ideas but it does give researchers reason to reconsider long-refuted notions about the inheritance of acquired characteristics.

Recently the US Office of Environmental Health Hazard Assessment strengthened scientific evidence that "Certain environmental factors have been linked to abnormal changes in epigenetic pathways in experimental and epidemiological studies. However, because these epigenetic changes are subtle and cumulative and they manifest over time, it is often difficult to establish clear-cut causal relationships between an environmental factor, the epigenetic change and the disease".

These findings enforce the need for scientists in many ecological areas, such as evolutionary biology, environmental protection and environmental health to follow and consider developments in the area of epigenetics.

Austria has a long history in research on ecology and evolutionary theory. Already in the 1990s Rupert Riedl, founder of the Society of Austrian Scientists for Environmental Protection and an active member of the Club of Rome, addressed epigenetic concepts in his work on system biology (discussed in this book).

The ecophysiologist Wolfgang Wieser, proposed a parallel concept in evolution with importance to epigenetics: He argued, in 1997, [2] that "the focus of evolutionary biology shifts from explaining the origin of species to the modelling of processes by which autonomous entities cooperate to form systems of greater complexity. Whereas the evolutionary theory is still dominated by the ideas of competition, the concept of transitions is dominated by the ideas of cooperation, control and conflicts on a different level of organization". Transitions include the

Epigenetics and Human Health
Edited by Alexander G. Haslberger, Co-edited by Sabine Gressler
Copyright © 2010 WILEY-VCH Verlag GmbH & Co. KGaA, Weinheim
ISBN: 978-3-527-32427-9

replication of molecules forming populations of molecules in compartments and the transition of solitary individuals forming integrated societies. "The common feature of these transitions is that entities capable of independent replication before the transition can replicate only as a part of a larger whole after it. Each of the major transitions represents an organizational level occupied by a certain type of biological system that evolved and created phylogenetic relationships". Conflicts between entities and systems are inevitable, constituting just as constructive an element of the evolutionary process as competition between entities. The quality that increases with each transition involves specialization and differentiation, allowing the exploitation of new sources of energy and materials. "On the concept of genes as selfish particles rests the study of populations and genetics, on the concept of genes as systemic components rests the science of developmental and other branches of organismic biology. Organismic function is the result of the extreme interdependence of its parts, and the dominant strategy in this game is the near absolute epigenetic control of gene activity". In the multicellular organism the dominant mechanism behind "division of labor" is the epigenetic control of gene activity by molecular inhibition. The major engineering feat behind this selective process is the shutdown of genes by means of molecular inhibitors. However, the act of gene inhibition is only the final step in chains of reactions that tie each gene into an information network of great complexity [3]. "Epigenetic modifications construct those cellular and physiological niches, in which genes are selected" [4]. Considering this view on evolution "Nothing in evolutionary biology makes sense except in the light of conflicts between parts and systems."

These concepts of my colleagues and friends, Wolfgang Wieser and Rupert Riedl, have demonstrated the importance of physiological and evolutionary control mechanisms for an understanding of physiological adaptations to our natural environment and environmental changes. This understanding should guide our research priorities in the understanding of interactions between human societies and their environment.

I hope that this book will contribute to encouraging and strengthening further research on links between hereditary, environmental and nutritional aspects as such interdisciplinary aspects will not only stimulate progress in the understanding of mechanisms of evolution but also establish ways to protect environmental safety and human health. As the present president of the Forum of Austrian Scientists for Environmental Protection I am glad that our society started to consider the consequences and interactions between epigenetic research and ecology and initiated conferences as a starting point for the present book.

References

1 Pray, L. (2004) Epigenetics: genome, meet your environment. *The Scientist*, **18**, 14.
2 Wieser, W. (1997) A major transition in Darwinism. *Tree*, **12**, 367–370.
3 Wieser, W. (2001) Private and collective interests; conflicts and solutions: the central theme of current thinking in evolutionary biology. *Zoology (Jena)*, **104**, 184–191.
4 Wieser, W. (2007) *Gehirn und Genom*, C.H. Beck, München.

4
Interaction of Hereditary and Epigenetic Mechanisms in the Regulation of Gene Expression

Thaler Roman, Eva Aumüller, Carolin Berner, and Alexander G. Haslberger

Abstract

Hereditary dispositions and environmental factors such as nutrition and the natural and societal environment interact with human health. Diet compounds raise increasing interest due to their influence in epigenetic gene expression. Nutrition and specific food ingredients have been shown to alter epigenetic marks such as DNA methylation or histone acetylation involving regulation of genes with relevance for fundamental mechanisms such as antioxidative control, cell cycle regulation or expression of immune mediators.

4.1
Hereditary Dispositions

Interactions between genes and environment are not linear and often include direct and indirect cause of events. Many complex diseases are linked to various heritable dispositions like single nucleotide polymorphisms (SNP) or allelic translocations. Consequently they have become a main focus in modern biomedical research and have also started to raise public interest. Single nucleotide polymorphisms are common rather than exceptions. SNPs may determine the efficiency of gene transcription, gene translation or protein structure, leading to an altered amount of enzyme and/or enzyme activity, thus influencing further metabolic pathways. SNPs can occur within coding sequences of genes, non-coding regions of genes or in intergenic regions. SNPs within a coding sequence do not necessarily change the functional efficiency of the protein (silent mutations) [1, 2].

SNPs occur on average somewhere between every 1 and 100 in 1,000 base pairs in the euchromatic human genome. SNPs and copy number variations determine human genetic variation and are assumed to influence peoples' variable responses to toxins or pharmaceuticals and to contribute to different penetrance of diseases. The International HapMap (haplotype map) Project aimed to study the scope of

SNP variations in different population groups. Clusters of SNPs located on the same chromosome tend to be inherited in blocks. It is expected that the outcomes of this project will provide crucial tools that will allow researchers to detect genetic variations with effects on health and disease. In so-called genome-wide association studies, researchers compare genomes of individuals with known diseases to a control group in order to detect suspicious tag SNPs which might play a role in development, genesis or course of disease (HapMap project, snp.cshl.org/).

A broad variety of functional SNPs on several genes, covering almost all aspects for cell viability, is well described in the literature. In superoxide metabolism, for example, an increase in superoxide radicals is given by the SNP (rs4880) regarding the mitochondrial superoxide dismutase (MnSOD) gene [3]. Several kinds of cancer [4], Alzheimer's disease [5], and an accelerated aging process are suggested to be related to this SNP. In contrast, three SNPs in the human forkhead box O3A gene (FOXO3A) were statistically significantly associated with longevity. Polymorphisms in this gene were indeed associated with the ability to attain exceptional old age. In nutritional sciences, the methylenetetrahydrofolatereductase (MTHFR) gene is well known as it is essential for folic acid metabolism [6]. The polymorphism C677T in the MTHFR gene is suggested to have an influence in several methylation pathways as for example in the DNA-methylation [7, 8].

So far, a multitude of ailments have been correlated with respective SNPs. Yet, for individuals the predictability of the development of diseases from the analysis of SNPs is still difficult because often the functional relevance of SNPs is missing. The mostly small and variable penetrance of single SNPs leads to statistical limitations and many clinical studies associating SNPs and diseases show poor reproducibility [9].

The individual combination of relevant SNPs in addition to environmental influences may define risks for developing diseases. Experiences with sets of candidate SNPs and the work of biobanks still need to be evaluated to understand gene–environment interactions [10].

4.2
The Epigenome

Additional to the genetic code, mammalian cells contain an additional regulatory level which predominates over the DNA code: modification of gene expression by altering chromatin without changing the DNA sequence. Thus, due to different chromatin status, the same genetic variants might be, for example, associated with different phenotypes, depending on environmental influences. New insights in research clarify the molecular pathways by which, among others, nutrition and lifestyle factors influence chromatin packaging, where here these epigenetic changes then strongly correlate with the development of multifactorial chronic diseases like cancer, diabetes or obesity [11–14].

A rapidly growing number of genes with epigenetic regulation altering their expression by remodeling chromatin have been identified. Methylation of cytosines in DNA, histone modifications as well as alterations in the expression level of micro RNA (miRNA) and short interference RNA (siRNA) are the mechanisms involved in chromatin remodelling. The term "epigenome" is used to define a cell's overall epigenetic state. Epigenetic modifications can be stably passed over numerous cycles of cell division. Some epigenetic alterations can even be inherited from one generation to the next [15, 16]. Studies conducted by the Department of Community Medicine and Rehabilitation of the Umeå University in Sweden showed transgenerational effects due to nutritional habits during a child's slow growth period (SGP). Evident correlations with descendants' risk of death from cardiovascular disease and diabetes were also seen [17]. The finding that monozygotic twins are epigenetically indistinguishable early in life but, with age, exhibit substantial differences in the epigenome, indicates that environmentally determined alterations in a cell's epigenetic marks are responsible [18].

During the development of germ cells and during early embryogenesis DNA is specifically methylated and these marks confer genome stability, imprinting of genes, totipotency, correct initiation of embryonic gene expression and early lineage development of the embryo [19]. According to experiments in the Agouti mouse model, early epigenetic programming is alterable through the mother's diet during pregnancy, leading to lifelong modification of selected genes in the offspring [20].

4.3
Epigenetic Mechanisms

Epigenetic regulation includes DNA methylation, histone modifications and post-transcriptional alteration of gene expression based on microRNA interference.

4.3.1
Methylation

Basic biological properties of DNA-segments such as gene density, replication timing and recombination are tightly linked to their guanine–cytosine (CG) content. Therefore, isochors (DNA fractions >300 kb on average) of the genome can be classified accordingly [21]. CpG islands are defined as genomic regions with a GC percentage greater than 50% and with an observed CpG (cytosine base followed by a guanine base) ratio greater than 60%. In mammals, CpG islands typically are 200–3000 base pairs long. CpGs are rare in vertebrate DNA due to the tendency of such arrangements to be methylated to 5-methylcytosines then converted into thymines by spontaneous deamination.

The consequences are large DNA-regions low in GC and gene density, clearly visible on isochor maps as "genome deserts". However, some regions escaped large scale methylation during evolution and, therefore, show a high amount of GCs, which is generally parallel to a high CpG island and gene density [16]. CpG islands mostly occur in these isochors, at or near the gene's transcriptional start site. Promoters of tissue-specific genes that are situated within CpG islands are, normally, largely unmethylated in expressing as well as non-expressing tissues [22]. There are three known ways by which cytosine methylation can regulate gene expression: (i) 5-methylcytosine can inhibit or hinder the association of some transcriptional factors with their cognate DNA recognition sequences, (ii) methyl-CpG-binding proteins (MBPs) can bind to methylated cytosines mediating a repressive signal and (iii) MBPs can interact with chromatin forming proteins modifying the surrounding chromatin, linking DNA methylation with chromatin modification [23]. Mostly, DNA methylation causes a repression of mRNA gene expression, however, when CpG methylation blocks a repressor binding site within a gene promoter, this may induce a transcriptional activation, as shown for Interleukin-8 in breast cancer [24].

DNA-methylation at position five of CpG-cytosines is conducted by DNA methyltransferases (DNMTs), which are expressed in most dividing cells [25]. DNMT1 enzyme is responsible for the maintenance of global methylation patterns on DNA. It preferentially methylates CpGs on hemimethylated DNA (CpG methylation on one site of both DNA strands), therefore guaranteeing transfer of methylation marks through the cell cycle in eukaryotic cells. The DNMT1 enzyme is directly incorporated in the DNA replication complex. The de novo methyltransferases DNMT3a and DNMT3b establish methylation patterns at previously unmethylated CpGs. DNMT3L is a DNMT-related enzyme which associates with DNMT3a/3b. It influences its enzymatic activity while lacking one of its own. Finally, for the DNMT2 enzyme, in mammals a biological function remains to be demonstrated [22, 25].

Most DNMTs contain a sex-specific germline promoter which is activated at specific stages during gametogenesis. Genomic methylation patterns are largely erased during proliferation and migration of primordial germ cells and re-established in a sex-specific manner during gametogenesis, resulting in a high methylation of the genome. Close regulation of the DNMT genes during these stages and during early embryogenesis is needed [25]. After fertilization, a second phase of large epigenetic reprogramming takes place. Upon fertilization, a strong, presumably active DNA demethylation can be observed in the male pronucleus while the maternal genome is slowly and passively demethylated. Imprinting by methylation is maintained for both the paternal and the maternal genome. DNA-demethylation occurs until the morula stage, followed by de novo methylation [26]. See Figure 4.1, adapted from Morgan *et al.* 2005 and Dean *et al.* 2003 [19, 26].

For targeting DNA de novo methylation, three mechanisms are described. (i) DNMT3 enzymes themselves might recognize DNA or chromatin via their conserved PWWP (relatively well-conserved Pro-Trp-Trp-Pro residues, present in all eukaryotes) domain. (ii) By interaction of DNMTs with site-specific transcriptional repressor proteins DNMTs can be targeted to gene promoter regions. (iii) *In vitro* studies have shown that the introduction of double-stranded RNA corresponding to the promoter region of the target gene leads to its de novo DNA methylation and decreased gene expression, suggesting the existence of an RNAi-mediated

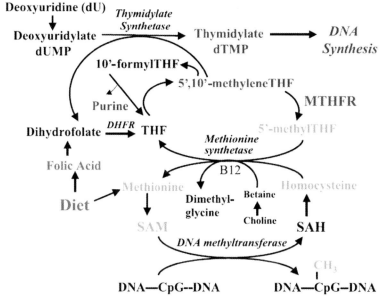
Figure 4.1 DNA methylation pathway.

DNA methylation mechanism. However, further efforts are needed to clarify these pathways [23].

DNMTs also seem to mediate gene silencing by modifying chromatin via protein–protein interactions. They biochemically interact with histone methyltransferases (HAT) and histone deacetylases (HDACs). As mentioned above, binding of MBPs to methylated CpGs mediates silencing of gene expression by the associaton with chromatin remodeling co-repressor complexes. Thus, under certain circumstances, gene silencing by DNA methylation may be attributed directly to chromatin modifications. So far, six different methyl-CpG-binding domains (MBD) are known for the MBPs: MBD1 to 4, MeCP2 and Kaiso. All of them mediate silencing of gene expression by chromatin remodeling [27]. For example, the MBD1 and a histone H3 methyltransferase enzyme (SetDB1) interact during the cell cycle, linking DNA methylation to rearrangement of chromatin by histone methylation [23]. Even if mechanisms for an active demethylation for DNA are not well understood [28], gene silencing by DNA methylation is reversible. For example, Il-4 expression in undifferentiated T cells is silenced via binding of MBD2 on the methylated promoter of the gene [29]. After differentiation, TH2 cells express the transcription factor GATA-3 which competes with MBD2 for binding to the IL-4 promoter. In this case epigenetic factors impose a threshold to be overcome in order to achieve efficient gene expression. The binding of MBDs to DNA seems to be sequence specific for most domains.

In order to achieve DNA methylation, often alterations at the chromatin level must first occur. As previously discussed, DNMTs interact with HATs and HDACs. Small modifications of histones by acetylation, methylation, phosphorylation and

Table 4.1 Consequence of histone modifications on chromatin.

Histone	Modification	Effect
H1	Phosphorylation	Chromatin condensation, gene specific condensation and repression.
	Ubiquitination	Transcriptional activation
H2A	Acetylation	Transcriptional activation
	Ubiquitination	Elusive
H2B	Ubiquitination	Prerequisite of H3 methylation
	Phosphorylation	Chromatin condensation
	Acetylation	Chromatin remodeling
	Methylation	Chromatin stabilization
H3	Methylation (H3-K4, R17)	Transcriptional activation
	Methylation (H3-K9, K79)	Transcriptional repression
	Acetylation	Transcriptional activation
	Phosphorylation	Chromatin condensation; transcriptional activation
	Ubiquitination	Nucleosome loosening
H4	Acetylation	Transcriptional activation
	Methylation (H4-K20)	Transcriptional repression
	Methylation (H4-R3)	Transcriptional activation

ubiquitation on certain amino acids can alter gene expression by chromatin remodeling. The effect of these small modifications on chromatin remodeling depends on the type, amount and site of modification as well as the interactions, as shown in Table 4.1 (adapted from He and Lehming 2003 [30]).

4.3.2
Histone Modifications

Some modifications, including acetylation and phosphorylation, are reversible and dynamic, others, such as methylation, are found to be more stable and involved in long-term alterations [31, 32].

As an attempt to organize the complexity of the different possible modifications, the establishment of a "histone code" is a major focus of epigenetic research. We know 24 methylation sites on a histone where the possibility of mono-, di-, or trimethylated lysine residues and mono- or dimethylated arginine side chains lead to 3×10^{11} different histone methylation states. The need to distinguish between short-term changes in histone modification associated with ongoing processes and changes that have long-term effects still poses a problem to be solved [33, 34]. The significance of some histone modifications for gene expression has been identified, especially regarding acetylation and deacetylation of histones [31]. Histone acetylation is the major factor regulating the degree of chromatin folding. Loss of acetylation at Lys16 and trimethylation at Lys20 of histone H4 are commonly seen in human tumor cells [35].

Several factors are involved in the process of gene silencing by DNA-methylation. One prerequisite is attributed to the recruitment of HDACs through the methyl-DNA binding motifs, while methylation at Lys9 on H3 is another prerequisite. Methylation of both DNA and histones seems to have a reciprocative reinforcing effect in gene silencing. Histone variants like macro H2A, accumulated on the inactive X chromosome, have been reported to play a role in gene expression regulation. Their presence in the IL-8 promoter-region. for example, is connected to tissue-related gene silencing of the IL-8 gene [30]. The comprehension of a histone code will be an important step toward understanding further mechanisms for epigenetic gene expression. Over time, the relationship between DNA-methylation and histone modifications will become clearer [23, 32].

Through alteration of gene expression and destabilization of chromatin, histone modifications can have an impact on the risk of cancer. In higher eukaryotes, HDACs are grouped into four classes (I–IV). Class I HDACs are found exclusively in the nucleus, whereas class II HDACs shuttle between the nucleus and the cytoplasm. Modified cancer pathogenesis is linked to varying activities of both HATs and HDACs. This mechanism is best described for acute promyelocytic leukemias, where a chromosomal translocation leads to inappropriate HDACs activity. In gastric cancer, esophageal squamous cell carcinoma, and prostate cancer, an increased HADC1 expression is related to pathogenesis. In contrast, in colon cancer an overexpression of HDAC2 causes a decreased expression of the APC (adenomatous polyposis coli) tumor suppressor gene. Another way of influencing the expression of tumor suppressor genes is abnormal functions of HDACs, for instance by atypical targeting. These observations led to the current exploration of HDAC inhibitors as anticancer therapeutics. The examined substances aim to shift several cell functions which are known to be down regulated in cancer cells. Among other presumed advantages of the approach cancer cells show an increased sensitivity for HDAC inhibitors compared to normal cells. Thus there may be a possibility to prevent or to decelerate the development of tumors by such substances.

Not only HDACs but also HATs influence the risk of developing cancer. Especially in cancer of epithelial origin an overexpression or mutation of HAT genes has been detected. Some lines of lung, breast, and colorectal cancer have in common a mutation which inactivates a specific HAT.

Changes of the methylation status of histones have also been observed in some types of cancer. The loss of the trimethylated form of the lysine 20 residue of the H4 histone is characteristic for many cancer cells. The demethylation of lysine 9 on H3 histones also increases significantly the formation of B-cell lymphoma in mice because of its link to chromatin silencing [36].

4.3.3
Micro RNAs

In addition to DNA and histone modifications, micro RNAs are part of the epigenetic gene expression regulatory complex. Micro RNAs are small non-coding RNAs that posttranscriptionally regulate the expression of complementary messenger RNAs and function as key controllers in a countless number of cellular processes, including proliferation, differentiation and apoptosis. Over the last few years, increasing evidence has indicated that a substantial number of micro RNA genes have been subjected to epigenetic alterations, resulting in aberrant patterns of expression upon the occurrence of cancer [37].

4.4
Environmental Influences

Influences from the environment on human health do not only include direct effects from the natural or the social environment but also involve indirect effects such as changes in ecosystems induced by developments in societies. The reports of the United Nations- Millennium Ecosystem Assessment (www.millenniumassessment.org/) demonstrate the complexity of influences as well as concepts for analysis of their effects on human health. Many of these factors may be of relevance for an analysis of possible epigenetic consequences.

4.4.1
Nutritional and Environmental Effects in Early Life Conditions

Environmental factors with importance in early life include not only food compounds, such as vitamin B_{12} or genistein, but also toxins, such as endogene disruptors. A transgenerational effect by an environmental toxin has been described for the endocrine disruptor vinclozolin, a common fungicide. Transient embryonic exposure to the endocrine disruptor vinclozolin in rats resulted in a number of disease states or tissue abnormalities, including prostate disease, kidney disease, immune system abnormalities, testis abnormalities, and tumor development in adult animals from the F1 generation and all subsequent generations examined (F1–F4). The incidence or prevalence of these transgenerational disease states was high and consistent across all generations. The effects were transferred through the male germ line. The described symptoms appear to be attributable to a heritable alteration of epigenetic programming of DNA-methylation in the germ line, which alters the transcriptomes of developing organs [38].

Early mammalian development is a crucial period for establishing and maintaining epigenetic marks. Modifications of the epigenome are not limited to the fetal period but extended to the plastic phase of early life. Broad epigenetic reprogramming can be seen after fertilization achieving totipotency of developing embryo cells, while methylation patterns associated with imprinting are sustained. Epigenetic modifications settled during fetal development are generally stable and passed through cell division processes throughout the lifetime.

Prenatal exposure to famine for instance is also likely to change the epigenetic status. Observed individuals showed less DNA methylation of imprinted IGF-2 gene than their unexposed same sex siblings [39]. A further, nutrition-related epigenetic alteration has been demonstrated in the viable yellow Agouti (= Avy) mouse model [16]. Maternal diet containing bisphenol A (BPA), an estrogenic monomer used in polycarbonate plastic production, significantly decreases the offspring's methylation of the Avy gene promoter which induces a different phenotype. Maternal nutritional supplementation with methyl-donors counteracts the effects from BPA, showing that simple dietary changes may protect against harmful epigenetical effects caused by environmental toxins.

Modified promoter methylation and, accordingly, modified gene expression of the hepatic glucocorticoid receptor and the peroxisome proliferator–activated receptor alpha (PPAR alpha), both important elements in carbohydrate and lipid metabolism regulation, can be seen in the offspring of rats fed a protein restricted diet. The selective methylation of PPAR alpha in the liver without consequences for the related transcription factor PPAR gamma demonstrates that maternal nutrition and behavior can also influence specific promoter regions rather than being associated with global DNA-methylation alteration. Such changes in gene expression and promoter methylation were also seen to be transmitted to the next generation without further nutritional challenge for the first generation [40].

Finally, the elucidation of the impact of epigenetic modifications on behavior and psychic health presents an intriguing challenge. Rodent experiments show that some epigenetic changes can be induced promptly after birth through the mother's physical behavior toward her newborn [41–43]. Licked and groomed newborns appear to grow up to be relatively brave and calm. In contrast, neglected newborns grow up to be nervous and hyperactive. The difference in their behavior can be explained by analyzing specific regions in the brain. The hippocampus of both groups reveals different DNA-methylation patterns for specific genes, agreeing with the difference in behavior. This entails a better development of the hippocampus in the licked newborns, possibly by releasing less of the stress hormone cortisol. Furthermore, recent research results have demonstrated that complex epigenetic mechanisms have long-lasting effects in mature neurons and that they possibly play a vital role in the etiology of major psychoses, such as schizophrenia or bipolar disorder.

Paternal age at conception is a strong risk factor for schizophrenia, explaining about a quarter of all cases. The possible mechanisms for the elevated risk may be *de novo* point mutations or defective epigenetic regulation of paternal genes. The risk might also be related to paternal toxic exposures, nutritional deficiencies, suboptimal DNA repair enzymes or other factors that influence the reliability of the transfer of genetic information in the constantly replicating male germ line [44, 45].

4.4.2
Environmental Pollution and Toxins

Air pollution, for instance, particular matters/small particles in the air and cigarette smoke, appear to have omnipresent toxicological influences on humans. Promoter hypermethylation in early tumorgenesis is likely to have a clinical importance, because dissentient promoter methylation in tumor suppressor genes has been detected in a large percentage of human lung cancers. Noticeably the p16 gene, which expresses an inhibitor of cyclin-dependent kinase 4 and 6, consequently interrupting the cell cycle progression, is methylated at its promoter region in 20–65% of lung tumors [46]. Smoking habits alter the extent of promoter methylation patterns. Increasing methylation of p16 could be seen with increasing smoking duration, packets per year and smoking when juvenile, whereas methylation decreases gradually over time when a person quits smoking.

Similar effects could be seen in rodent lung cancer, induced by the particulate carcinogens carbon black and diesel exhaust, where the tumor cells showed an overmethylation in the p16 promoter region. Similar observations could be found in murine lung tumors, caused by cigarette smoke. Remarkable improvements could be achieved by treating rodents with histone deacetylation inhibitors in a lung cancer mouse model. A greater than 50% reduction of tumor growth was seen in treated mice.

In addition to well-known carcinogens, air pollution components and particulate matter (PM10), nickel and beryllium compounds have also been shown to have an impact on histone acetylation and/or altered DNA methylation patterns [47]. These examples already show that epigenetic mechanisms will help and strongly develop toxicogenomic approaches [48].

4.5
Dietary Effects

Dietary habits as well as a sedentary lifestyle clearly contribute to today's increasing number of chronic diseases. The influence played by numerous food compounds on the epigenetic machinery is of growing interest. The large number of established and probable epigenetic active compounds found in food will challenge the understanding of how diet may influence epigenetic gene expression. Studies of dietary effects on epigenetic gene regulation are still in their infancy, but first results have been reported [49]. Food contains compounds influencing both DNA methylation and histone modifications. However, the consumption of some of these compounds can vary by season, dietary habits, age and environment.

4.5 Dietary Effects

Dietary compounds like vitamin B_{12} or folic acid are implicated in the regulation of the cytosine methylation pathway [50–52]. Vitamin B_{12} plays a central role as it acts as a co-factor of the methionine synthetase, remethylating homocysteine to methionine. Methionine is further activated to S-adenosylmethionine (SAM), the methyl donor for DNA methylation. SAM converts to S-adenosylhomocysteine (SAH) after DNA-methylation. Reversible hydrolysis of SAH to homocysteine completes the cycle (Figure 4.2, adapted from Kim [50]). Folate, cholin or betaine are potent methyl-donors directly implicated in the DNA-methylation pathway.

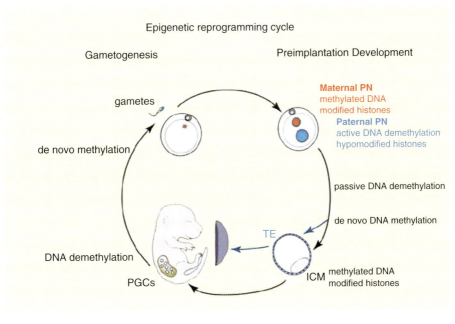

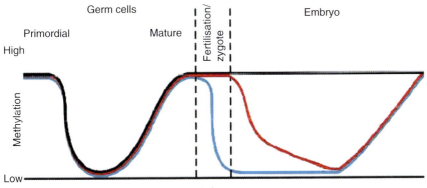

Figure 4.2 Epigenetic reprogramming cycle. Methylation on maternal and paternal DNA during gametogenesis and first steps of embryogenesis.

However, under conditions of vitamin B_{12} depletion, cellular folate accumulates in the methylfolate form due to the activity of methionine synthetase being blocked, thereby creating a conditional form of folate deficiency. Deficiency in vitamin B_{12} leads also to an accumulation of serum homocysteine. Under conditions of elevated homocysteine concentrations, levels of the potent SAM–inhibitor SAH repress DNA methylation. The role of folate in DNA-methylation seems to be rather complex. Evidence from animal, human, and *in vitro* studies suggest that folate-dependent DNA methylation is highly complex, gene and site specific [50, 53–55]. These studies have shown that the extent and direction of changes in SAM and SAH in cell lines in response to folate deficiency are cell specific and that genomic-site and gene-specific DNA demethylation are not affected by the changes in SAM and SAH induced by folate depletion.

Furthermore, transgenerational studies on a rat model demonstrated that folate deficiency during pregnancy impacts on methyl metabolism, but does not affect global DNA methylation in rat foeti [56]. A study analyzing the expression of the antioxidative enzyme mitochondrial superoxide dismutase (MnSOD) showed that this gene is expressed differently in vegetarians and omnivores due to different DNA-methylation patterns in the MnSOD promoter region: vegetarians expressed a significantly higher amount of MnSOD, showing a low promoter methylation for this gene. Inversely, a lower MnSOD expression by a higher MnSOD promoter methylation was seen in omnivores [49]. Several factors which may reduce DNA-methylation machinery in vegetarians are discussed. Studies analyzing the B-vitamin status of vegans and vegetarians compared with omnivores, showed generally a higher dietary supply of folate and a lower dietary supply of vitamin B_{12} (since vitamin B_{12} intake comes predominantly from animal food products) and methionine for vegetarians and vegans [57, 58]. Correlations between plasma folate and vitamin B_{12} values were found and study findings showed that vegetarians and vegans show higher plasma homocysteine concentrations [57]. Furthermore, DNA hypomethylation due to high homocysteine levels has been reported *in vitro* and *in vivo* [59, 60]. Inhibition of DNA methyltransferase 1 activity by 30% and reduced binding of methyl CpG binding protein 2 have also been seen in this context. A further study [61] comparing vegetarians and omnivores observed a reverse correlation between SAH concentrations and DNA global methylation levels in blood. However, this study was unable to correlate homocysteine concentrations with the degree of DNA global methylation and found no correlation between the degree of CpG methylation of the promoter of the p66Shc gene (involved in oxidative stress) and homocysteine, or SAM or SAH levels.

Other diet-derived factors such as diallyl sulfide [62], an organosulfur compound found in garlic, genisteine, the main flavanoid in soy [16], vitamin D_3 or all-*trans*-retinoic acids [63] have been shown to influence DNA methylation by altering histones and chromatin structure. On the histone level, many more food compounds show effects on epigenetic gene regulation. Food compounds acting as histone modifiers are generally weak enzyme ligands and thus are needed in high concentrations to generate a consistent effect and, therefore, might subtly regulate gene expression [64]. A number of dietary agents are considered to have a role in epigenetics as HDAC inhibitors. In the literature, well described agents in this context are butyrate, diallyl disulfide or sulforaphane. Intriguing examples of an HDAC inhibitor are conjugated linoleic acids (CLAs), which have anticancerogenic and antiatherogenic properties. Interestingly, CLAs decrease Bcl-2 and induce p21, a known HDAC target. The effects of CLAs on HDACs are discussed. Regarding the short-chain fatty acids (SCFs) resulting from the microbial fermentation of dietary fibers, acetate, propionate and butyrate are taken up by the colonic mucosa. Butyrate is the smallest known HDAC inhibitor, and supresses histone acetylation at high micromolar or low millimolar concentrations *in vitro*, levels nonetheless considered to be achievable in the gastro intestinal tract by bacterial metabolism [64]. Butyrate is the preferred energy source for colonocytes and is transported across the epithelium. Butyrate, and to a lesser extent other SCFs, influence gene expression and inflammation. . Therefore, diet compounds, toxins or medications interfering with the colonic fermentation may also modulate potential epigenetic effects generated by the microbial flora. There is now much work to be done to clarify the influence on histone modifications of a great number of dietary factors, alone or in combination, and their tissue specific characteristics. A list of discussed dietary HDAC activity modulators is shown in Table 4.2.

Evidence shows that a diet rich in vegetables and fruits may prevent some kinds of cancers [65]. In this context, the role of flavonoids is often discussed. Flavonoids can act either in a pro- or antioxidative manner depending on their structure and characters. Because of the antioxidative properties of some flavonoids, oxidative

Table 4.2 Effect of different dietary compounds on HDAC.

Dietary agent	To be found in	Effect
Butyrate	GI, generated by microbiom	Attenuation of HDACs
Diallyl disulfide	Garlic	Attenuation of HDACs
Sulforaphane	Cruciferous vegetables such as brussel sprouts, broccoli, cabbage, cauliflower	Attenuation of HDACs
CLA	Especially in meat and dairy products derived from ruminants	Attenuation of HDACs
Isothiocyanates	Mustard oil, rocket plant	Attenuation of HDACs
All-trans retinoic acid (derived from vitamin A)	Meat, liver, butter (sources of vitamin A)	Attenuation of HDACs
Theophylline	Tea	Enhance HDAC activity
Resveratrol	Red grapes, peanuts, raspberries	Enhance HDAC activity

damage of DNA can be prevented and cancerogenesis altered. Guarrera *et al.* postulated recently that flavonoids might influence the gene expression of DNA repair genes, which could be a possible explanation for the decrease in tumor development [66].

In many animal models caloric restriction has been shown to prevent the development of cancer, thereby sirtuins become of increasing interest. Resveratrol activates the HDAC sirtuin 1 (SIRT1), an HDAC belonging to class III [36]. SIRT is an abbreviation for silent mating type information regulation two and there are seven groups in humans [36]. Sirtuins are nicotineamide adenine dinucleotide (NAD(+))-dependent deacetylases which exhibit a well-defined regression during aging that is dramatically reverted in transformed cells.

4.5.1
Nutrition and the Immune System

Expression of Interleukin 8 and the IL23/IL17 pathway is pivotal in the development of chronic inflammation such as Crohn's disease (CD) or Inflammatory Bowel disease [67, 68].

IL-8, a member of the CXC chemokine family, is an important activator and chemoattractant for neutrophils and is primarily regulated at the transcriptional level. IL-8 was shown to be affected by methylation of two CpGs as well as by histone acetylation. Compounds taking part in DNA methylation and histone acetylation pathways such as folic acid, vitamin B_{12}, genistein, zebularine and valproate may influence epigenetic modification and gene expression of the IL-8 gene in human colon cancer cells.

By contrast to leukemic blasts, where IL-8 is activated by differentiation [69], immune-activated transcription of IL-8 gene may be silenced after differentiation of cells from solid tumors such as breast cancer [70]. Mechanisms may involve histone deacetylation by elements located outside the immediate 5'flanking region.

The commensal microflora regulates the local expansion of CD4 T cells producing proinflammatory cytokines including IL-17 (Th17 cells) in the colonic lamina propria [71].TH cells, which produce IL-17 and IL-17F, two highly homologous cytokines whose genes are located in the same chromosomal region have been recently identified to promote tissue inflammation. In these cells Histone H3 acetylation and Lysine 4 tri-methylation were specifically associated with the activity of the IL-17 and IL-17F gene promoters [72].

As natural food compounds and diets have been shown to modify the epigenetic regulation of the expression of the discussed mediators [73, 74] dietary strategies might be a possibility in modulating inflammatory diseases.

4.5.2
Nutrition and Aging

Hereditary disposition, environmental elements, individual lifestyle and nutritional factors are known to interact with the aging process.

The balance between genetic and epigenetic impacts in the development of malignancy seems to change from childhood to later age. Whereas, for example, the majority of childhood tumors are associated with an inherited genetic or epigenetic (e.g., imprinted) burden, this balance shifts in favor of acquired epigenetic and genetic hits in tumors of adults and the elderly [75] (Figure 4.3).

The impacts on the aging-process can be divided into intrinsic and extrinsic factors. Intrinsic aging (cellular aging) mainly depends on the individual hereditary background. Extrinsic aging is generated by external factors like smoking, excessive alcohol consumption and poor nutrition. During aging, epigenetic changes occur, for instance global DNA methylation decreases in contrast to a

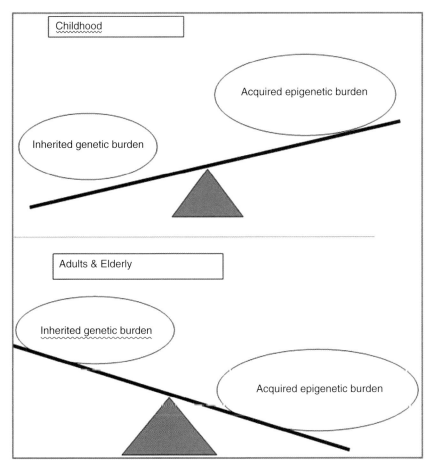

Figure 4.3 The balance between genetic and epigenetic impacts in development of malignancy is changing from childhood to later age. Whereas the majority of childhood tumors are associated with an inherited genetic or epigenetic (e.g., imprinted) burden, this balance shifts in favor of acquired epigenetic and genetic hits in tumors of adults and elderly.

CpG island hypermethylation. Progressive loss of global methylation during aging caused by passive demethylation is probably due to increasing DNMT1 enzyme inefficiency. Epigenetic changes can be seen as important determinants of cellular senescence and organism aging. Sirtuins take part in various cellular developments like chromatin remodelling, mitosis, and life-span duration and are, furthermore, involved in the regulation of gene expression, insulin secretion, and DNA repair. Since sirtuins control the activity of many proteins involved in cell growth, it was suggested that sirtuins are involved in the elongation of life-span mediated by caloric restriction. Indeed, in mammals, caloric restriction decreases the incidence or delays the onset of age-associated diseases including cardiovascular diseases, cancer, and osteoporosis, as well as neurodegenerative diseases. SIRT1 is known to be down regulated in senescent cells and during aging. Transcription factors like p53, p73, NFκB, E2F1, p300 and others are also shown to be SIRT1 substrates, suggesting an oncogenetic potential for SIRT 1. In an ongoing process more and more sirtuin inhibitors are being discovered. Recent detections are sirtinol, cambinol, dihydrocoumarin and indole, coumarin and indoles being natural food compounds. Coumarin can be found in dates, dihydrocoumarin naturally in sweet clover and synthetically manufactured in foods and cosmetics. Indole is produced by intestinal bacteria as a product of tryptophan degradation [76, 77]. Early findings about this novel class of HDACs are encouraging. Even more intensive research will be necessary to clarify the connections between sirtuins, nutrition, aging and cancer genesis.

4.6
Inheritance and Evolutionary Aspects

Epigenetic mechanisms contribute to the control of gene expression as a result of environmental signals due to their inherent malleability. Considerable evidence suggests that nutritional imbalance and metabolic disturbances during critical time windows of development may have a persistent effect on the health and may even be transmitted to the next generation. For example, in addition to a "thrifty genotype" inheritance, individuals with obesity, type 2 diabetes, and metabolic syndrome may have suffered improper "epigenetic programming" during their fetal/postnatal development due to maternal inadequate nutrition and metabolic disturbances and also during their lifetime. This epigenetic misprogramming can even be transmitted to the next generation(s), (Asim K Duttaroy, Evolution, Epigenetics, and Maternal Nutrition, Darwins Day Celebration, 2006).

A model for transgenerational epigenetic inheritance explains that epigenetic marks at alleles seem to be resistant to demethylation during gametogenesis, and therefore the epigenetic state of the allele in the gamete correlates, to a considerable extent, with the phenotype of the individual. Then, after fertilization, the marks are not cleared completely, and after stochastic reestablishment there would be some memory of the epigenetic state that existed in the gametes of the parent [78–80].

Presently more than 100 examples with transgenerational inheritance are known, where these examples may presently just represent the tip of an iceberg (Jablonka, Vienna, Darwin lecture, 2009 and in print).

Increased knowledge of these epigenetic mechanisms will also improve our understanding of the mechanisms of evolution. As predicted by Rupert Riedl working with the so-called "General Systems Theory" the evolution of complex adaptation requires a match between the functional relationships of the phenotypic characters and their genetic representation [61]. According to Riedl "evolvability" results from such a match. If the epigenetic regulation of gene expression "imitates" the functional organization of the traits then the improvement by mutation and selection is facilitated. Riedl predicts that the evolution of the genetic representation of phenotypic characters tends to favor those representations which imitate the functional organization of the characters. Imitation means that complexes of functionally related characters shall be "coded" as developmentally integrated characters but coded independently of functionally distinct character complexes [81]. Epigenetic mechanisms thus appear as a machinery of recursive feedback which determines the average rate of change of characters at all levels from molecules to cells, organisms and their environment [75]. The newly established existence of inherited epigenetic variation may alter our thinking about soft inheritance and evolutionary change [82]. In synergy with other elements of modern synthetic theory of evolution such as EvoDevo, epigenetic inheritance might further help to explain environmental influences, speed and discontinuity in evolution.

4.7
Conclusion

The WHO report "Genomics and World Health (WHO, Geneva, 2002) underlined that "Except for genetic diseases that result from a single defective gene, most common diseases result from environmental factors, together with variations in individual susceptibility, which reflect the action of several genes. Further research should lead to the discovery of specific molecular targets for therapy, provide information that will allow treatment to be tailored to individual needs, and, in the longer-term, generate a new approach to preventive medicine based on genetic susceptibility to environmental hazards." The translation of public health genomics and genetic testing into individualized preventive health care will be an extraordinarily important aspect of modern developments in biomedicine. However, these developments will need to include epigenetic aspects in tests and concepts.

Because of the increasing evidence for epigenetic consequences of many pollutants and toxins, epigenetic evaluations will also need to be taken seriously in toxicology and environmental health. In contrast, diets and food ingredients with better defined consequences on epigenetic mechanisms will become an attractive strategy for chemoprevention, or even therapy, in many fields.

We thank the Austrian Federal Ministry for Health for support and Dr. Judith Schuster for critical reading.

References

1 Nothnagel, M., Ellinghaus, D., Schreiber, S., Krawczak, M., and Franke, A. (2008) A comprehensive evaluation of SNP genotype imputation. *Hum. Genet.*, **151**(2), 163–171.

2 Haiman, C.A., and Stram, D.O. (2008) Utilizing HapMap and tagging SNPs. *Methods Mol. Med.*, **141**, 37–54.

3 Taufer, M., Peres, A., de Andrade, V.M., de Oliveira, G., Sa, G., do Canto, M.E., dos Santos, A.R., Bauer, M.E., and da Cruz, I.B. (2005) Is the Val16Ala manganese superoxide dismutase polymorphism associated with the aging process? *J. Gerontol. A Biol. Sci. Med. Sci.*, **60**, 432–438.

4 Kang, D., Lee, K.M., Park, S.K., Berndt, S.I., Peters, U., Reding, D., Chatterjee, N., Welch, R., Chanock, S., Huang, W.Y., and Hayes, R.B. (2007) Functional variant of manganese superoxide dismutase (SOD2 V16A) polymorphism is associated with prostate cancer risk in the prostate, lung, colorectal, and ovarian cancer study. *Cancer Epidemiol. Biomarkers Prev.*, **16**, 1581–1586.

5 Wiener, H.W., Perry, R.T., Chen, Z., Harrell, L.E., and Go, R.C. (2007) A polymorphism in SOD2 is associated with development of Alzheimer's disease. *Genes Brain Behav.*, **6**(8), 770–775.

6 Miyaki, K., Takahashi, Y., Song, Y., Zhang, L., Muramatsu, M., and Nakayama, T. (2008) Increasing the number of SNP loci does not necessarily improve prediction power at least in the comparison of MTHFR SNP and haplotypes. *J. Epidemiol.*, **18**, 243–250.

7 Boccia, S., Boffetta, P., Brennan, P., Ricciardi, G., Gianfagna, F., Matsuo, K., van Duijn, C.M., and Hung, R.J. (2009) Meta-analyses of the methylenetetrahydrofolate reductase C677T and A1298C polymorphisms and risk of head and neck and lung cancer. *Cancer Lett.*, **273**, 55–61.

8 Boccia, S., Hung, R., Ricciardi, G., Gianfagna, F., Ebert, M.P., Fang, J.Y., Gao, C.M., Gotze, T., Graziano, F., Lacasana-Navarro, M., Lin, D., Lopez-Carrillo, L., Qiao, Y.L., Shen, H., Stolzenberg-Solomon, R., Takezaki, T., Weng, Y.R., Zhang, F.F., van Duijn, C.M., Boffetta, P., and Taioli, E. (2008) Meta- and pooled analyses of the methylenetetrahydrofolate reductase C677T and A1298C polymorphisms and gastric cancer risk: a huge-GSEC review. *Am. J. Epidemiol.*, **167**, 505–516.

9 Janssens, A.C., Gwinn, M., Bradley, L.A., Oostra, B.A., van Duijn, C.M., and Khoury, M.J. (2008) A critical appraisal of the scientific basis of commercial genomic profiles used to assess health risks and personalize health interventions. *Am. J. Hum. Genet.*, **82**, 593–599.

10 Vineis, P., Veglia, F., Garte, S., Malaveille, C., Matullo, G., Dunning, A., Peluso, M., Airoldi, L., Overvad, K., Raaschou-Nielsen, O., Clavel-Chapelon, F., Linseisen, J.P., Kaaks, R., Boeing, H., Trichopoulou, A., Palli, D., Crosignani, P., Tumino, R., Panico, S., Bueno-De-Mesquita, H.B., Peeters, P.H., Lund, E., Gonzalez, C.A., Martinez, C., Dorronsoro, M., Barricarte, A., Navarro, C., Quiros, J.R., Berglund, G., Jarvholm, B., Day, N.E., Key, T.J., Saracci, R., Riboli, E., and Autrup, H. (2007) Genetic susceptibility according to three metabolic pathways in cancers of the lung and bladder and in myeloid leukemias in nonsmokers. *Ann. Oncol.*, **18**, 1230–1242.

11 Tsugane, S., and Sasazuki, S. (2007) Diet and the risk of gastric cancer: review of epidemiological evidence. *Gastric Cancer*, **10**, 75–83.

12 Johanning, G.L., Heimburger, D.C., and Piyathilake, C.J. (2002) DNA methylation and diet in cancer. *J. Nutr.*, **132**, 3814S–3818S.

13 Mann, N.J., Li, D., Sinclair, A.J., Dudman, N.P., Guo, X.W., Elsworth, G.R., Wilson,

A.K., and Kelly, F.D. (1999) The effect of diet on plasma homocysteine concentrations in healthy male subjects. *Eur. J. Clin. Nutr.*, **53**, 895–899.

14 Jirtle, R.L., and Skinner, M.K. (2007) Environmental epigenomics and disease susceptibility. *Nat. Rev. Genet.*, **8**, 253–262.

15 Anway, M.D., Memon, M.A., Uzumcu, M., and Skinner, M.K. (2006) Transgenerational effect of the endocrine disruptor vinclozolin on male spermatogenesis. *J. Androl.*, **27**, 868–879.

16 Dolinoy, D.C., Weidman, J.R., Waterland, R.A., and Jirtle, R.L. (2006) Maternal genistein alters coat color and protects Avy mouse offspring from obesity by modifying the fetal epigenome. *Environ Health Perspect.*, **114**, 567–572.

17 Kaati, G., Bygren, L.O., and Edvinsson, S. (2002) Cardiovascular and diabetes mortality determined by nutrition during parents' and grandparents' slow growth period. *Eur. J. Hum. Genet.*, **10**, 682–688.

18 Fraga, M.F., Ballestar, E., Paz, M.F., Ropero, S., Setien, F., Ballestar, M.L., Heine-Suner, D., Cigudosa, J.C., Urioste, M., Benitez, J., Boix-Chornet, M., Sanchez-Aguilera, A., Ling, C., Carlsson, E., Poulsen, P., Vaag, A., Stephan, Z., Spector, T.D., Wu, Y.Z., Plass, C., and Esteller, M. (2005) Epigenetic differences arise during the lifetime of monozygotic twins. *Proc. Natl. Acad. Sci. U S A*, **102**, 10604–10609.

19 Morgan, H.D., Santos, F., Green, K., Dean, W., and Reik, W. (2005) Epigenetic reprogramming in mammals. *Hum. Mol. Genet.*, **14** Spec No 1, R47–R58.

20 Dolinoy, D.C., Das, R., Weidman, J.R., and Jirtle, R.L. (2007) Metastable epialleles, imprinting, and the fetal origins of adult diseases. *Pediatr. Res.*, **61**, R30–R37.

21 Costantini, M., Clay, O., Auletta, F., and Bernardi, G. (2006) An isochore map of human chromosomes. *Genome. Res.*, **16**, 536–541.

22 Bestor, T.H. (2000) The DNA methyltransferases of mammals. *Hum. Mol. Genet.*, **9**, 2395–2402.

23 Klose, R.J., and Bird, A.P. (2006) Genomic DNA methylation: the mark and its mediators. *Trends Biochem. Sci.*, **31**, 89–97.

24 De Larco, J.E., Wuertz, B.R., Yee, D., Rickert, B.L., and Furcht, L.T. (2003) Atypical methylation of the interleukin-8 gene correlates strongly with the metastatic potential of breast carcinoma cells. *Proc. Natl. Acad. Sci. U S A*, **100**, 13988–13993.

25 Schaefer, C.B., Ooi, S.K., Bestor, T.H., and Bourc'his, D. (2007) Epigenetic decisions in mammalian germ cells. *Science*, **316**, 398–399.

26 Dean, W., Santos, F., and Reik, W. (2003) Epigenetic reprogramming in early mammalian development and following somatic nuclear transfer. *Semin. Cell Dev. Biol.*, **14**, 93–100.

27 Clouaire, T., and Stancheva, I. (2008) Methyl-CpG binding proteins: specialized transcriptional repressors or structural components of chromatin? *Cell Mol. Life Sci.*, **65**, 1509–1522.

28 Simonsson, S., and Gurdon, J. (2004) DNA demethylation is necessary for the epigenetic reprogramming of somatic cell nuclei. *Nat. Cell Biol.*, **6**, 984–990.

29 Hutchins, A.S., Mullen, A.C., Lee, H.W., Sykes, K.J., High, F.A., Hendrich, B.D., Bird, A.P., and Reiner, S.L. (2002) Gene silencing quantitatively controls the function of a developmental trans-activator. *Mol. Cell*, **10**, 81–91.

30 He, H., and Lehming, N. (2003) Global effects of histone modifications. *Brief Funct. Genomic Proteomic*, **2**, 234–243.

31 Shahbazian, M.D., and Grunstein, M. (2007) Functions of site-specific histone acetylation and deacetylation. *Annu. Rev. Biochem.*, **76**, 75–100.

32 Cheung, P., and Lau, P. (2005) Epigenetic regulation by histone methylation and histone variants. *Mol. Endocrinol.*, **19**, 563–573.

33 Turner, B.M. (2000) Histone acetylation and an epigenetic code. *Bioessays.*, **22**, 836–845.

34 Turner, B.M. (2002) Cellular memory and the histone code. *Cell*, **111**, 285–291.

35 Fraga, M.F., Ballestar, E., Villar-Garea, A., Boix-Chornet, M., Espada, J., Schotta, G., Bonaldi, T., Haydon, C., Ropero, S.,

Petrie, K., Iyer, N.G., Perez-Rosado, A., Calvo, E., Lopez, J.A., Cano, A., Calasanz, M.J., Colomer, D., Piris, M.A., Ahn, N., Imhof, A., Caldas, C., Jenuwein, T., and Esteller, M. (2005) Loss of acetylation at Lys16 and trimethylation at Lys20 of histone H4 is a common hallmark of human cancer. *Nat. Genet.*, **37**, 391–400.

36 Davis, C.D., and Ross, S.A. (2007) Dietary components impact histone modifications and cancer risk. *Nutr. Rev.*, **65**, 88–94.

37 Guil, S., and Esteller, M. (2009) DNA methylomes, histone codes and miRNAs: tying it all together. *Int. J. Biochem. Cell Biol.*, **41**, 87–95.

38 Anway, M.D., Memon, M.A., Uzumcu, M., and Skinner, M.K. (2006) Transgenerational effect of the endocrine disruptor vinclozolin on male spermatogenesis. *J. Androl.*, **27**, 868–879.

39 Heijmans, B.T., Tobi, E.W., Stein, A.D., Putter, H., Blauw, G.J., Susser, E.S., Slagboom, P.E., and Lumey, L.H. (2008) Persistent epigenetic differences associated with prenatal exposure to famine in humans. *Proc. Natl. Acad. Sci. U S A*, **105**, 17046–17049.

40 Gluckman, P.D., Lillycrop, K.A., Vickers, M.H., Pleasants, A.B., Phillips, E.S., Beedle, A.S., Burdge, G.C., and Hanson, M.A. (2007) Metabolic plasticity during mammalian development is directionally dependent on early nutritional status. *Proc. Natl. Acad. Sci. U S A*, **104**, 12796–12800.

41 Darnaudery, M., and Maccari, S. (2008) Epigenetic programming of the stress response in male and female rats by prenatal restraint stress. *Brain Res. Rev.*, **57**, 571–585.

42 Henry, C., Kabbaj, M., Simon, H., Le Moal, M., and Maccari, S. (1994) Prenatal stress increases the hypothalamo-pituitary-adrenal axis response in young and adult rats. *J. Neuroendocrinol.*, **6**, 341–345.

43 Zuena, A.R., Mairesse, J., Casolini, P., Cinque, C., Alema, G.S., Morley-Fletcher, S., Chiodi, V., Spagnoli, L.G., Gradini, R., Catalani, A., Nicoletti, F., and Maccari, S. (2008) Prenatal restraint stress generates two distinct behavioral and neurochemical profiles in male and female rats. *PLoS ONE*, **3**, e2170.

44 Mill, J., Tang, T., Kaminsky, Z., Khare, T., Yazdanpanah, S., Bouchard, L., Jia, P., Assadzadeh, A., Flanagan, J., Schumacher, A., Wang, S.C., and Petronis, A. (2008) Epigenomic profiling reveals DNA-methylation changes associated with major psychosis. *Am. J. Hum. Genet.*, **82**, 696–711.

45 Satta, R., Maloku, E., Zhubi, A., Pibiri, F., Hajos, M., Costa, E., and Guidotti, A. (2008) Nicotine decreases DNA methyltransferase 1 expression and glutamic acid decarboxylase 67 promoter methylation in GABAergic interneurons. *Proc. Natl. Acad. Sci. U S A*, **105**, 16356–16361.

46 Hayslip, J., and Montero, A. (2006) Tumor suppressor gene methylation in follicular lymphoma: a comprehensive review. *Mol. Cancer*, **5**, 44.

47 Vineis, P., Hoek, G., Krzyzanowski, M., Vigna-Taglianti, F., Veglia, F., Airoldi, L., Autrup, H., Dunning, A., Garte, S., Hainaut, P., Malaveille, C., Matullo, G., Overvad, K., Raaschou-Nielsen, O., Clavel-Chapelon, F., Linseisen, J., Boeing, H., Trichopoulou, A., Palli, D., Peluso, M., Krogh, V., Tumino, R., Panico, S., Bueno-De-Mesquita, H.B., Peeters, P.H., Lund, E.E., Gonzalez, C.A., Martinez, C., Dorronsoro, M., Barricarte, A., Cirera, L., Quiros, J.R., Berglund, G., Forsberg, B., Day, N.E., Key, T.J., Saracci, R., Kaaks, R., and Riboli, E. (2006) Air pollution and risk of lung cancer in a prospective study in Europe. *Int. J. Cancer*, **119**, 169–174.

48 Reamon-Buettner, S.M., Mutschler, V., and Borlak, J. (2008) The next innovation cycle in toxicogenomics: environmental epigenetics. *Mutat. Res.*, **659**, 158–165.

49 Thaler, R., Karlic, H., Rust, P., and Haslberger, A.G. (2009) Epigenetic regulation of human buccal mucosa mitochondrial superoxide dismutase gene expression by diet. *Br. J. Nutr.*, **101**(5), 743–749.

50 Kim, Y.I. (2004) Folate and DNA methylation: a mechanistic link between folate deficiency and colorectal cancer? *Cancer Epidemiol. Biomarkers Prev.*, **13**, 511–519.

51 Hayashi, I., Sohn, K.J., Stempak, J.M., Croxford, R., and Kim, Y.I. (2007) Folate deficiency induces cell-specific changes in the steady-state transcript levels of genes involved in folate metabolism and 1-carbon transfer reactions in human colonic epithelial cells. *J. Nutr.*, **137**(3), 607–613.

52 Choi, S.W., Friso, S., Ghandour, H., Bagley, P.J., Selhub, J., and Mason, J.B. (2004) Vitamin B-12 deficiency induces anomalies of base substitution and methylation in the DNA of rat colonic epithelium. *J. Nutr.*, **134**, 750–755.

53 Cravo, M., Fidalgo, P., Pereira, A.D., Gouveia-Oliveira, A., Chaves, P., Selhub, J., Mason, J.B., Mira, F.C., and Leitao, C.N. (1994) DNA methylation as an intermediate biomarker in colorectal cancer: modulation by folic acid supplementation. *Eur. J. Cancer Prev.*, **3**, 473–479.

54 Kim, Y.I., Baik, H.W., Fawaz, K., Knox, T., Lee, Y.M., Norton, R., Libby, E., and Mason, J.B. (2001) Effects of folate supplementation on two provisional molecular markers of colon cancer: a prospective, randomized trial. *Am. J. Gastroenterol.*, **96**, 184–195.

55 Rampersaud, G.C., Kauwell, G.P., Hutson, A.D., Cerda, J.J., and Bailey, L.B. (2000) Genomic DNA methylation decreases in response to moderate folate depletion in elderly women. *Am. J. Clin. Nutr.*, **72**, 998–1003.

56 Maloney, C.A., Hay, S.M., and Rees, W.D. (2007) Folate deficiency during pregnancy impacts on methyl metabolism without affecting global DNA methylation in the rat fetus. *Br. J. Nutr.*, **97**, 1090–1098.

57 Majchrzak, D., Singer, I., Manner, M., Rust, P., Genser, D., Wagner, K.H., and Elmadfa, I. (2006) B-vitamin status and concentrations of homocysteine in Austrian omnivores, vegetarians and vegans. *Ann. Nutr. Metab.*, **50**, 485–491.

58 Millet, P., Guilland, J.C., Fuchs, F., and Klepping, J. (1989) Nutrient intake and vitamin status of healthy French vegetarians and nonvegetarians. *Am. J. Clin. Nutr.*, **50**, 718–727.

59 Jamaluddin, M.S., Chen, I., Yang, F., Jiang, X., Jan, M., Liu, X., Schafer, A.I., Durante, W., Yang, X., and Wang, H. (2007) Homocysteine inhibits endothelial cell growth via DNA hypomethylation of the cyclin A gene. *Blood*, **110**(10), 3648–3655.

60 Jiang, Y., Sun, T., Xiong, J., Cao, J., Li, G., and Wang, S. (2007) Hyperhomocysteinemia-mediated DNA hypomethylation and its potential epigenetic role in rats. *Acta. Biochim. Biophys. Sin. (Shanghai)*, **39**, 657–667.

61 Geisel, J., Schorr, H., Bodis, M., Isber, S., Hubner, U., Knapp, J.P., Obeid, R., and Herrmann, W. (2005) The vegetarian lifestyle and DNA methylation. *Clin. Chem. Lab. Med.*, **43**, 1164–1169.

62 Mathers, J.C. (2006) Nutritional modulation of ageing: genomic and epigenetic approaches. *Mech. Ageing. Dev.*, **127**, 584–589.

63 Krawczyk, B., and Fabianowska-Majewska, K. (2006) Alteration of DNA methylation status in K562 and MCF-7 cancer cell lines by nucleoside analogues. *Nucleosides Nucleotides Nucleic Acids*, **25**, 1029–1032.

64 Dashwood, R.H., Myzak, M.C., and Ho, E. (2006) Dietary HDAC inhibitors: time to rethink weak ligands in cancer chemoprevention? *Carcinogenesis*, **27**, 344–349.

65 World Cancer Research Fund/American Institute for Cancer Research (2007) *Food, Nutrition, Physical Activity, and the Prevention of Cancer: A Global Perspective*, AICR, Washington DC.

66 Guarrera, S., Sacerdote, C., Fiorini, L., Marsala, R., Polidoro, S., Gamberini, S., Saletta, F., Malaveille, C., Talaska, G., Vineis, P., and Matullo, G. (2007) Expression of DNA repair and metabolic genes in response to a flavonoid-rich diet. *Br. J. Nutr.*, **98**, 525–533.

67 Li, J., Moran, T., Swanson, E., Julian, C., Harris, J., Bonen, D.K., Hedl, M., Nicolae, D.L., Abraham, C., and Cho, J.H. (2004) Regulation of IL-8 and IL-1beta expression in Crohn's disease associated NOD2/CARD15 mutations. *Hum. Mol. Genet*, **13**, 1715–1725.

68 McGovern, D.P., Rotter, J.I., Mei, L., Haritunians, T., Landers, C., Derkowski,

C., Dutridge, D., Dubinsky, M., Ippoliti, A., Vasiliauskas, E., Mengesha, E., King, L., Pressman, S., Targan, S.R., and Taylor, K.D. (2009) Genetic epistasis of IL23/IL17 pathway genes in Crohn's disease. *Inflamm. Bowel. Dis.*, **15**(6), 883–889.

69 Delaunay, J., Lecomte, N., Bourcier, S., Qi, J., Gadhoum, Z., Durand, L., Chomienne, C., Robert-Lezenes, J., and Smadja-Joffe, F. (2008) Contribution of GM-CSF and IL-8 to the CD44-induced differentiation of acute monoblastic leukemia. *Leukemia*, **22**, 873–876.

70 Chavey, C., Muhlbauer, M., Bossard, C., Freund, A., Durand, S., Jorgensen, C., Jobin, C., and Lazennec, G. (2008) Interleukin-8 expression is regulated by histone deacetylases through the nuclear factor-kappaB pathway in breast cancer. *Mol. Pharmacol.*, **74**, 1359–1366.

71 Niess, J.H., Leithauser, F., Adler, G., and Reimann, J. (2008) Commensal gut flora drives the expansion of proinflammatory CD4 T cells in the colonic lamina propria under normal and inflammatory conditions. *J. Immunol.*, **180**, 559–568.

72 Akimzhanov, A.M., Yang, X.O., and Dong, C. (2007) Chromatin remodeling of interleukin-17 (IL-17)-IL-17F cytokine gene locus during inflammatory helper T cell differentiation. *J. Biol. Chem.*, **282**, 5969–5972.

73 Romier, B., Van De Walle, J., During, A., Larondelle, Y., and Schneider, Y.J. (2008) Modulation of signalling nuclear factor-kappaB activation pathway by polyphenols in human intestinal Caco-2 cells. *Br. J. Nutr.*, **100**, 542–551.

74 Kikuno, N., Shiina, H., Urakami, S., Kawamoto, K., Hirata, H., Tanaka, Y., Majid, S., Igawa, M., and Dahiya, R. (2008) Genistein mediated histone acetylation and demethylation activates tumor suppressor genes in prostate cancer cells. *Int. J. Cancer*, **123**, 552–560.

75 Haslberger, A., Varga, F., and Karlic, H. (2006) Recursive causality in evolution: a model for epigenetic mechanisms in cancer development. *Med. Hypotheses*, **67**, 1448–1454.

76 Akkermans, L.M.A., Kleerebezem, M., Reid, G., Rijkers, G., Timmerman, H., and Vos, W.M.D. (2007) Gut microbiota in health and disease. Report of the 2nd International Workshop in Amsterdam, pp. 1–14.

77 Kwon, H.J., Owa, T., Hassig, C.A., Shimada, J., and Schreiber, S.L. (1998) Depudecin induces morphological reversion of transformed fibroblasts via the inhibition of histone deacetylase. *Proc. Natl. Acad. Sci. U S A*, **95**, 3356–3361.

78 Jablonka, E. (2004) Epigenetic epidemiology. *Int. J. Epidemiol.*, **33**, 929–935.

79 Rakyan, V.K., and Beck, S. (2006) Epigenetic variation and inheritance in mammals. *Curr. Opin. Genet. Dev.*, **16**, 573–577.

80 Rakyan, V.K., Chong, S., Champ, M.E., Cuthbert, P.C., Morgan, H.D., Luu, K.V., and Whitelaw, E. (2003) Transgenerational inheritance of epigenetic states at the murine Axin(Fu) allele occurs after maternal and paternal transmission. *Proc. Natl. Acad. Sci. U S A*, **100**, 2538–2543.

81 Wagner, G.P., and Laubichler, M.D. (2004) Rupert Riedl and the re-synthesis of evolutionary and developmental biology: body plans and evolvability. *J. Exp. Zoolog. B. Mol. Dev. Evol.*, **302**, 92–102.

82 Richards, E.J. (2006) Inherited epigenetic variation – revisiting soft inheritance. *Nat. Rev. Genet.*, **7**, 395–401.

Part II
Hereditary Aspects

5
Methylenetetrahydrofolate Reductase C677T and A1298C Polymorphisms and Cancer Risk: A Review of the Published Meta-Analyses

Stefania Boccia

Abstract

The present chapter aims to illustrate some key concepts of genetic association studies, with an emphasis on meta-analysis as a powerful study design to summarize quantitatively the results of the scientific literature. Additionally, a systematic review of the published meta-analyses of the effect of methylenetetrahydrofolate reductase (*MTHFR*) gene polymorphisms on cancer risk is reported. MTHFR plays a central role in folate metabolism by irreversibly catalyzing the conversion of 5,10-methylenetetrahydrofolate to 5-methyltetrahydrofolate, the primary circulating form of folate and a cosubstrate for homocysteine methylation to methionine. The C677T polymorphism has been investigated in relation to cancer, as the *MTHFR* TT genotype is related to DNA hypomethylation, particularly in individuals with reduced plasma folate concentrations. The results of the present review of the published meta-analyses support a protective effect of the 677 TT genotype on colorectal cancer and adult lymphoblastic leukemia, while an increased risk for gastric cancer, with the former probably resulting from the increased levels of the MTHFR substrate (which is essential for DNA synthesis) due to the variant allele, and the latter from impaired folate levels (acid folic protects from cancer by limiting aberrant DNA methylation). Additional large studies collecting data on folate serum levels might help to clarify the result of the complex interaction of dietary folate intake and MTHFR genotype on cancer risk.

5.1
Key Concepts of Population-Based Genetic Association Studies

5.1.1
Definition and Goals of Genetic Epidemiology

Genetic epidemiology faces the challenge of understanding and compiling evidence for the contribution of genetic risk in common human diseases. It is a discipline closely related to traditional epidemiology, however, it focuses on the genetic determinants of diseases and the joint effects of genes and non-genetic determinants [1]. As Shpilberg and colleagues optimistically stated, "The sequencing of the human genome offers the greatest opportunity for epidemiology since John Snow discovered the Broad Street pump" [2]. Although in the past great successes have been obtained for monogenic disorders following the laws of Mendelian inheritance by classic linkage studies, genetic epidemiology today focuses increasingly on complex disorders such as cancer, ischemic heart disease, asthma and diabetes mellitus [3–9]. In this context, population-based genetic association studies aim to quantify the magnitude of the association between one or more genetic polymorphisms and a disease trait in an identified population. Currently, it is believed that an individual genetic susceptibility to common complex disorders probably involves many genes, although their effects may only be small. Thus, in order to detect these small effects large sample sizes are required. However, the combination of even a few small effects [Odds Ratio (OR) <2.0] could account for a sizeable population attributable fraction of common diseases, shedding new light on disease etiology and environmental determinants. This aspect, together with the identification of large numbers of single nucleotide polymorphisms (SNPs) throughout the genome, have led to acknowledgment of the importance of association studies in genetic epidemiology [10].

5.1.2
Study Designs in Genetic Epidemiology

Study designs adopted for population-based genetic association research are identical to those used in traditional epidemiology [11]. The most commonly used study design is the case-control. One of the main reasons for the general use of the case-control study design in genetic association studies is that "recall bias", which usually afflicts poorly designed case-control studies, does not affect the main variable of interest (the genotype) in genetic studies. In fact, the genotype is not supposed to change over time, therefore, unlike observational studies, genetic studies are not affected by "reverse causation". The primary aim of many genetic association studies, however, is to evaluate the interaction between a gene and an environmental exposure, which means that cohort studies usually have the advantage in that they are less likely to misclassify environmental exposure. Lastly, population stratification, which results from the mixing of different ethnic groups (with different allele frequencies), usually represents the main source of bias in genetic association studies. In fact, in this situation confounding arises since the ethnicity itself might be related to a specific disease as well as the allele frequencies, however, it is easily controlled by stratification for ethnicity [10].

5.1.3
The Human Genome

The haploid genome is about 3.3 billion bp, with just 30 000–40 000 protein-coding genes [12]. Unrelated individuals share about 99.99% of their genome, however the DNA sequence may vary between two versions of the same chromosome in many ways. Among the most important are the SNPs, which represent a variation in a single nucleotide with a frequency in the population higher than 1% [13]. With several million SNPs present in the human genome, the number of possible genetic associations that can be tested is almost unlimited [8]. Considering the possible combinations of SNPs, disease and outcomes, it has been estimated that there are around a trillion association analyses that could be performed [14]. In this scenario, it is clear that the high chance of an initial "statistically significant" finding turns out to be a false-positive finding, even for a large and well-designed study, so that replication in this context is demanded. Finally, emerging technologies with high-throughput genotyping allow researchers to study hundreds of thousands of SNPs simultaneously, so that the task of understanding the role played by human genetic variation in complex diseases is becoming daunting [15].

5.1.4
Meta-Analysis in Genetic Epidemiology

The current scenario of published studies in genetic epidemiology, however, is characterized by the prevalence of underpowered studies with limited sample sizes, often with flawed study design and biased analyses; the selective reporting of "positive" findings (publication bias); the lack of standardization among studies and difficulties in assessing interactions with environmental risk factors [16]. In this context, the creation of networks of researchers studying the same disease, in which they share both their knowledge and their genetic data as well as use the same methodology, is warranted [16]. In the meantime, the use of meta-analyses might help in clarifying the weight of a specific polymorphism on a certain disease. Currently, meta-analyses are the most cited study design in health sciences [17] and are widely accepted as the highest level of evidence in medicine. They can be defined as the systematic and rigorous quantitative integration of information on the same research question [18]. A meta-analysis goes beyond a literature review, by synthesizing the results of the individual studies into a new result [19]. It also differs from a "pooled analysis" because in the analysis they summarize the results of the previous studies, not the results from individual subjects. Initially adopted for summarizing results from clinical trials, meta-analyses were then widely applied to observational studies including, more recently, their utilization in the field of genetic epidemiology. Since the first publication appeared in 1990 [20], more than 400 meta-analyses of genetic epidemiology have been published, particularly in the last few years, as reported in Figure 5.1.

However, the growth of meta-analysis has not gone unchallenged, even in the field of genetic medicine. Combining data could improve statistical power when several small studies on the same question are present, but simply putting problematic data together will not overcome their problems [22]. Studies might reach different conclusions depending on their quality, and "negative" studies might remain unpublished, so that the result of a meta-analysis might be as strongly biased as a single study. In this context, as Ioannidis states "beside getting summary estimates, meta-analysis is probably more useful for listing and dissect-

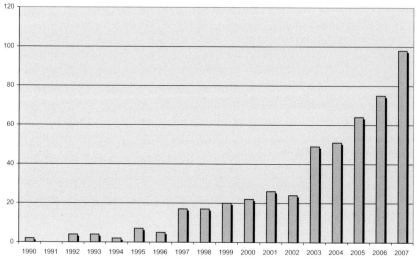

Figure 5.1 Trend of published meta-analysis in genetic epidemiology since the first, published in 1990, until 2007 [21].

ing sources of bias, quantifying heterogeneity, and proposing some potential explanations for dissecting genuine heterogeneity from bias" [22]. The reliability of the results from meta-analysis, however, depends also on rigorous methodology. Published meta-analyses should: clearly define the research question; state the sources and literature search strategy; state the studies inclusion/exclusion criteria adopted; assess and explain any heterogeneity among the studies; pool the data according to appropriate statistical methods; perform subgroup analysis (mainly by ethnicity and quality); and investigate publication bias. A recently published paper reviews all the published meta-analyses in the genetic epidemiology field, and highlights their poor methodological quality [23]. The authors conclude the paper by encouraging adherence to traditional meta-analyses guidelines for the meta-analyst of population-based genetic association studies and show that, besides the clear problem of publication bias, population stratification and the need to pool results in a way that reflects the underlying biology of the genetic effects deserve special attention [24].

5.1.5
Human Genome Epidemiology Network

From this introduction, it should be clear that genetic epidemiology benefits from a large-scale population based approach to identify genes underlying common diseases, to assess associations between genetic variants and disease susceptibility and to examine potential gene–environment interactions [25]. In this context, networks can support studies with sample sizes large enough to achieve "definitive" results and yield results that can be translated into public health application [26]. The Human Genome Epidemiology Network (HuGENet), established in 1998 as an open and global collaboration of organizations from different field of science [27], recently launched a global network of consortia working on human genome epidemiology [28]. This Network of Investigator Networks aims to create resources for sharing information, offering methodological support and facilitating rapid confirmation of findings.

5.1.6
"Mendelian Randomization"

One of the most important contributions of genetic epidemiology could be the ability to overcome limitations of classical epidemiological techniques, through the "Mendelian randomization" approach. The basic principle is that if a genetic variant alters the level of, or mirrors the biological effects of, a modifiable environmental exposure that itself alters disease risk, then this genetic variant should be related to disease risk to the extent predicted by its effect on exposure to the risk factor [29, 30]. In fact, since alleles are allocated essentially at random, such an association would not be subject either to confounding or reverse causation. Thus, common genetic polymorphisms that have a well characterized biological function, for example, by influencing exposure propensities or by modifying the biological effect of an environmental exposure, can be used as a surrogate to measure the effect of a suspected environmental exposure on disease risk. For instance, as later detailed in this chapter, the homozygosity for the C677T variant of the *MTHFR* gene is associated with reduced folate-dependent enzyme activity that can be partly reversed by folate supplementation. A case-control study of the relation between the TT genotype (associated with decreased folate blood levels) and risk of neural tube defects [31], can be interpreted as equivalent to a randomized trial of the effect on disease risk of alteration of the availability of folate. Despite this approach having some theoretical drawbacks, "Mendelian randomization" provides a promising means of examining the effects of modifiable exposures on disease risk.

5.2
Methylenetetrahydrofolate Reductase Gene Polymorphisms (C677T and A1298C) and Its Association with Cancer Risk

5.2.1
Gene and Function

The 5,10-methylenetetrahydrofolate reductase gene maps to chromosome 1p36.3 [32]. The complementary DNA sequence is 2.2 kb long, containing 11 exons, and the gene product is a 77-kD protein [32]. MTHFR plays a central role in folate metabolism by irreversibly catalyzing the conversion of 5,10-methylenetetrahydro-

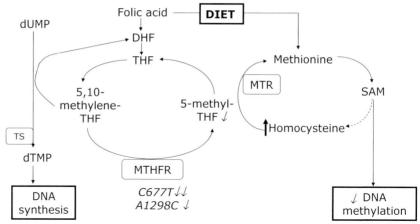

Figure 5.2 Folate pathway [33]. dUMP, deoxyuridine monophosphate; dTMP, deoxythymidine monophosphate; DHF, dihydrofolate; MTHFR, methylenetetrahydrofolate reductase; MTR, methionine synthase; SAM, s-adenosylmethionine; THF, tetrahydrofolate; TS, thymidylate synthase.

folate to 5-methyltetrahydrofolate, the primary circulating form of folate and a cosubstrate for homocysteine methylation to methionine (Figure 5.2). In humans, folate plays the fundamental role of providing methyl groups for *de novo* deoxynucleotide synthesis and for intracellular methylation reactions [34].

MTHFR enzyme function might influence cancer risk in two ways. The substrate of the MTHFR enzyme, 5,10-methylenetetrahydrofolate, is involved in the conversion of deoxyuridylate monophosphate (dUMP) to deoxythymidylate monophosphate (dTMP), and low levels of 5,10-methylenetetrahydrofolate would lead to an increased dUMP/dTMP ratio. In this situation an increased incorporation of uracil into DNA in place of thymine may follow, resulting in an increased chance of point mutations and DNA/chromosome breakage [34]. A less active form of MTHFR would lead, all other factors being equal, to an accumulation of 5,10-methylenetetrahydrofolate, thus a lower dUMP/dTMP ratio, and a presumably lower cancer risk [34]. The second way in which an impaired MTHFR activity might influence cancer risk is determined by the level of s-adenosylmethionine (SAM), the common donor of methyl that is necessary for the maintenance of the methylation patterns in DNA. Changes in methylation modify DNA conformation and gene expression. A less active form of MTHFR leads to lower SAM and consequently to hypomethylation; this phenomenon would be expected to increase the risk of some cancers [35] (Figure 5.1). Similarly, low folate intake may modify cancer risk by inducing uracil misincorporation during DNA synthesis, leading to chromosomal damage, DNA strand breaks and impaired DNA repair, and DNA hypomethylation [36].

5.2.2
C677T and A1298C Gene Variants

Twenty-nine rare mutations of *MTHFR* have been described in homocystinuric patients resulting in very low enzymatic activity [37], while two common polymorphisms are present in healthy individuals with reduced enzyme activity: C→T in exon 4 at nucleotide 677, leading to Ala222Val [38]; and A→C in exon 7 at nucleotide 1298, leading to Glu429Ala [39]. These polymorphisms are located 2.1 kb apart, and have been investigated in association with the risk of gastric and other cancers [40]. The T allele frequency of *MTHFR* 677 polymorphism is reported to be 0.36–0.44 in Europeans, 0.35–0.53 in Asians, 0.33–0.35 in US, and 0.10–0.24 in African Americans [41]. For *A1298C*, the variant allele frequency is reported to be 0.14–0.35 in Europeans and 0.11–0.17 in Asians. Individuals who are homozygous for the *MTHFR* 677 less frequent variant (TT) have 30% of the expected enzyme activity *in vitro*, compared with those who are homozygous for the common variant (CC), while heterozygous carriers have 65% activity. It has been reported that subjects who are TT homozygous for the *MTHFR* 677 exhibit reduced folate status and higher serum homocysteine levels compared to those who carry at least one 677C allele [42]. The evidence on the association of the 1298 variant allele with increased folate levels is less consistent [42]. Recent studies reported that the *MTHFR* TT genotype is related to DNA hypomethylation [43], particularly in individuals with reduced plasma folate concentrations [44]. Inconsistent results derive from studies on *A1298C* polymorphism, plasma folate and homocysteine levels. Both *MTHFR* C677T and *A1298C* can be detected by means of polymerase chain reaction (PCR) followed by restriction fragment-length polymorphism (RFLP) analysis with *HinfI* and *MboII* for *C677T* and *A1298C*, respectively [38]. Other methods include direct DNA sequencing or TaqMan assays.

5.2.3
Gene–Environment Interaction

Some nutrients involved in the folate metabolic pathway (e.g., vitamin B_6 and B_{12}, and methionine), alcohol (a folate antagonist) and smoking (which impairs the folate level) may interact with plasma folate levels and the *MTHFR* polymorphisms in determining cancer risk [45]. It has been reported that alcohol perturbs folate metabolism by reducing folate absorption, increasing folate excretion, or by inhibiting methionine synthase [46]. The inverse association between folate intake and plasma homocysteine levels can be modified by alcohol intake and by the *MTHFR* 677, but not the 1298 polymorphism [47]. The inverse effect of smoking on folate status might be confounded by alcohol intake or dietary habits [48], even though the association persists after adjusting for dietary folate intake and alcohol [48]. Additional studies reported that elevated folate turnover in response to rapid tissue proliferation or DNA repair in aerodigestive tissues among people exposed to tobacco smoke might partially explain this phenomenon.

5.3
Meta-Analyses of Methylenetetrahydrofolate Reductase C677T and A1298C Polymorphisms and Cancer

Identification of the meta-analyses of genetic association studies on MTHFR and cancer was carried out through a search of Medline and Embase, up to March 2008. The following terms were used: (("Methylenetetrahydrofolate Reductase (NADPH2)" [Mesh] OR MTHFR OR Methylenetetrahydrofolate Reductase)) AND (("Neoplasms" [Mesh]) OR cancer OR carcinoma)) AND (("Meta-Analysis as Topic" [Mesh] OR meta-analyses OR meta-analysis)), without any restriction on language. All studies included in the meta-analysis and their results are summarized in Figures 5.3 and 5.4.

Fourteen studies were identified, of which four were on breast cancer [52–55] four on colorectal cancer [49, 50, 56, 57], three on gastric cancer [58–60], and others concerned different tumor sites [51, 61, 62]. One paper included also a pooled analysis on *MTHFR* and gastric cancer (data not shown) [58]. All the published studies reported meta-ORs of *MTHFR C677T* and cancer, while only seven reported them on *A1298C* (Figures 5.1 and 5.2). As shown in Figure 5.3, the vast majority of the studies reported the absence of a significant association with the *C677T* variant genotype and cancer, with the exception of those on gastric and colorectal cancer (Figure 5.3). Here, a significant increased risk of gastric cancer

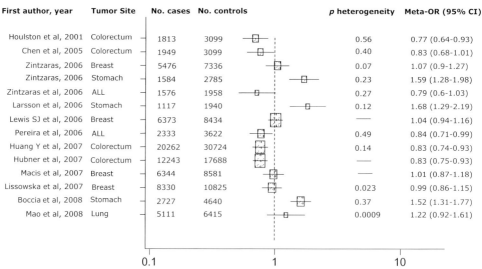

Figure 5.3 Forest plot depicting the results of meta-analyses of methylenetetrahydrofolate reductase (MTHFR) C677T polymorphism and cancer. OR, odds ratio; CI, confidence interval. The OR is from the random effect model except Houlston et al. [49], Chen et al. [50] and Pereira et al. [51] where the fixed effect model was used. The comparison is MTHFR 677TT versus CC except Macis et al. [52] and Pereira et al. [51] who compared TT versus C carriers. ALL, acute lymphoblastic leukemia.

associated with the *C677T* homozygous variant has been reported from three meta-analyses, of which the one published in 2008 considered more than 6000 individuals (Figure 5.3, [58]). Absence of statistical heterogeneity and publication bias strengthens these results. For breast and lung cancer, none of the published meta-analyses reported a significantly increased risk related to the *MTHFR* C677T variant (Figure 5.3). Among them, however, Macis *et al.* [52] reported that stratifying meta-analysis results according to the age at breast cancer, the C677T homozygous variant increased the risk of breast cancer in premenopausal woman (OR = 1.42 for TT versus TC + CC; 95% CI: 1.02–1.98). On the other hand, the homozygous variant of C677T was found, from all the four published meta-analyses, to confer a significant protective effect on colorectal cancer (Figure 5.3), among these four was one including more than 50 000 individuals [57]. Lastly, a borderline significant protective effect of the homozygous variant of *C677T* was reported from the 2 meta-analyses on acute lymphoblastic leukemia, especially among adults (OR = 0.45 for TT versus TC + CC; 95% CI: 0.26–0.77) [51].

For *MTHFR* A1298C, the results show the absence of significant association with cancer (Figure 5.4), however a borderline protective effect for colorectal cancer was noted by Huang *et al.* [57].

MTHFR plays a central role in balancing DNA synthesis (which involves 5,10-methylenetetrahydrofolate), and DNA methylation (which involves 5,10-methyltetrahydrofolate). Specifically, the 677T allele contributes to DNA hypomethylation, which in turn may lead to altered gene expression; at the same time this polymorphism might exert a protective effect, as observed for colorectal cancer

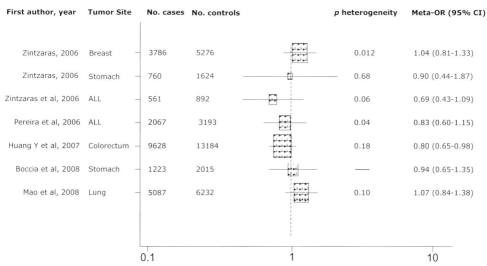

Figure 5.4 Forest plot depicting the results of meta-analyses of methylenetetrahydrofolate reductase (MTHFR) A1298C polymorphism and cancer. OR, odds ratio; CI, confidence interval. The OR is from the random effect model except for Pereira *et al.* [51] who used the fixed effect model. The comparison is MTHFR 677TT versus CC except for Pereira *et al.* [51] who compared TT versus C carriers. ALL, acute lymphoblastic leukemia.

and adult lymphoblastic leukemia, by increasing the levels of the MTHFR substrate, essential for DNA synthesis. Therefore, the exact interpretation of MTHFR-cancer association is not straightforward; although the observed increased risk for gastric cancer associated with the *MTHFR* 677 homozygous variant suggests that dietary folate might be protective in gastric carcinogenesis, mainly by limiting aberrant DNA methylation in the situation of impaired folate status.

As previously discussed, whether the C677T and A1298C variants act as beneficial or deleterious might depend on the subjects' folate status, alcohol consumption and possibly smoking habits. In this sense, it would have been valuable to explore the overall effect of *MTHFR* variants on cancer risk by stratifying meta-ORs on folate status (dietary folate intake or serum folate levels), alcohol intake and smoking habits. In order to do this, however, individual studies included in each meta-analysis should have enough power to explore such interactions, while in many instances the small sample sizes clearly prevent it. Additionally, even if gene–environment interactions were explored from the individual studies, often authors do not report genotype data stratified according to environmental covariates, thus preventing the possibility of performing stratified meta-analyses. In this sense, pooled analyses are preferable to meta-analyses because they are based on individual-level data, which also allow one to adjust for potential confounders.

The meta-analyses on *MTHFR* C677T and A1298C included in this chapter did not report subgroup meta-analyses according to the previously mentioned covariates, however, most of them discussed the results of the individual paper exploring gene–environment interactions in the discussion section. The pooled analysis of four studies on C677T *MTHFR* and gastric cancer [58], however, stratified results according to folate levels and showed an increased risk among individuals with low levels (OR = 2.05; 95% CI: 1.13–3.72) compared to those with high folate levels (OR = 0.95; 95% CI: 0.54–1.67). This result strengthens the previous observation of a potential role of folate in gastric carcinogenesis, suggesting that concomitant inadequate folate intake and impaired MTHFR activity might be important susceptibility factors for gastric cancer. In circumstances of folate deficiency, a decrease in downstream MTHFR-products results in a lower global DNA methylation status. Recently, aberrant methylation of proto-oncogenes has been explored as both a mechanism and a marker of carcinoma progression [63], with some studies reporting an altered methylation pattern, particularly for diffuse gastric cancer. Additionally, it has been recently reported that a significant global DNA hypomethylation occurs in MTHFR 677 TT subjects when compared with those with the wild type genotype [43], especially when plasma folate level is reduced. Taken together, these results suggest that the increased risk for gastric cancer associated with the homozygous *MTHFR* 677 variant might be linked to the subsequent impaired folate levels affecting DNA methylation status. Therefore, the negative association between the homozygous variant *MTHFR* genotype and gastric cancer might be counterbalanced to some extent by an adequate folate intake.

In conclusion, results from the published meta-analyses on *MTHFR* C677T and cancer support a protective effect of the 677 TT genotype on colorectal cancer and adult lymphoblastic leukemia, while there is an increased risk for gastric cancer.

This dual effect might be related to the mentioned dual activity of the MTHFR enzyme, so that the protective effect for colorectal cancer might result from the increased levels of the MTHFR substrate due the variant allele, which is essential for DNA synthesis, while the increased risk for gastric cancer suggests that dietary folate might be protective in gastric carcinogenesis mainly by limiting aberrant DNA methylation. Other genes involved in folate metabolism, however, should be considered for a more comprehensive understanding of the exact role of the folate pathway in cancer susceptibility. Large prospective cohort studies based on serological dosage of folate levels and/or detailed and repeated nutritional data would further clarify the role of folate in carcinogenesis.

References

1 Burton, P.R., Tobin, M.D., and Hopper, J.L. (2005) Key concepts in genetic epidemiology. *Lancet*, **366** (9489), 941–951.

2 Shpilberg, O., Dorman, J.S., Ferrell, R.E., Trucco, M., Shahar, A., and Kuller, L.H. (1997) The next stage: molecular epidemiology. *J. Clin. Epidemiol.*, **50**, 633–638.

3 Botstein, D., and Risch, N. (2003) Discovering genotypes underlying human phenotypes: past successes for Mendelian disease, future approaches for complex disease. *Nat. Genet.*, **33** (Suppl.), 228–237.

4 Risch, N. (1990) Linkage strategies for genetically complex traits. II. The power of affected relative pairs. *Am. J. Hum. Genet.*, **46**, 229–241.

5 Lander, E., and Kruglyak, L. (1995) Genetic dissection of complex traits: guidelines for interpreting and reporting linkage results. *Nat. Genet.*, **11**, 241–247.

6 Todd, J.A. (1999) Interpretation of results from genetic studies of multifactorial diseases. *Lancet*, **354** (Suppl. 1), SI15–SI16.

7 Risch, N.J. (2000) Searching for genetic determinants in the new millennium. *Nature*, **405**, 847–856.

8 Colhoun, H.M., McKeigue, P.M., and Davey Smith, D. (2003) Problems of reporting genetic associations with complex outcomes. *Lancet*, **361**, 865–872.

9 Zondervan, K.T., and Cardon, L.R. (2004) The complex interplay among factors that influence allelic association. *Nat. Rev. Genet.*, **5**, 89–100.

10 Cordell, H.J., and Clayton, D.G. (2005) Genetic association studies. *Lancet*, **366** (9491), 1121–1131.

11 Hattersley, A.T., and McCarthy, M.I. (2005) What makes a good genetic association study? *Lancet*, **366**, 1315–1523.

12 Strachan, T., and Read, A.P. (2003) *Human Molecular Genetics*, vol. **3**, Garland Science Publishers, Oxford.

13 Suh, Y., and Vijg, J. (2005) SNP discovery in associating genetic variation with human disease phenotypes. *Mutat. Res.*, **573**, 41–53.

14 Ioannidis, J.P. (2003) Genetic associations: false or true? *Trends Mol. Med.*, **9**, 135–138.

15 Ioannidis, J.P., Bernstein, J., Boffetta, P., Danesh, J., Dolan, S., Hartge, P. et al. (2005) A network of investigator networks in human genome epidemiology. *Am. J. Epidemiol.*, **162**, 302–304.

16 Ioannidis, J.P., Gwinn, M., Little, J., Higgins, J.P., Bernstein, J.L., Boffetta, P. et al. (2006) A road map for efficient and reliable human genome epidemiology. *Nat. Genet.*, **38**, 3–5.

17 Patsopoulos, N.A., Analatos, A.A., and Ioannidis, J.P. (2005) Relative citation impact of various study designs in the health sciences. *JAMA*, **293**, 2362–2366.

18 West, S., King, V., Carey, T.S., Lohr, K.N., McKoy, N., Sutton, S.F., and Lux, L. (2002) Systems to rate the strength of scientific evidence. *Evid. Rep. Technol. Assess.*, **47**, 1–11.

19 Berman, N.G., and Parker, R.A. (2002) Meta-analysis: neither quick nor easy. *BMC Med. Res. Methodol.*, **2**, 10.
20 Alvan, G., Bechtel, P., Iselius, L., and Gundert-Remy, U. (1990) Hydroxylation polymorphisms of debrisoquine and mephenytoin in European populations. *Eur. J. Clin. Pharmacol.*, **39**, 533–537.
21 Boccia, S., Gianfagna, F., La Torre, G., Persiani, R., D'Ugo, D., van Duijn, C.M., and Ricciardi, G. (2008) Genetic susceptibility to gastric cancer: a review of the published meta-analyses, in *Research Focus on Gastric Cancer* (ed. D.C. Cardinini), Nova Publisher, pp. 137–164.
22 Ioannidis, J.P. (2006) Meta-analysis in public health: potentials and problems. *Ital JPH*, **3** (2), 9–14.
23 Attia, J., Thakkinstian, A., and D'Este, C. (2003) Meta-analyses of molecular association studies: methodologic lessons for genetic epidemiology. *J. Clin. Epidemiol.*, **56**, 297–303.
24 Stroup, D.F., Berlin, J.A., Morton, S.C. et al. (2000) Meta-analysis of observational studies in epidemiology: a proposal for reporting. *JAMA*, **283**, 2008–2011.
25 Khoury, M.J. (2004) The case for a global human genome epidemiology initiative. *Nat. Genet*, **36**, 1027–1028.
26 Seminara, D., Khoury, M.J., O'Brien, T.R., Manolio, T., Gwinn, M.L., Little, J., Higgins, J.P., Bernstein, J.L., Boffetta, P., Bondy, M., Bray, M.S., Brenchley, P.E., Buffler, P.A., Casas, J.P., Chokkalingam, A.P., Danesh, J., Davey Smith, G., Dolan, S., Duncan, R., Gruis, N.A., Hashibe, M., Hunter, D., Jarvelin, M.R., Malmer, B., Maraganore, D.M., Newton-Bishop, J.A., Rioli, E., Salanti, G., Taioli, E., Timpson, N., Uitterlinden, A.G., Vineis, P., Wareham, N., Winn, D.M., Zimmern, R., Ioannidis, J.P., Human Genome Epidemiology Network and the Network of Investigator Networks (2007) The emergence of networks in human genome epidemiology: challenges and opportunities. *Epidemiology*, **18**, 1–8.
27 Khoury, M.J., and Dorman, J.S. (1998) The human genome epidemiology network. *Am. J. Epidemiol.*, **148** (1), 1–3.
28 Ioannidis, J.P., Bernstein, J., Boffetta, P., Danesh, J., Dolan, S., Hartge, P., Hunter, D., Inskip, P., Jarvelin, M.R., Little, J., Maraganore, D.M., Bishop, J.A., O'Brien, T.R., Petersen, G., Rioli, E., Seminara, D., Taioli, E., Uitterlinden, A.G., Vineis, P., Winn, D.M., Salanti, G., Higgins, J.P., and Khoury, M.J. (2005) A network of investigator networks in human genome epidemiology. *Am. J. Epidemiol.*, **162** (4), 302–304.
29 Davey Smith, G., and Ebrahim, S. (2003) "Mendelian randomization": can genetic epidemiology contribute to understanding environmental determinants of disease? *Int. J. Epidemiol.*, **32**, 1–22.
30 Chen, L., Davey Smith, G., Harbord, R.M. et al. (2008) Alcohol intake and blood pressure: a systematic review implementing a Mendelian randomization approach. *PLoS Med.*, **5** (3), e52.
31 Shields, D.C., Kirke, P.N., Mills, J.L., Ramsbottom, D., Molloy, A.M., Burke, H., Weir, D.G., Scott, J.M., and Whitehead, A.S. (1999) The "thermolabile" variant of methylenetetrahydrofolate reductase and neural tube defects: an evaluation of genetic risk and the relative importance of the genotypes of the embryo and the mother. *Am. J. Hum. Genet.*, **64**, 1045–1055.
32 Goyette, P., Sumner, J.S., Milos, R. et al. (1994) Human methylenetetrahydrofolate reductase: isolation of cDNA mapping and mutation identification. *Nat. Genet.*, **7**, 195–200.
33 Boccia, S., Brand, A., Brand, H., and Ricciardi, G. (2009) The integration of gene-environment interactions into healthcare and disease prevention as a major challenge for Public Health Genomics. *Mutat. Res.*, **667**, 27–34.
34 Blount, B.C., Mack, M.M., Wehr, C.M. et al. (1997) Folate deficiency causes uracil misincorporation into human DNA and chromosome breakage: implications for cancer and neuronal damage. *Proc. Natl. Acad. Sci. U S A*, **94**, 3290–3295.
35 Stern, L.L., Mason, J.B., Selhub, J. et al. (2000) Genomic DNA hypomethylation, a characteristic of most cancers, is present in peripheral leukocytes of individuals who are homozygous for the C677T polymorphism in the

methylenetetrahydrofolate reductase gene. *Cancer Epidemiol. Biomarkers Prev.*, **9**, 849–853.

36 Duthie, S.J. (1999) Folic acid deficiency and cancer: mechanisms of DNA instability. *Br. Med. Bull.*, **55**, 578–592.

37 Sibani, S., Leclerc, D., Weisberg, I.S. et al. (2003) Characterization of mutations in severe methylenetetrahydrofolate reductase deficiency reveals an FAD-responsive mutation. *Hum. Mutat.*, **21**, 509–520.

38 Frosst, P., Blom, H.J., Milos, R. et al. (1995) A candidate genetic risk factor for vascular disease: a common mutation in methylenetetrahydrofolate reductase. *Nat. Genet1*, **10**, 111–113.

39 van der Put, N.M., Gabreels, F., Stevens, E.M. et al. (1998) A second common mutation in the methylenetetrahydrofolate reductase gene: an additional risk factor for neural-tube defects? *Am. J. Hum. Genet.*, **62**, 1044–1051.

40 Kim, Y.I. (2000) Methylenetetrahydrofolate reductase polymorphisms, folate, and cancer risk: a paradigm of gene-nutrient interactions in carcinogenesis. *Nutr. Rev.*, **58**, 205–209.

41 Botto, L.D., and Yang, Q. (2000) 5,10-Methylenetetrahydrofolate reductase gene variants and congenital anomalies: a HuGE review. *Am. J. Epidemiol.*, **151**, 862–877.

42 Parle-McDermott, A., Mills, J.L., Molloy, A.M. et al. (2006) The MTHFR 1298CC and 677TT genotypes have opposite associations with red cell folate levels. *Mol. Genet Metab.*, **88**, 290–294.

43 Castro, R., Rivera, I., Ravasco, P. et al. (2004) 5,10-methylenetetrahydrofolate reductase (MTHFR) 677C–>T and 1298A–>C mutations are associated with DNA hypomethylation. *J. Med. Genet.*, **41**, 454–458.

44 Friso, S., Choi, S.W., Girelli, D. et al. (2002) A common mutation in the 5,10-methylenetetrahydrofolate reductase gene affects genomic DNA methylation through an interaction with folate status. *Proc. Natl. Acad. Sci. U. S. A.*, **99**, 5606–5611.

45 Bailey, L.B. (2003) Folate, methyl-related nutrients, alcohol, and the MTHFR 677C–>T polymorphism affect cancer risk: intake recommendations. *J. Nutr.*, **133**, S3748–S3753.

46 Mason, J.B., and Choi, S.W. (2005) Effects of alcohol on folate metabolism: implications for carcinogenesis. *Alcohol*, **35**, 235–241.

47 Chiuve, S.E., Giovannucci, E.L., Hankinson, S.E. et al. (2005) Alcohol intake and methylenetetrahydrofolate reductase polymorphism modify the relation of folate intake to plasma homocysteine. *Am. J. Clin. Nutr.*, **82**, 155–162.

48 Stark, K.D., Pawlosky, R.J., Beblo, S. et al. (2005) Status of plasma folate after folic acid fortification of the food supply in pregnant African American women and the influences of diet, smoking, and alcohol consumption. *Am. J. Clin. Nutr.*, **81**, 669–667.

49 Houlston, R.S., and Tomlinson, I.P. (2001) Polymorphisms and colorectal tumor risk. *Gastroenterology*, **121**, 282–301.

50 Chen, K., Jiang, Q.T., and He, H.Q. (2005) Relationship between metabolic enzyme polymorphism and colorectal cancer. *World J. Gastroenterol.*, **11**, 331–335.

51 Pereira, T.V., Rudnicki, M., Pereira, A.C., Pombo-de-Oliveira, M.S., and Franco, R.F. (2006) Do polymorphisms of 5,10-methylenetetrahydrofolate reductase (MTHFR) gene affect the risk of childhood acute lymphoblastic leukemia? *Eur. J. Epidemiol.*, **21**, 885–886.

52 Macis, D., Maisonneuve, P., Johansson, H., Bonanni, B., Botteri, E., Iodice, S., Santillo, B., Penco, S., Gucciardo, G., D'Aiuto, G., Rosselli Del Turco, M., Amadori, M., Costa, A., and Decensi, A. (2007) Methylenetetrahydrofolate reductase (MTHFR) and breast cancer risk: a nested-case-control study and a pooled meta-analysis. *Breast. Cancer Res. Treat*, **106**, 263–271.

53 Lissowska, J., Gaudet, M.M., Brinton, L.A., Chanock, S.J., Peplonska, B., Welch, R., Zatonski, W., Szeszenia-Dabrowska, N., Park, S., Sherman, M., and Garcia-Closas, M. (2007) Genetic polymorphisms in the one-carbon metabolism pathway and breast cancer risk: a population-based

case-control study and meta-analyses. *Int. J. Cancer*, **15** (120), 2696–2703.

54 Lewis, S.J., Harbord, R.M., Harris, R., and Smith, G.D. (2006) Meta-analyses of observational and genetic association studies of folate intakes or levels and breast cancer risk. *J. Natl. Cancer Inst.*, **98**, 1607–1022.

55 Zintzaras, E. (2006) Methylenetetrahydrofolate reductase gene and susceptibility to breast cancer: a meta-analysis. *Clin. Genet.*, **69**, 327–336.

56 Hubner, R.A., and Houlston, R.S. (2007) MTHFR C677T and colorectal cancer risk: a meta-analysis of 25 populations. *Int. J. Cancer*, **120**, 1027–1035.

57 Huang, Y., Han, S., Li, Y., Mao, Y., and Xie, Y. (2007) Different roles of MTHFR C677T and A1298C polymorphisms in colorectal adenoma and colorectal cancer: a meta-analysis. *J. Hum. Genet.*, **52**, 73–85.

58 Boccia, S., Hung, R., Ricciardi, G., Gianfagna, F., Ebert, M.P., Fang, J.Y., Gao, C.M., Götze, T., Graziano, F., Lacasaña-Navarro, M., Lin, D., López-Carrillo, L., Qiao, Y.L., Shen, H., Stolzenberg-Solomon, R., Takezaki, T., Weng, Y.R., Zhang, F.F., van Duijn, C.M., Boffetta, P., and Taioli, E. (2008) Meta- and pooled analyses of the methylenetetrahydrofolate reductase C677T and A1298C polymorphisms and gastric cancer risk: a huge-GSEC review. *Am. J. Epidemiol.*, **167**, 505–516.

59 Larsson, S.C., Giovannucci, E., and Wolk, A. (2006) Folate intake, MTHFR polymorphisms, and risk of esophageal, gastric, and pancreatic cancer: a meta-analysis. *Gastroenterology*, **131**, 1271–1283.

60 Zintzaras, E. (2006) Association of methylenetetrahydrofolate reductase (MTHFR) polymorphisms with genetic susceptibility to gastric cancer: a meta-analysis. *J. Hum. Genet.*, **51**, 618–624.

61 Mao, R., Fan, Y., Jin, Y., Bai, J., and Fu, S. (2008) Methylenetetrahydrofolate reductase gene polymorphisms and lung cancer: a meta-analysis. *J. Hum. Genet.*, **53**, 340–348.

62 Zintzaras, E., Koufakis, T., Ziakas, P.D., Rodopoulou, P., Giannouli, S., and Voulgarelis, M. (2006) A meta-analysis of genotypes and haplotypes of methylenetetrahydrofolate reductase gene polymorphisms in acute lymphoblastic leukemia. *Eur. J. Epidemiol.*, **21**, 501–510.

63 Dunn, B.K. (2003) Hypomethylation: one side of a larger picture. *Ann. N. Y. Acad. Sci.*, **983**, 28–42.

6
The Role of Biobanks for the Understanding of Gene–Environment Interactions

Christian Viertler, Michaela Theresia Mayrhofer, and Kurt Zatloukal

Abstract

The identification of susceptibility genes and genetic modifiers is of great medical relevance, not only because it provides insight into the complex network of biological processes but also because it provides the basis for new diagnostic tests. Moreover, genetic variants which confer susceptibility or resistance also serve as indicators of targets for therapeutic interventions. The bottleneck to finding environmental and genetic associations that are robust enough to guide targeted treatment and prevention strategies is access to both collections of large numbers of high quality biological samples and associated medical data. Consequently, networks of biorepositories for research are initiated to benefit from the added value of a larger pool of sample and data sets which can provide the number of samples required to achieve statistical significance. These networks promise to greatly advance our understanding of disease development, prevention and therapy. In this chapter, we take the complex disease NAFLD (non-alcoholic fatty liver disease) as an example to discuss the challenges faced by (networks of) biobanks in biomedical research and for investigation of gene–environment interactions.

6.1
Background

Over the last two decades, collections of biological samples such as blood, tissue, cells, or DNA have gained a new significance for biomedical research. Today, biological samples with associated medical and research data from large numbers of patients and healthy persons, as well as biomolecular research tools, are recognized as key resources to unravel genetic and environmental factors causing diseases and influence their outcome. Furthermore, they are required for the identification of new targets for therapy and may help to reduce attrition in drug discovery and development [1–3].

In order to understand the causal pathway of a chronic metabolic disease, it is essential to acknowledge the role of gene–environment interactions as determinants of disease risk and progression. Therefore, these studies should consider the effects of expressed genes and intermediary phenotypes. This requires access to longitudinal medical records, follow-up assessment and repeated sample collection from individuals [4]. Studying the etiology of complex diseases is a challenging task because they are influenced by a large variety of additive effects. These effects represent the sum of consequences of genetic predisposition, lifestyle and environmental factors, including exposure to the pathogens. Consequently, revealing these complex interactions depends critically on the study of large sets of well-documented samples and data provided by biobanks [5]. Among a plethora of biobank formats, population-based biobanks and disease-oriented biobanks can be identified as the major formats.

6.1.1
What Purpose Do Different Biobank Formats Serve?

Population-based biobanks typically follow a longitudinal approach and are often established in the context of national or regional medical research initiatives (e.g., UK, Estonia, Iceland). The specific strength of the population-based approach is that it allows one to assess the natural frequency in occurrence and progression of common diseases in an *a priori* healthy population. Here, the emphasis is laid on the study of genetic variants and environmental risk factors. The study of complex diseases requires comparison of large numbers of affected and unaffected individuals ("cases" and "controls"). Population-based biobanks have thus become indispensable to elucidate molecular processes and causal pathways, whether genetic or environmental, and to translate biomedical research into improvements in healthcare. A major drawback of the population-based approach is that for most study designs a sufficient number of even the most common diseases in healthy individuals can only be detected after a long period of time [3, 6]. For example, a cohort of 500 000 middle-aged participants may be expected to generate 10 000 incident cases of wide spread diseases, such as diabetes or coronary artery disease, within 7–8 years. This will take up to 20 years or more in the case of common cancers and even longer for rare diseases [7].

In contrast, disease-oriented biobanks have usually emerged from routine medical services. Both, the high number of represented diseases and the different stages of diseases collected as a "by-product" of health care make them a valuable resource for modern research. Over the last few years, these collections have been subsequently adapted to emerging biomedical developments [8].

Disease-oriented biobanks are important for evaluation of the human disease relevance of discoveries made in various model organisms, such as mice, Drosophila or yeast [6]. Furthermore, for the discovery of disease triggering effects, cancer research and drug development depend critically on the study of large sample and data sets from both patients and healthy individuals. To understand the link between molecular targets for drug intervention and the molecular pathol-

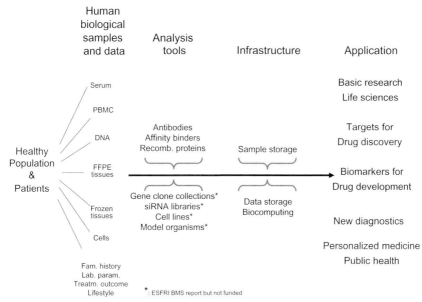

Figure 6.1 Key components and applications of biobanks [10].

ogy of disease would help us translate the results of clinical trials into a molecular understanding of desired responsiveness and adverse side-effects [9]. Key results would be the identification of pathways involved in disease progression and the development of more effective drugs (Figure 6.1). This, in combination with corresponding biomarkers – defined by the NIH as an objectively measured and evaluated indicator of normal biological or pathogenic processes and pharmacological response to a therapeutic intervention [11] – would accelerate the progress in personalized medicine for specific patient groups [12]. Furthermore, these biobanks are valuable resources for biomedical research to explore the function and medical relevance of human genes and their products. However, to be able to use these resources accordingly, there is a need for transnational coordination and collaboration.

6.1.2
Why Do We Need Networks of Biobanks?

Although currently established biobanks have their unique strengths, even large biobanks are often unable to provide the quantity of samples that are required by studies in order to achieve statistical significance. The exploration of genomic associations typically requires several thousand cases to study main effects. For the investigation of gene–environment interactions tens of thousands are needed [7]. So it is not surprising that many biobanks initiate and/or join networks in order to benefit from the added value of access to a larger pool of sample and data sets.

Moreover, biobanks, especially those collaborating across national borders, are facing the challenges of the heterogeneous ethical and legal frameworks in which they are operating. Furthermore, biobanks typically suffer from fragmentation of the biobanking-related research community, differently structured and organized collections, variable access rules and a lack of commonly applied standards [2, 8]. Harmonized standard operating procedures (SOPs) are essential for the whole process from sample collection, data acquisition and storage to technical issues of different analyses.

Furthermore, a main factor limiting high-quality molecular genetic epidemiology studies concerns the resources required to obtain detailed, accurately measured phenotypic data. The phenotypic data in existing biobanks is often variable in content, format, depth and vocabulary. This shortcoming can be remedied by collection of additional data, or by retrospective harmonization of phenotypes that have already been collected in the various biobanks. Thus, there is an urgent need for improved assessment and classification of multivariate phenotypes associated with a complex disease. Indeed, the lack of a common language and standardized vocabulary to describe phenotypic characteristics in sufficient detail represents a major barrier to both national and trans-national research collaboration.

In order to face these heterogeneous legal, ethical and scientific challenges, concerted actions are necessary in order to devise procedures for collecting, exchanging and linking samples and data [4]. At the trans-national level, many efforts are currently under way to unite the fragmented biobanking community and to work toward a set of homogenous rules through joint and networking activities.

The Public Population Project in Genomics (P^3G), for instance, promotes international collaboration between researchers in the field of population genomics and biobanking to ensure public access to population genomics data according to prevailing ethical and legal norms. This platform has been launched in order to provide the international population genomics community with the resources, tools and know-how to facilitate data management for improved methods of knowledge transfer. One of its major goals is sharing research tools for effective collaboration between biobanks. This will enable the international research community to share expertise and resources and facilitate knowledge transfer. Along these lines, P^3G fosters the harmonization of nomenclature of biological, medical, demographic and social data collected from participants, mainly in the context of population-based studies (e.g., the Swedish Twin Study of Adults: Genes and Environments (STAGE), for further details see http://www.p3gobservatory.org/catalogue.htm?questionnaireId=73). A particular tool to facilitate harmonization is the Data Schema and Harmonization Platform for Epidemiological Research (DataSHaPER). It supports the construction of cross-sectional baseline questionnaires to define a core set of information that is of particular scientific relevance for a specific type of biobank (http://www.p3gobservatory.org/datashaper/presentation.htm). More than 25 international biobanks have contributed to the conception of the DataSHaPER.

The European Community also places much emphasis on improving transnational collaboration in the field of biobanking. Already under Framework Program 5, a significant number of networking activities have been supported. A policy which was continued under both Framework Program 6 and 7.[1] Moreover, the European roadmap for research infrastructures foresees a pan-European research infrastructure for biobanking and biomolecular resources (BBMRI) to further develop these resources and to provide access to academia and industry. The preparation phase BBMRI is funded within the Framework Program 7. BBMRI builds on existing sample collections, resources, technologies, and expertise, which will be specifically complemented with innovative components [13]. In particular, "BBMRI will comprise (i) all major population-based and disease-oriented biobank formats, (ii) biomolecular resources, such as collections of antibodies and other affinity binders and a variety of molecular tools to decipher protein interactions and function, (iii) bio-computing and sample storage infrastructure. All resources will be integrated into a pan-European distributed hub-structure-like network, and will be properly embedded into European scientific, ethical, legal and societal frameworks" [10]. Specific tasks in the planning of BBMRI are the preparation of an inventory of existing resources, implementation of common standards and access rules, establishment of incentives for resource providers, and development of solutions for international exchange of biological samples and data which properly consider the heterogeneity of pertinent national legislation and ethical principles.

Today, the focus is on building sustainable research infrastructures rather than fostering short-term research collaborations. Consequently the emphasis has shifted toward strengthening trans-national networking activities to support sustainable long-term research collaborations and infrastructures.

6.2
The Investigation of Gene–Environment Interactions as a Challenge for Biobanks

Population-based and disease-oriented biobanks are essential to establish the disease relevance of human genes and provide opportunities to evaluate their interaction with lifestyle and environment. Since the sequence of the human genome was determined some years ago, biomedical research has progressed from the study of rare monogenic diseases to the study of common multifactorial diseases and places much emphasis on an individual's disease risks.

The non-alcoholic fatty liver disease (NAFLD) is a typical example of a complex disease that results from interactions between the environment and several different genetic factors. In fact NAFLD is one of the most common liver disorders seen

---

1) For further information, please consider the Report and Recommendations of the Networking Meeting for EU-funded Biobanking Projects held in Brussels on 20–21 November 2008; ftp://ftp.cordis.europa.eu/pub/fp7/docs/report-meeting-eu-funded-biobanks_en.pdf

by hepatologists [14]. It has been estimated that up to one out of three Western adults has NAFLD, a disease closely associated with the metabolic syndrome. The looming pandemic of type 2 diabetes mellitus and obesity as key components of the metabolic syndrome, reaching now eastern countries as well as children, suggests that this prevalence is likely to increase further [15].

The disease spectrum of NAFLD ranges from simple steatosis (fatty liver) through non-alcoholic steatohepatitis (NASH) – characterized by liver cell injury, inflammation, and fibrosis – to cirrhosis, liver failure and hepatocellular carcinoma.

Steatohepatitis can develop in alcoholics, obese and type II diabetics (BMI > 30) and can also result from drug toxicity (e.g., amiodarone, perhexilin maleate, tamoxifen, HIV antiretroviral drugs). Cirrhosis due to NASH may now account for up to 20% of cirrhosis cases and may also play a major role in the category of cryptogenic cirrhosis; furthermore, up to 10% of hepatocellular carcinomas may result from NASH [16–19]. This development makes the metabolic syndrome and its manifestations in the liver a major health problem with an estimated world-wide prevalence of steatohepatitis of 3–5%. Moreover, NAFLD is emerging as an important independent risk factor for cardiovascular morbidity and mortality [20–22].

Whereas most patients with risk factors for NAFLD develop hepatic steatosis, only a minority will ever develop progressive disease: for example, only about 10–20% of even morbidly obese patients develop more than simple steatosis [23, 24], 20% of heavy drinkers or 50% of obese type II diabetic patients develop steatohepatitis (Figure 6.2). In contrast to simple steatosis, patients with NASH are at significant risk of developing cirrhosis (up to 25% within 10 years) and dying from liver disease (up to 10% within 10 years).

These marked differences in the individual risk have led to the question: what factors determine whether a patient develops advanced NAFLD? Environmental and genetic risk factors may play a role in etiology and progression of the disease.

Over the years, potential risk factors for the development of NAFLD have been detected in numerous studies (Table 6.1). In single studies for example, dietary saturated fat, antioxidant intake and small bowel overgrowth are suggested to play a role in disease progression [25, 26]. Family studies and interethnic variations in

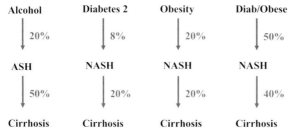

Figure 6.2 Differences in individual risk for development of steatohepatitis and progression to cirrhosis depending on etiology.

Table 6.1 Environmental and genetic factors determining the risk for the development of NAFLD [15].

Potential lifestyle factors and accompanying diseases influencing susceptibility to NAFLD:
  Alcohol (low intake protective)
  Physical activity (protective)
  Dietary factors (low antioxidant vitamins C/E, higher intake of saturated fat)
  Obesity
  Type 2 diabetes mellitus
  Small bowel intestinal overgrowth
  Sleep apnea syndrome

Potential candidate genes in fatty liver disease:
  Genes determining the magnitude and pattern of fat deposition
  Genes determining insulin sensitivity
  Genes involved in hepatic lipid synthesis, storage and export
  Genes involved in hepatic fatty acid oxidation
  Genes influencing the generation of oxidant species
  Genes encoding proteins involved in the response to oxidant stress
  Cytokine genes and receptors
  Genes encoding endotoxin receptors
  Immune response genes
  NAFLD-related fibrosis genes
  General fibrosis genes

susceptibility suggest that genetic factors are important in determining the risk of progressive NAFLD: for example, the prevalence of cryptogenic cirrhosis is higher in Hispanics and lower in Afro-Americans compared to European-American patients, despite a similar prevalence of type 2 diabetes mellitus [27–30]. Genetic risk factors for NAFLD most likely include genes that influence hepatic free fatty acid supply, the magnitude of oxidative stress, the release and effect of cytokines and/or the severity of fibrosis.

The identification of susceptibility genes and genetic modifiers is of great medical relevance, not only because it provides insights into the complex network of biological processes affected in steatohepatitis but also because it provides the basis for new diagnostic tests for identification of high risk patients. Current diagnosis (grading, staging) of steatohepatitis relies on liver biopsy as a diagnostic gold standard for differentiation between "simple" steatosis and (N)ASH [16]. Routine biochemical serum tests (liver function tests) underestimate in many cases the severity/activity of steatohepatitis. Although the degree of liver fibrosis may be a crude predictor for the development of liver cirrhosis [31], more sophisticated individual risk markers/profiles for the development of cirrhosis and hepatocellular cancer are required. Moreover, the relative contribution of cardiovascular versus liver-related morbidity/risk may vary significantly among individual NAFLD/NASH patients. Given the high prevalence of steatosis and steatohepatitis in the general population, there is an urgent need for non-invasive diagnostic tests and prognostic biomarkers which predict the individual disease course and need for and response to therapy. Furthermore, genetic variants which confer susceptibility or resistance to steatohepatitis also serve as indicators of targets for therapeutic interventions.

6.2.1
How to Evaluate Risk Factors for Metabolic Syndrome and Steatohepatitis?

Data for the evaluation of environmental risk factors whether collected by general questionnaires (e.g., through P³G) or by individual anamnesis of the patient must be interpreted with caution. For instance, in contrast to HbA1C that is commonly used for blood glucose long-term monitoring of diabetic patients, no objective markers have been found reliable enough to indicate alcohol intake or physical activity.

The conclusions drawn from classical case-control, candidate gene, and allele association studies are also subject to common pitfalls, for example, false positive by chance findings or false negative by small, underpowered studies [32]. The traditional hypothesis-driven approaches rely on selecting genes on the basis of their assumed role in disease pathogenesis, often derived from studies in animal models. More recently, several hypothesis-free or generating approaches have been developed; for example, microarray and proteomic studies in tissues from patients and animal models, QTL (quantitative trait loci) mapping and mouse mutagenesis studies [15].

6.2.2
Why Are Biobanks Needed in This Context and What Challenges Do They Have to Face?

High quality human biological samples from biobanks with associated well documented clinical and research data are key resources to evaluate the interaction of genetic background and environmental risk factors. Data generated from model systems need to be evaluated for their relevance in human tissue to identify the corresponding human susceptibility and modifier genes. New prognostic biomarkers, which predict the individual risk of a patient as well as a possible response to therapy, can only be identified in large studies with human biological samples enabled by biobanks. Depending on the study design and properties of the markers investigated, the discovery of statistically significant new biomarkers requires 5000–50000 cases. For example, in contrast to the readily measurable body mass index (BMI),[2] measurements for many other lifestyle-environmental determinants, such as nutritional components, have a much lower test-retest reliability [7].

2) $BMI = weight/height^2$

Moreover, high quality samples are essential for emerging techniques well suited to investigate the molecular basis of complex diseases such as chronic liver disease, diabetes or obesity. Microarray analysis, for instance, is an appropriate method for assessing simultaneously the expression of a large number of genes in a tissue sample. In combination with "proteomics" approaches, this technique can be employed to compare patterns of gene expression in liver tissue samples with corresponding protein profiles in the serum of NAFLD patients.

To tap the potential of modern "-omics" technologies, the development of improved cryo-preservation and storage procedures based on a systematic experimental approach is required. Ample experience and best practice protocols exist for pre-analytical processing for DNA and RNA analysis, but there is much less knowledge on the biobanking requirements for proteins or metabolites. For example, there are several factors whose impact on metabolome analysis has not yet been further evaluated, such as the metabolic state of the donor at the time of sample acquisition and the pre-analytical procedures from acquisition, processing, and stabilization to storage of samples [33].

Establishing a metabolic disease cohort that is big enough to address key questions on individual's susceptibilities can hardly be achieved by a single institution. In many cases, the different samples (e.g., liver, muscle, subcutaneous and visceral adipose tissue and blood) needed to investigate molecular mechanisms underlying these diseases and the variations in organ manifestations, are collected separately. Furthermore, detailed disease phenotype characterization and highly specified sample collection procedures are expensive and laborious (Table 6.2).

It is an emerging challenge for biobanks to fulfill the growing demands, especially of new "-omics" technologies (genome, transcriptome, proteome and metabolome analyses), concerning both quantity *and* quality of biological specimens. On the one hand, the potential to gain further insight in gene–environment interactions has increased due to the enormous analytical capacities provided by new technologies and reduced costs per data generated. On the other hand, the access to high quality biological samples and associated data becomes increasingly the limiting factor for translating the technological achievements into medical progress. Consequently, environmental and genetic association studies robust enough to guide targeted treatment and prevention strategies can only be achieved through multinational cooperation and collaboration of biobanks.

Acknowledgments

This work was supported by the funds of the pan-European biobanks and biomolecular resources research infrastructure (BBMRI, Grant Agreement Nr. 212111) and the project Standardization and Improvement of Generic Pre-Analytical Tools and Procedures for *in vitro* Diagnostics (SPIDIA, Grant Agreement. Nr. 222916).

Table 6.2 Investigation of gene–environment interactions: major challenges for biobanks.

Issue	Challenge
Statistical significance	Tens of thousands of cases required (especially for biomarkers and gene–environment studies)
Quantity and quality of samples	Growing demands of emerging techniques and specific study designs
Data acquisition and management	Standardized vocabulary and questionnaires
Trans-national collaboration and coordination	Heterogeneous ethical, legal, societal frameworks
Accurate phenotype data for high-quality molecular genetic epidemiology studies	Data variability in content, format, depth, vocabulary
"-omics" technologies	Development of SOPs for biobanking practices (e.g., improved pre-analytical processing and storage)
Environmental exposure	Lack of specific biomarkers (variability vs. reliability)
Studying the etiology of a complex disease	Large variety of additive effects (e.g., genetic predisposition, life-style, environmental factors, exposure to pathogens)

References

1 Kaiser, J. (2002) Population databases boom, from Iceland to the U.S. *Science*, **298**, 1158–1161.
2 Hans-E, H., and Carlstedt-Duke, J. (2004) Building global networks for human diseases: genes and populations. *Nat. Med.*, **10**(7), 665–667.
3 Manolio, T.A., Bailey-Wilson, J.E., and Collins, F.S. (2006) Genes, environment and the value of prospective cohort studies. *Nat. Rev. Genet.*, **7**, 812–820.
4 ESF (2008) Population Surveys and Biobanking. ESF Science Policy Briefing, 32.
5 Viertler, C., and Zatloukal, K. (2008) Biobanking and biomolecular resources research infrastructure (BBMRI): implications for pathology. *Pathologe*, **29** (Suppl. 2), 210–213.
6 Asslaber, M., and Zatloukal, K. (2007) Biobanks: transnational, European and global networks. *Brief Funct. Genomic Proteomic*, **6**, 193–201.
7 Burton, P.R., Hansell, A.L., Fortier, I., Manolio, T.A., Khoury, M.J., Little, J., and Elliott, P. (2009) Size matters: just how big is BIG?: quantifying realistic sample size requirements for human genome epidemiology. *Int. J. Epidemiol.*, **38**, 263–273.
8 Hirtzlin, I., Dubreuil, C., Preaubert, N., Duchier, J., Jansen, B., Simon, J., Lobato De Faria, P., Perez-Lezaun, A., Visser, B., Williams, G.D., and Cambon-Thomsen, A. (2003) An empirical survey on biobanking of human genetic material and data in six EU countries. *Eur. J. Hum. Genet.*, **11**, 475–488.

9 The Innovative Medicines Initiative (IMI) (2006) Creating biomedical R&D leadership for Europe to benefit patients and society. IMI: Stategic Research Agenda

10 BBMRI (2007) Grant agreement for Combination of Collaborative Project and Coordination and Support Actions. 212111.

11 Atkinson, A.J., Colburn, W.A., Degruttola, V.G., Demets, D.L., Downing, G.J., Hoth, D.F., Oates, J.A., Peck, C.C., Schooley, R.T., Spilker, B.A., Woodcock, J., and Zeger, S.L. (2001) Biomarkers and surrogate endpoints: preferred definitions and conceptual framework. *Clin. Pharmacol. Ther.*, **69**, 89–95.

12 Sotiriou, C., and Piccart, M.J. (2007) Taking gene-expression profiling to the clinic: when will molecular signatures become relevant to patient care? *Nat. Rev. Cancer*, **7**, 545–553.

13 Yuille, M., van Ommen, G.J., Brechot, C., Cambon-Thomsen, A., Dagher, G., Landegren, U., Litton, J.E., Pasterk, M., Peltonen, L., Taussig, M., Wichmann, H.E., and Zatloukal, K. (2008) Biobanking for Europe. *Brief Bioinform* **9**, 14–24.

14 Neuschwander-Tetri, B.A., and Caldwell, S.H. (2003) Nonalcoholic steatohepatitis: summary of an AASLD Single Topic Conference. *Hepatology*, **37**, 1202–1219.

15 Wilfred de Alwis, N.M., and Day, C.P. (2008) Genes and nonalcoholic fatty liver disease. *Curr. Diab. Rep.*, **8**, 156–163.

16 Angulo, P. (2002) Nonalcoholic fatty liver disease. *N. Engl. J. Med.*, **346**, 1221–1231.

17 Bugianesi, E., Leone, N., Vanni, E., Marchesini, G., Brunello, F., Carucci, P., Musso, A., De Paolis, P., Capussotti, L., Salizzoni, M., and Rizzetto, M. (2002) Expanding the natural history of nonalcoholic steatohepatitis: from cryptogenic cirrhosis to hepatocellular carcinoma. *Gastroenterology*, **123**, 134–140.

18 Farrell, G.C., and Larter, C.Z. (2006) Nonalcoholic fatty liver disease: from steatosis to cirrhosis. *Hepatology*, **43**, S99–S112.

19 Brea, A., Mosquera, D., Martin, E., Arizti, A., Cordero, J.L., and Ros, E. (2005) Nonalcoholic fatty liver disease is associated with carotid atherosclerosis: a case-control study. *Arterioscler Thromb. Vasc. Biol.*, **25**, 1045–1050.

20 Schindhelm, R.K., Diamant, M., Dekker, J.M., Tushuizen, M.E., Teerlink, T., and Heine, R.J. (2006) Alanine aminotransferase as a marker of non-alcoholic fatty liver disease in relation to type 2 diabetes mellitus and cardiovascular disease. *Diabetes. Metab. Res. Rev.*, **22**, 437–443.

21 Targher, G., Bertolini, L., Padovani, R., Poli, F., Scala, L., Tessari, R., Zenari, L., and Falezza, G. (2006) Increased prevalence of cardiovascular disease in Type 2 diabetic patients with non-alcoholic fatty liver disease. *Diabet. Med.*, **23**, 403–409.

22 Villanova, N., Moscatiello, S., Ramilli, S., Bugianesi, E., Magalotti, D., Vanni, E., Zoli, M., and Marchesini, G. (2005) Endothelial dysfunction and cardiovascular risk profile in nonalcoholic fatty liver disease. *Hepatology*, **42**, 473–480.

23 Wanless, I.R., and Lentz, J.S. (1990) Fatty liver hepatitis (steatohepatitis) and obesity: an autopsy study with analysis of risk factors. *Hepatology*, **12**, 1106–1110.

24 Dixon, J.B., Bhathal, P.S., and O'Brien, P.E. (2001) Nonalcoholic fatty liver disease: predictors of nonalcoholic steatohepatitis and liver fibrosis in the severely obese. *Gastroenterology*, **121**, 91–100.

25 Musso, G., Gambino, R., De Michieli, F., Cassader, M., Rizzetto, M., Durazzo, M., Faga, E., Silli, B., and Pagano, G. (2003) Dietary habits and their relations to insulin resistance and postprandial lipemia in nonalcoholic steatohepatitis. *Hepatology*, **37**, 909–916.

26 Wigg, A.J., Roberts-Thomson, I.C., Dymock, R.B., McCarthy, P.J., Grose, R.H., and Cummins, A.G. (2001) The role of small intestinal bacterial overgrowth, intestinal permeability, endotoxaemia, and tumour necrosis factor alpha in the pathogenesis of non-alcoholic steatohepatitis. *Gut*, **48**, 206–211.

27 Browning, J.D., Kumar, K.S., Saboorian, M.H., and Thiele, D.L. (2004) Ethnic differences in the prevalence of cryptogenic cirrhosis. *Am. J. Gastroenterol.*, **99**, 292–298.

28 Cossrow, N., and Falkner, B. (2004) Race/ethnic issues in obesity and obesity-related comorbidities. *J. Clin. Endocrinol. Metab.*, **89**, 2590–2594.

29 Struben, V.M., Hespenheide, E.E., and Caldwell, S.H. (2000) Nonalcoholic steatohepatitis and cryptogenic cirrhosis within kindreds. *Am. J. Med.*, **108**, 9–13.

30 Willner, I.R., Waters, B., Patil, S.R., Reuben, A., Morelli, J., and Riely, C.A. (2001) Ninety patients with nonalcoholic steatohepatitis: insulin resistance, familial tendency, and severity of disease. *Am. J. Gastroenterol.*, **96**, 2957–2961.

31 Matteoni, C.A., Younossi, Z.M., Gramlich, T., Boparai, N., Liu, Y.C., and McCullough, A.J. (1999) Nonalcoholic fatty liver disease: a spectrum of clinical and pathological severity. *Gastroenterology*, **116**, 1413–1419.

32 Daly, A.K., and Day, C.P. (2001) Candidate gene case-control association studies: advantages and potential pitfalls. *Br. J. Clin. Pharmacol.*, **52**, 489–499.

33 Griffin, J.L., and Nicholls, A.W. (2006) Metabolomics as a functional genomic tool for understanding lipid dysfunction in diabetes, obesity and related disorders. *Pharmacogenomics*, **7**, 1095–1107.

7
Case Studies on Epigenetic Inheritance
Gunnar Kaati

Abstract

The effects in adulthood of nutrition during adolescence, childhood, infancy, and the fetal and embryonic stages of development have attracted much attention in research. It is argued that the effects (adult diseases) are caused by nutritional constraints at critical phases of key fetal organ development. Here it is asked whether there are transgenerational effects. It is suggested that various nutrients transiently influence the expression of specific subsets of genes during the slow growth period just before the prepubertal peak in growth velocity of children.

Food availability during the slow growth period (SGP) just before the prepubertal peak in growth velocity of the paternal grandfather exerted an effect on the longevity of his grandchildren. Scarcity of food in the paternal grandfather's SGP was associated with significant extended survival of his grandchildren, while food abundance was associated with shortened life span of the grandchildren. One explanation was genomic imprinting, an intergenerational "feedforward" control loop linking grandparental nutrition with the grandchild's growth.

Cardiovascular mortality was reduced with poor availability of food in the father's SGP, but also with good availability in the mother's SGP. If the paternal grandfather was exposed to a surfeit of food during his SGP, the proband had a fourfold excess mortality related to diabetes. A father's exposure to a surfeit of food during his SGP, on the other hand, tended to protect the proband from diabetes.

The effects were sex-specific; the paternal grandfather's food supply was only linked to the mortality of grandsons, while the paternal grandmother's food supply was only associated with the granddaughter's mortality. This indicates the existence of a direct biological transgenerational effect, an epigenetic inheritance with the sex-specific patterns of transmission suggesting a direct role for the Y chromosome and possible the X chromosome.

The effects were not the result of the proband's own early life during childhood, and/or effects of genetic selection.

In conclusion, the studies explored the possible transgenerational effects from exposures occurring during the SGP prior to the prepubertal peak growth velocity. Such exposures seemed to influence gene expression in the next generation(s).

Epigenetics and Human Health
Edited by Alexander G. Haslberger, Co-edited by Sabine Gressler
Copyright © 2010 WILEY-VCH Verlag GmbH & Co. KGaA, Weinheim
ISBN: 978-3-527-32427-9

7.1
Introduction

In this chapter a research project, initiated in the early 1980s, that sought transgenerational explanations of adult disease is presented. At the overall level, the theoretical goal of the project was to link social medicine (now largely a discarded subject in medical schools) to the emerging field of epigenetics; the ambition was to integrate different systems in an expanded theoretical approach. This objective is not pursued here. Instead the focus is on the problems delved into, the data used, and of course the results. The data and the empirical assumptions of the project are rather unique, and will be discussed in some detail.

During the last 40–50 years, the search for the developmental origins of adult disease has proceeded along very disparate paths. A wide variety of explanatory factors (i.e., concerning coronary heart disease) has been advanced. In a survey in 1981 Hopkins and Williams examined 246 putative risk factors and the number has continued to increase [86].

Currently the etiology of coronary heart disease (and many other diseases, as well as longevity) is explained by (i) risk factors (immediate), and (ii) fetal growth impairment or fetal programming. In this chapter a different line of research is presented namely (iii) epigenetic inheritance. This approach ought to be interesting since it (a) probes the etiology of premature disease in humans from a transgenerational perspective, (b) shows what can be done with historical data, and (c) discusses the viability of epigenetic research in humans.

From the 1960s onwards the identification and modification of risk factors has been the main target of a large number of prevention programs attempting to reduce mortality and morbidity due to coronary heart disease by modifying the risks [1]. The promises of the prevention of thousands of deaths a year have not been fulfilled; the value of the knowledge underpinning these interventions has been questioned [2, 3]. The whole theory that underpins the interventions is problematic but a bigger problem is the possibility of thinking and acting rationally in a comprehensive sense, not a topic to be pursued here [4].

The first steps in a novel approach to the etiology of coronary heart diseases can, with hindsight, be said to have been taken with the publication of Anders Forsdahls *Are poor living conditions during childhood an important risk factor for arteriosclerotic disease?* [5]. Forsdahl had been working as a general practitioner in an impoverished region in the far north of Norway; his experiences from that work convinced him that childhood conditions impact on health status in later life.

David Barker and his associates [6, 83] carried Forsdahl's question forward in hypothesizing that undernutrition *in utero* permanently changes the body's structure, function and metabolism in ways that lead to coronary heart disease in later life [6]. Since then the "fetal programming" of later disease has spawned an ever increasing volume of research dealing with a wide range of other outcomes from diabetes [7], breast cancer [8, 9], asthma [10], low IQ [11], infant mortality [6, 12], hypertension [13], impairment of hearing or vision [14] schizophrenia [15], to lifestyle choices [16]. Recently, the perspective has been renamed "developmental origins of adult disease" and anchored in an evolutionary biology perspective [17–20].

In the "developmental origins" approach undernutrition *in utero* is seen as the main determining factor in the development of coronary heart disease [84]. But what if the causes of, or the risk factors for, coronary heart disease have their origin in the lives of the ancestors? And not in a sense of Mendelian inheritance, but as heritable changes in gene expression that are not due to changes in DNA sequence. The epigenetic approach promises to open up new avenues to the explanation of heart and coronary heart disease, type 2 diabetes, obesity, cancer, schizophrenia and other disorders in later life [85, 87]. Equally promising is that the epigenetic paradigm also provides a link to social factors, or internal and external environmental clues or exposures, in disease. It also provides a paradigm within which explanatory factors, reviewed, for example, by Hopkins and Williams, can be judged as to their biological plausibility.

Developmental experience and environmental influences are known to modify gene expression through epigenetic mechanisms, but it has been widely assumed that acquired epigenetic changes are "wiped clean" between generations and not transmitted to offspring. However, the phenomenon of transient environmental exposures producing transgenerational phenotypic effects has been documented in experimental rodents for a long time [21–23], but the mediating molecular mechanism(s) are not known.

To date the field relies on data from controlled animal studies where molecular mechanisms can be dissected. Although the biological plausibility of the proposed underlying epigenetic mechanisms are established [24–31, 88] the biological plausibility of the findings of transgenerational genetic/epigenetic effects on germ cells needs further corroboration.

7.2
Methodology

7.2.1
On the Study of Epigenetic Inheritance

To advance the field it is necessary to ascertain whether the transgenerational epigenetic effects/ mechanisms uncovered in animals can be confirmed in humans. Does environmental change or exposure in one generation influence important outcomes, gene expression, in the next or subsequent generations? The main problem to be confronted today is how to study epigenetic inheritance in humans.

Jablonka seems to be very optimistic about the possibility of carrying out long-term, transgenerational studies. She considers that the new molecular methods that allow the expression profiles of thousands of genes to be simultaneously observed and studied will improve the ability to assess phenotypes and relate them

7 Case Studies on Epigenetic Inheritance

to present and past environmental factors as well as to genotypes [27]. One line of research, she suggests, is to expose individuals in an isogenic line (lines where genetic differences between individuals are minimal) to a variety of treatments over several generations (such as extreme diets), and then test and compare gene expression profiles in them and their offspring. Thereby it should be possible to evaluate the effects that ancestral conditions have on descendants. Of course this would maybe be the ultimate test because if the "treated" parent is male, and the offspring inherits its pattern of gene expression, it is likely that the variation was transmitted through the sperm. Whether this design is realistic is an open question. At this stage, however, a more down to earth strategy seems more realistic.

7.2.2
The Ideal Study Design

In Figure 7.1 the design that underlay the planning in the early 1980s of the work presented here is outlined. As it stands, it was, and still is, an ideal design. It was recognized at the planning stage that the design could not be implemented owing primarily to resource limitations but it was also realized that a realistic approach had to combine historical and prospective data. In the end, resource limitations limited the data collection to one community (the original plan included four different types of communities), and to the proband generations and two parental

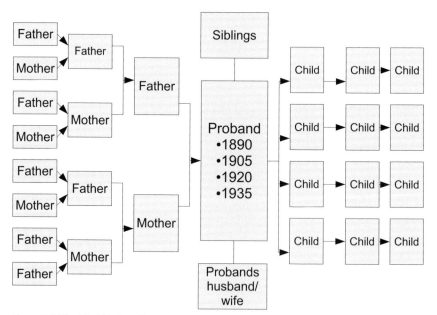

Figure 7.1 The Ideal Design: Generations.

generations (and no post-proband generations, several post-proband generations were included in the original design).

The study of transgenerational effects requires information from across generations, which is not easy to obtain in humans. The reasons are obvious. First, to collect new information over several generations is hardly possible in the current structure of research; secondly, the costs of such studies are very high; and thirdly, the problems of controlling all factors but those to be studied are rather overwhelming in these kinds of studies.

In the short run the only way to study transgenerational effects in health in humans is to combine historical data with data obtained from living and future generations of individuals related to the probands, but even this approach seems out of reach owing above all to the cost.

As regards historical data, the situation in Sweden is good. From early times a rather complete documentation of the population and the social and economic conditions in the parishes was carried out. The quality of the information is generally very high [32, 33].

The collection of data (DNA samples) from current and future generations is mainly a question of resources.

Two sets of data provide the empirical base for the work presented here; three Överkalix cohorts and the Avon Longitudinal Study of Parents and Children (ALSPAC) data set. The two data sets, and the construction of measures and indicators, and the statistical analyses performed are described in detail in papers by Bygren *et al.* [34], Kaati *et al.* [35], and Pembrey *et al.* [89].

7.2.3
The Överkalix Cohorts of 1890, 1905 and 1920

The cohorts used in this project consisted of 320 probands from three 50% random samples from the birth cohorts of 1890, 1905, and 1920 (data were also collected from the 1936 cohort but have not been used yet owing to some uncertainties about some secrecy issues). The cohorts and their ancestors were followed as they moved about in local parish registers of the Swedish and Finnish communities, and most could be traced to their death, emigration, or current residence. A detailed description of the cohorts is provided in the papers [34, 35].

The probands' parents and grandparents were traced back to their birth. In total, 1818 persons.

Birth dates, the dates May 1st and November 1st and historical data on harvests were used to assess ancestors' access to food during the slow growth period (SGP) of their childhood. Birth years only were retrieved for 13 of the grandparents. May 1st of the subsequent year was used as the worst time following a crop failure; the best time following a good harvest was represented by November 1st while intermediate availability was represented by July 1st.

Sensitivity analysis was performed where they were excluded and included, and denoted as living through years of a surfeit of food or of a poor availability. All but one of the significant results was unchanged.

Table 7.1 Överkalix: the samples, the attrition and the cohorts.

	Birth year			
	1890	1905	1920	All
Sample	108	99	113	320
Not traced	1	0	2	3
Emigrated	4	3	2	9
Alive at the end of the follow-up	0	2	42	44
Cause of death unknown	10	3	1	14

The causes of death were taken from the parish registers and coded at Statistics of Sweden according to the 9th International Revision of the Classification of Diseases (ICD-9). A cardiovascular death was defined as caused by hypertension, coronary artery disease, or stroke; that is, at least one ICD-9 number in the class 400–439 coded as direct, intermediate, underlying, contributory cause of death. A diabetes death was also noted when the ICD-9 number 250 was one of the causes. Age at death was used as a covariate (Table 7.1).

7.2.4
The ALSPAC Data Set

The ALSPAC (Avon Longitudinal Study of Parents and Children) data set contains all pregnancies in three districts around Bristol, UK. It was designed with particular attention to pregnancy and early development. Women with due dates between April 1st 1991 and December 31st 1992 were enrolled. For 10 000 of the 14 024 pregnancies surviving 5 months, a questionnaire was completed by the father containing questions about, among other things, the onset of smoking.

7.2.5
The Proband's Childhood

The childhood environment was partly described by the family's social circumstances, especially landownership. The father's death during the proband's intrauterine life and up to the age of 13, and correspondingly the mother's death when the proband was 0–12 years old (if the death occurred before the proband's death) were used as indicator variables. The literacy level of the father and the mother according to the clergy's mark in five grades when the parents were confirmed at the age of 14 were used as a continuous variable. The occupation of the father was dichotomized into blue-collar (mostly farmhands) and other. The siblings were grouped into firstborn, middle- and lastborn.

7.2.6
Food Availability

The availability of food in the county of Norrbotten during any year has been classified based on regional harvest statistics (unprinted), grain prices [36], the estimates of a 19th century statistician [37] and general historical facts. Food availability was classified as poor, moderate, or good. Most harvests were completed in September, with the slaughtering of pigs and cattle occurring later in the fall. We used the 1st of May in the year after a crop failure as the time when food was least available. The period following a good harvest, when food was most available, was represented by the 1st of November. The age of a parent or grandparent on those dates was used to determine the availability of food during his or her SGP.

7.2.7
Ancestors' Experience of Crises

There is some information indicating the quality of harvests during the end of the 18th and the 19th century available from various sources. In the statistical tables kept for every parish since 1749 the parish clergy was required to make quantitative estimates of the harvests until 1820 (Tabellverket). Thereafter, county governors reported harvests to the central authorities. This sort of information was used in the national statistics in 1855 to give an indication of national harvest results and development. The statistician Hellstenius presented harvest estimates at county level. For that purpose, he presumably used both the statistical tables and the reports from county governors. He graded the harvests from 0 (total crop failure), to 6 (very good harvests) [37]. Others have used information about price fluctuations. Jörberg [36] found that there is a correlation between harvest results and price fluctuations during the 18th century. Poverty fluctuations are another method of calculation that could be used. In the end it proved possible to use Hellstenius' estimates for Norrbotten for the period 1816–1849 as the basis of our analysis. These crop failure identifications were consolidated with sources of price fluctuations and governmental aid, and with protocols from local parish meetings, communications from the county governor and other qualitative sources. For the study-period 1803–1815 and 1850–1900 estimates from the statistical tables were used as a primary source.

7.2.8
Growth Velocity

It was hypothesized that there are two sensitive periods in children's growth, which might be crucial in the transmission of information between generations. In particular, the period just before the prepubertal peak in growth velocity, when the requirements of nutrition are rather low and the child might therefore be overfed, may result in epigenetic manipulations. The concept of a slow growth period (SGP) was thus born.

The SGP was determined from the growth velocity during childhood, measured in centimeter increases per year. The ages for SGP were set at 8–10 years for girls and 9–12 years for boys. The period had to be estimated from a modern cohort, and turned out to be shorter for girls than for boys [38]. Furthermore, the successive decrease in age at puberty during the 19th and 20th centuries had to be taken into account [39]. The delay of the onset of puberty in parents and grandparents and hence the SGP compared to the modern cohort was set at 1 year.

The growth velocity during childhood has undergone a change from the end of the 19th century up to the present. A peak occurs among children in the years before puberty, which is preceded by a slow growth velocity period [38]. The average age of puberty has decreased and the distribution around the average age has become less skewed to the right since the end of the 19th century. During the 19th century, the age at puberty did not change much [39]. For parents and grandparents born in the 19th century the decrease in the age of puberty has been set at 1 year. The SGP was determined from the modern growth velocity estimates adjusted for the decrease of pubertal age. Food availability during the SGP was used as a proxy for nutritional intake.

7.3
Patterns of Transgenerational Responses

7.3.1
The Social Context

The study area was one of isolation and impoverishment. The landscape was barren, suitable for farming only in some areas and in smallholdings with 1–2 cows. The capacity of the land was thus restricted owing to the limitations of the arable land, although this did not seem to have affected the growth of the population. In 1750 the population was 792, fifty years later the population was 1414, almost a two-fold increase. It reached 2500 in 1850, an increase of 70%. The expansion of the population continued unabated, reaching 3165 persons in 1859 (Tabellverket), 5547 in 1900 (Folkräkningen 1900) [40], and peaked in 1950 to a population of 9214. Then the decline began and in 2004 the population was 4063, back at the 1880 level (Folkmängd 1810–1990) [41]. The social composition of the population was very simple. Only 4–6 adults could be classified as being educated and well off during the period up to modern times.

The population growth in the parish was chiefly a result of an increased number of births: 31 live births in 1750, 45 in 1800, 104 in 1850, and 139 in 1859. The rate of infant deaths remained almost constant during the same time, between 5–12 deaths per year in the 19th century. That was a remarkable decline in relation to the number of births and the poor socio-economic circumstances in the parish.

No systematic differences in child mortality between the sexes during this period can be detected, it seems to vary only by chance.

The population in the parish was culturally and genetically extremely homogeneous.

The grim socio-economic circumstances confronting, in particular, the children and the mothers emerges from the parish priest's reports of the child-bearing women's situation during the crisis year. It is noted that, for example, in reports covering the crisis years 1831 to 1836, in 1831 of the 75 mothers giving birth only 30 lived under "good" circumstances; a year later 21 out of 91; in 1833, 7 out of 60; in 1834, 5 out of 60; in 1835, 12 out of 75; and in 1836, 8 out of 67. The majority were classified as living under very poor or barely manageable circumstances (Tabellverket 1849–1859) [42].

7.3.2
The Ancestors' Nutrition

The crops were often poor during the 19th century in the study area, most of the arable land was already in use and almost nothing more was developable. So the provision of livelihood was basically dependent on the outcomes of harvests. The harvests of 1800, 1812, 1821 and 1829 were classified as total crop failures. In the period 1831–36, all years but 1834 were years of total crop failures and as 1834 was affected by lack of seed-corn it was also classified as a year of total crop failure. This was also true of 1809 because of the Swedish–Russian war when the armies laid their hands on much of the local stocks and made it difficult for the harvest to be brought in. In 1851 and 1856 there were again total crop failures and the harvests in 1867, 1877, 1881, 1888 and 1889 were poor. A surfeit of food was available after the harvest in 1799, 1801, 1813–1815, 1822, 1825–26, 1828, 1841, 1844, 1846, and 1853, 1860–61, 1863, 1870, 1876, 1879 and 1880. All other years had moderate harvests, neither very poor nor good. In 39% of cases the paternal grandmothers had experienced famine during their slow growth period, as had 50% of the paternal grandfathers. Good harvests during the SGP were experienced by 46 and 47%, respectively. Relief programs were not in place until the late 19th century.

7.3.3
Longevity and Paternal Ancestors' Nutrition

Were there intergenerational effects on survival from "overnutrition" during the slow growth period just before the prepubertal peak of growth velocity? There was indeed an effect of food availability during the SGP of the paternal grandfather on the longevity of the probands. Scarcity of food in the grandfather's SGP was associated with a significantly extended survival of his grandchildren, while food abundance was associated with a greatly shortened life span of the grandchildren.

The assumptions underlying the question were, building on McGill [43], Martyn et al. [44] and Signorelli and Trichopoulos [45], that the functioning mechanism was epigenetic. Whether imprints are initiated or erased is essential for normal development [46, 47]. An intergenerational "feedforward" control loop was proposed by Pembrey [48] linking grandparental nutrition with grandchildren's growth. The mechanism could be a specific response, for example, to their nutritional state directly modifying the setting of the gametic imprint on one or more genes.

Nutrition affects ovaries and testes from the moment they form during fetal life up to maturity [49]. An effect due to the fathers and mothers living in an environment with a surfeit of food or poor availability of food during their SGP on their children's longevity could not be detected.

A similar environment during SGP of most of their grandparents had no impact on their children's longevity and did not affect their grandchildren either, but for one important exception. The paternal grandfather living in an environment with a surfeit of food during his SGP did affect his grandchild's longevity negatively, and, conversely, affected it positively when the availability was poor.

It is interesting that abundance of nutrition seems to have affected the DNA directly when it affected the paternal grandfather and perhaps the paternal grandmother but not maternal ancestors. This was not seen for other childhood periods, but only, uniquely, for nutrition during the SGP.

It is also of interest that the impact skipped the parent generation which indicated an intergenerational "feedforward" control loop similar to the loop proposed by Pembrey [48] linking grandparental nutrition with the grandchild's growth.

7.3.4
The Influence of Nutrition During the Slow Growth Period on Cardiovascular and Diabetes Mortality

A question pursued was whether variations in childhood nutrition could influence the maturing gametes. It was hypothesized that exposure to a poor diet during a period of great need for energy at the prepubertal growth velocity peak and exposure to plentiful food during the slow growth period (SGP) compared to intermediate food availability would result in high cardiovascular mortality in subsequent generations. However, cardiovascular deaths had a dose-response to the food energy availability during the ancestor's SGP only. Especially, cardiovascular- and diabetes-related deaths were studied since imprinted genes have been implicated in cardiovascular and diabetes risk.

Overall they show that cardiovascular mortality was reduced with poor availability of food in the father's SGP, but also with good availability in the mother's SGP. This reciprocal effect of parental nutrition is intriguing in itself, but the most striking results come with diabetes. If the paternal grandfather was exposed to a surfeit of food during his SGP, then the proband had a fourfold excess mortality related to diabetes (OR 4.1, 95 % c.i. 1.33–12.93, $P = 0.01$) when age at death and the effects of possible over eating among parents and grandparents during their respective SGP were taken into account. A father's exposure to a surfeit of food during his SGP tended to protect the proband from diabetes (OR 0.13, 95% c.i 0.02–1.07, $P = 0.06$), hinting at some "see-saw" effect down the generations.

7.3.5
Is Human Epigenetic Inheritance Mediated by the Sex Chromosomes?

A question of considerable interest is whether nutritional exposure influences gene expression in the next generation(s). It was tested whether the male SGP is a specific time for triggering a transgenerational effect using the Överkalix data and the Avon Longitudinal Study of Parents and Children (ALSPAC), bearing in mind that the outcomes might be sex-specific. More specifically, it was tested whether paternal smoking habits during the fathers' SGP have transgenerational effects on the pregnancy outcomes of their future offspring, and also whether the effect of ancestral food supply on the proband mortality differs by sex.

Variations in response to nutritional "exposures" are determined by inherited differences, different developmental experience or both. Inherited differences are traditionally regarded as DNA sequence variations transmitted by parents, while environmental influences are usually thought of as only operating during the person's lifetime. Another possibility to be considered is a transgenerational response. This requires a mechanism for transmitting environmental exposure information that then alters gene expression in the next generation(s). This may be performed down the female line through the egg cytoplasm or transplacental route; a transgenerational response down the male line involves the carrying of ancestral exposure information by the sperm's chromosomes as a particular state, or by responsive DNA sequences. Viruses, prions or RNA molecules could also mediate transgenerational effects.

Were there sex-specific effects in linking ancestral food supply to mortality? Both parents' exposure to food availability during their SGP had significant influences on their daughter's mortality RR, with a possible effect of the mother's food supply on her son's mortality RR.

There were no significant effects of SGP food availability for either maternal grandparent on the mortality RR in the grandchildren. However, on the paternal side there were significant effects on the grandchildren's relative mortality RR and they are strikingly sex-specific. The male (but not the female) probands had an increased ($P = 0.009$) mortality RR of 1.67 when the paternal grandfather had experienced good food supply during his SGP (compared to the risk for male probands whose paternal grandfather had other experiences during his SGP).

The female (but not the male) probands had a twofold higher mortality RR when the paternal grandmother experienced good availability during her SGP compared to the mortality risk of probands whose paternal grandmothers had other experiences during the SGP ($P = 0.001$). Poor food supply during the SGPs of the paternal grandparents was followed by the opposite effect. The male probands whose paternal grandfather had such an experience had a reduced ($P = 0.025$) mortality RR of 0.65 and female probands had some tendency ($P = 0.116$) to reduced mortality RR (0.72) when the paternal grandmother had poor food supply.

The above described results are summarized in (Figure 7.2).

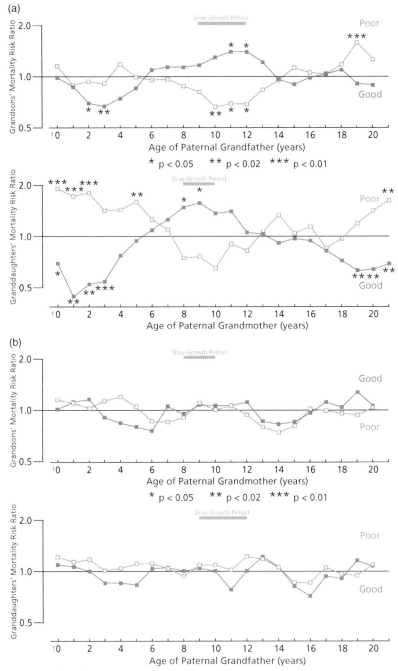

Figure 7.2 The effect of paternal grandparental food supply (good = filled squares, poor = open squares) at different times in their early life on the mortality rate of their grandchildren. (a) The mortality (y-axis) of the male proband as a response to his paternal grandfathers childhood (x-axis) food supply. (b) The corresponding mortality of the female proband as a response to her paternal grandmothers childhood food supply. Food supply at age 0 is the mean for the 33 month period from −267 days until the day before the second birthday and therefore includes fetal life.

7.3.5.1 Paternal Initiation of Smoking and Pregnancy Outcome

Almost half (5357 of 10 000) of the fathers responding to the question in the ALSPAC study smoked, mostly starting at the age of 16. But 166 reported smoking before 11 years of age when many would still be in the SGP. The effects on pregnancy outcome (birthweight and gestational length) of the onset of paternal smoking at <11 years, 11–12 years, 13–14 years, >14 years, and of not smoking were compared. In a regression analysis, the initially significant reduction in birth weight with the earliest paternal onset of smoking lessened on correction for gender, parity, gestation length, and maternal smoking ($P = 0.1$). Given that boys are born earlier than girls and in view of the possible tendency of the Y chromosome for epigenetic transmission, the data were stratified by the sex of the baby. This showed a significant association between the age of onset of paternal smoking and mean gestation length being restricted to boys (even when controlled for socio-economic status).

7.3.6
Epigenetic Inheritance, Early Life Circumstances and Longevity

A possible explanation for the associations uncovered is that they could be produced by the social circumstances of the people concerned. As has been described earlier the social circumstances of the probands and their ancestor's were pretty hard.

About half of the fathers in all three cohorts were owners of small holdings, which in this parish usually meant quite small units. The mortality rate among parents was rather high: 18 and 29 cases (6% and 10%) of the parents of the male and female probands, respectively, died before the probands reached their 13th birthday.

The family size was large and 72 of the probands had 10 or more siblings, the largest sibship was 15. The large family size put a considerable stress on the parents to provide the necessities for living. The probands' parents' literacy rate was not total since 22% of the fathers and 24% of the mothers did not pass the literacy test performed by the clergy.

The transmission of influence of the exposure of good nutrition from the paternal grandfather to the grandsons entailed higher mortality risks for the proband [34, 35, 50]. The male probands had an increased risk of mortality if their fathers had good nutrition during their SGP but this became visible only after the introduction of the proband's childhood circumstances. The influence of the paternal grandfather's exposure to good nutrition during SGP was as expected although slightly attenuated after the introduction of childhood circumstances. The same was true for the influence of poor nutrition. The childhood environment of the probands had profound influences on the male probands HR. For them the mother's death during childhood was detrimental, while the better the mother's literacy, the better it protected the male proband. For the female proband no such influence of the childhood environment could be detected.

7.3.7
How to Explain the Effects of Food Availability During SGP on Human Health?

The possible explanations boil down to four: chromosomal transmissions of nutritionally-induced epigenetic modifications, intense genetic selection through differential survival or fertility, a statistical quirk or hidden bias producing a false association, or some mechanism of inheritance yet to be discovered. Of the four the first two will be considered here, the other two, although not to be dismissed out of hand, do not pose any direct problem.

7.3.7.1 Genetic Selection Through Differential Survival or Fertility?

At a time of relatively high infant mortality and with the possibility of variations in food supply influencing fertility, genetic selection is a possibility, although in this case it would have to be very intense to sustain the effects over three generations. Controlling for longevity, food availability in infancy and sibship size excludes evidence of intense selection. The analysis of the social context clearly shows that the demographic profile of the community studied was stable and no particular trends could be detected that would hint at genetic or other forms of selection. In fact, the infant mortality rate remained almost constant during the whole 19th century at between 5–12 deaths per year. The swings to the upper level from the median value were caused by a measles epidemic. No systematic differences in child mortality between the sexes during this period could be detected – the death rate in relation to sex seemed to vary only by chance. This was also the case concerning child mortality.

Since the same combination of exposures down the male line could influence both sibship size and the diabetic risk in grandchildren it could be argued that the same transgenerational influence causes both subfertility and/or increased embryonic/fetal loss and a diabetic susceptibility in those offspring who survive. If so, the transgenerational response to variations in nutrition could have evolved primarily in relation to the regulation of reproduction, with the diabetic risk being a secondary consequence. The fact remains: those who either starved or over-ate during the SGP could not be demonstrated to have different early mortality and their ancestors could not be shown to have different reproductive fitness.

7.3.7.2 Chromosomal Transmission of Nutritionally Induced Epigenetic Modifications

The demonstration of the existence of a sex-specific transgenerational response system in humans that is triggered by exposures at very specific times during the ancestors' development has a number of implications, for example, for the understanding of variations in health.

It is the SGP but not puberty that seems to be an exposure-sensitive period for the sperm-mediated transgenerational effects of nutrition; probably a number of exposures such as tobacco are influential.

What mechanism might be mediating the direct biological transgenerational effects? Transmissible agents would not be expected to segregate down the generations in the way observed. The most likely explanation is epigenetic inheritance with the sex-specific patterns of transmission suggesting a direct role for the Y chromosome and possible the X chromosome.

One test of the causal nature of an association is biological plausibility. The possibility of human epigenetic inheritance has not been taken seriously, despite experimental evidence in mammals [25]. One reason for this reluctance to take account of the idea of epigenetic inheritance is probably a certain suspicion of a hypothesis with a Lamarckian flavor. Another important reason for the reluctance has been the lack of compelling human observations and plausible molecular mechanisms that could be investigated experimentally.

If imprinted genes are now involved, the INS-IGF2-H19 imprinted domain is a candidate. It is paternally imprinted and variations at the INS VNTR are associated with diabetes (Judson et al., 2002) [51, 52, 90]. It has been shown that the father's untransmitted INS VTR allele can influence the effect of the transmitted allele on the child's diabetes risk, indicating that a transgenerational mechanism can operate on the INS-IGF2-H19 imprinted domain. The question is: Does it operate in a boy's testes during this SGP? [50].

BORIS (brother of the regulator of imprinted sites), a novel male germ-line specific protein associated with epigenetic reprogramming events, shares the same DNA-binding domain as CTCF, the insulator protein involved in reading imprinting marks at the imprinting control region between IGF2 and H19, among many other functions [53, 54]. The role that CTCF performs for cell survival is most likely performed by BORIS in the primary spermatocytes of the testis, where CTCF is silenced. This switch takes place in association with erasure and re-establishment of methylation marks [50].

Although there is a paucity of information on the human prepubertal testis, the available studies indicate that prespermatocytes are present from the age of 5 years [55]. From about 8 years of age the proportion of boys' seminiferous tubules with primary spermatocytes increases until full spermatogenesis at puberty. An increasing number of primary spermatocytes survive to progress through meiosis to spermatids. The SGP is therefore a crucial stage in development in that a viable pool of spermatocytes emerges and the beginning of re-programming of methylation imprints occurs. It is therefore a dynamic state in which different sensing mechanisms may operate.

7.4
Epigenetic Inheritance

The studies presented here uncover evidence of transgenerational sex-specific effects specifically as a result of experiences occurring during the male SGP. It is, however, necessary to remember that the studies presented here have not formally demonstrated epigenetic inheritance by molecular analysis; the mediating mechanisms are called "epigenetic" influence or inheritance.

The specific genomic patterns and extent of plasticity of DNA methylation are not clearly understood [56]. The same can be said about the sensitivity of genomic methylation patterns to external influences. There are differences in the timing of the acquisition of genomic methylation patterns in the male growing testes (begun before birth) and female (begun postnatally in growing oocytes) germ lines, but how germ cell development methylation patterns are maintained, propagated, and repaired is unclear [24].

Of great interest in this context are the differences between genetically identical twins. One unclear issue in, for example, complex diseases, is the relatively high degree of discordance of monozygotic twins. Discordance of monozygotic twins reaches 30% for idiopathic epilepsy, 30–50% for diabetes, 50% for schizophrenia, 70% for multiple sclerosis and rheumatoid arthritis and 80% for breast cancer [57, 58]. How, when and why do these differences originate? It is clear that some changes are initiated in parental or earlier generations [59]. Geneticists have been baffled by these issues. Despite the same DNA sequence and the same environment "identical" twins can sometimes show striking differences. It is here that epigenetic inheritance comes in. May subtle modifications to the genome that do not alter its DNA sequence provide the answer? In the case of discordance of monozygotic twins there can exist a substantial degree of epigenetic dissimilarity which is accumulated over thousands of mitotic division of the cells in two genetically identical organisms. Although they carry identical DNA sequences they may exhibit numerous epigenetic differences [28, 58, 60].

In general, it implies that the health and physiology of people can also be affected by inherited effects of the interplay of genes and environment in their ancestors but neither by inheriting particular genes nor the persistence of the ancestral environment.

The epigenetic inheritance system provides a cohesive explanation for various features of complex diseases as well as the development of adult disease, and epigenetic mechanisms can be a common denominator for a wide variety of epidemiological, clinical and molecular findings in diseases [61].

DNA sequence variations within genes and epigenetic regulation of genes are of course closely related and should be investigated in parallel.

7.4.1
Fetal Programming and Epigenetic Inheritance

It has been pointed out that for a man falling into the lowest risk groups for plasma lipid concentrations, blood pressure, cigarette smoking, and the presence of preexisting symptoms of coronary heart disease, the commonest cause of death is coronary heart disease [17, 62]. A search for explanations has been pursued for decades. At first, the search focused on the possibility that early life experiences have a powerful impact on the emergence of cardiovascular disease – and adult disease in general; then the search moved on to the fetal life as the crucial environment. Still it has to be asked: Which is the most vulnerable period: fetal, embryonic, or postnatal life? Here a further dimension of the early life experiences has been explored by analyzing whether intergenerational environments link parental or grandparental nutrition with a child's or grandchild's disease susceptibility and survival.

The original Barker hypothesis has been transformed to the "developmental origins of adult disease" perspective and anchored in evolutionary biology [17–20]. At the same time, it seems, the nature of the explanation proposed has changed. The proponents of this complex of ideas, drawn from evolutionary ecology, behavioral development, life-history theory, molecular biology and medical epidemiology, state that these fields have converged on the key finding "that a given genotype can give rise to different phenotypes, depending on the environmental conditions" [18]. It is in these terms that the fetal programming hypothesis is framed. It is difficult to see how the inclusion of epidemiological observations on birthweight and adult disease in this evolutionary perspective relates to current empirical issues [63]. Obviously the intention of the proponents must have been to strive for a consolidation of proximate and ultimate explanations though it is uncertain how successful it is. As it is developed at such an abstract level it seems not to provide much of guidance for empirical scientists apart from providing an abstract language and an all-encompassing perspective.

The question is then how the concept of developmental plasticity as an explanation of the associations between, for example, coronary heart disease and diabetes, improves our understanding of causes of disease and particularly its utility in interventions.

The relationship between fetal programming and epigenetic programming is unclear. The relationship between the genetic component of adult diseases, metabolic programming (or fetal programming) and cellular epigenetic information systems has to be considered the general problem in the field. It is not clear that the Barker hypothesis assumes metabolic programming in the fetal period [91], but if so, it is not clear in what diseases and whether they are also assumed to show transgenerational effects. A number of questions have to be posed: Are there transgenerational effects and if so through what mechanisms? Does fetal programming involve epigenetic reprogramming of the germ cells? Are there germ-line epimutations and/or germ-line mutations?

Critical periods and environments "disturb" in one way or another the normal development from a seed or fertilized egg, through a series of cell divisions, differentiation, and movements of tissues to the complete organism [64–71]. Where along this path do the disturbances arise that in adult years lead to disease and early death? It is obvious that it is a complex process that encompasses a broad assembly of influences both inside and outside the organism [30].

There is a substantial body of research that supports the claim that experiences throughout one's life have profound effects on personality, character, and health. Attachment theory, for example, is the research basis for the idea that "the first years last forever"; that is, certain maternal or caregiver behavior ("inherited" over generations) causes different kinds of attachment relations, with varying long-term consequences [72].

7.5
Future Directions

It is worth noticing that more than 98% of the human disease burden consists of diseases that cannot be explained by independent variations in single genes, but has to be explained by complex etiology [30, 73]. Therefore, a theoretical approach encompassing the complex etiology of adult disease has to focus on unfolding over time at different levels of scale and organization. That is, causes operate at many levels of organization (system levels) and at many stages in the chains of causation [65, 74, 75].

The real comprehension will only come with the understanding of how all the parts interact to generate organismal traits [68, 73, 92]. The twentieth century notion that genes represent a privileged level of explanation of the development and evolution of organismal traits is questionable. On the other hand, knowledge about gene sequences, processing of gene products, and gene–gene interactions is undoubtedly essential in any systems approach to the understanding of development and evolution [92, 93].

Ideally a framework that incorporates historical, individual and societal-dimensions and the processes of evolving health conditions over several decades would be required in order to make sense of the evolution of generations. This would allow the search for causes at multiple levels of organization as well as at the level of individual and community history, and the processes linking an antecedent event and a development to the emergence of risk factors and diseases [69, 71, 76].

A fruitful framework has to depict sets of interacting systems of resources, entities and influences that produce the life cycle of an organism [77]. The developmental means or/and systems exist at different levels of organization which is difficult to depict in a one-dimensional figure. In this framework the organisms and their environments define the relevant aspects of, and can affect, each other; it is, then, the inter-penetration of organism and environment [74]; or the dialectic between the interactants in the process that characterizes the framework [66, 67, 78, 94].

The focus here is on the transmission of variation from one generation to the next and it is the whole developmental system with all its different and interacting inheritance systems that has to be considered. Jablonka and Lamb [27, 59] provide an extension by outlining what they call multiple inheritance systems; they include the genetic inheritance system, cellular or epigenetic inheritance systems, the system underlying the transmission of behavior patterns through social learning, and the communication systems employing symbolic language.

These views emphasize thinking in terms of systems that consist of heterogeneous arrays of processes, entities, and environments; chemical and mechanical, micro-and macroscopic, social and psychological. The forces are multiple and mobile, distributed and systemic [79, 80, 95].

The metaphor of development carries the implication of an unfolding or unrolling of an internal program that determines the organism's life history from its origin as a fertilized zygote to its death. This metaphor fails to take account of the interactive processes that link the inside with the outside. Individual development is not unfolding.

Heredity is not an explanation of this process, but a statement of that which must be explained [81]. Genetic research is important for the future health of the population but genes alone are only factors to explain development; environments – organic and inorganic, microscopic and macroscopic, internal and external – change over generations [82, 96].

Although an extended systems approach provides a fresh way to approach and analyze health problems, it is obvious that the framework has to be developed further in order for it to be possible to deal with problems that the systems approach has led us to, namely to explain how intergenerational transfer of "information" occurs. What we can be sure of is that a multitude of causal factors is needed for a reasonable treatment of the life courses of individuals. The model raises other question such as what other resources are inherited at what developmental stage. And what are development stages? How far back does or should the causal networks extend? The concept of inheritance is used to explain the stability of biological forms from one generation to the next although this rather straightforward concept is extended here; in the context of developmental theory the concept of inheritance may apply to any resource that is reliably present in successive generations [27, 59].

A systems perspective is not all-embracing: no scientific theory has ever explained, nor claimed to explain, all aspects of its field of discourse.

7.6
Conclusions

We are only beginning to understand the complexity of how developmental exposures influence gene expression in subsequent generation(s). For a long time it has been assumed that the essence of all human disease is related to DNA sequence variation but now the idea that epigenetic changes might be linked to disease is well established. The possibility that environmental influences on parents or grandparents could influence their grandchildren is intriguing. It was shown that many patients with heart disease and diabetes could be suffering from the after-effects of their ancestors living in an environment abundant in food.

The SGP before the prepubertal peak of growth velocity was found to be sensitive for nutrition and resulted in transgenerational influences on longevity. Scarcity of food in the grandfather's SGP was associated with a significantly extended survival of his grandchildren, while food abundance was associated with greatly shortened life span.

Cardiovascular disease and diabetes were found to be partly determined by ancestors' experience during the SGP prior to the prepubertal peak in growth velocity. Cardiovascular mortality was reduced with poor availability of food in the father's SGP, but also with good availability in the mother's SGP. With regard to diabetes, the exposure of the paternal grandfather to a surfeit of food during his SGP was associated with a fourfold excess mortality.

The effect on the proband's mortality of ancestors' experience of food availability was found to be sex-linked and most distinct going from the paternal grandfather to the grandson and from the paternal grandmother to the granddaughter.

An influence from the proband's own childhood was found to partly compete with the transgenerational effects of ancestors food supply but the inheritance from grandparents to grandchildren was still pronounced.

This research suggests a closer degree of integration between environment, including social and economic realities of life, and biological evolution, and an overall more significant role for environment in determining morbidity and mortality of life chances.

In an evolutionary perspective these studies resurrect the evolutionary debate over the role of the environment, and its specific interplay with biological evolution. The environment is assumed to determine only on what grounds selection takes place and what characteristics are necessary for better reproductive opportunities. It does, however, not suggest a re-evaluation of the Lamarckian theory that argued for the inheritance of acquired characteristics.

From the perspective of a controlled trial the work presented here has a number of shortcomings, the most obvious being the historical nature of the data, and the indirect nature of the indicators used.

In order to advance the field it is necessary to replicate the findings on other cohorts exposed to nutritional excess/deficits. The measurement should ideally be applied directly rather than indirectly although this seems very difficult to carry out, as pointed out above.

Another way to advance the field is to acquire data from controlled animal studies where molecular mechanisms can be dissected. The biological plausibility of the findings of transgenerational genetic/epigenetic effects on germ cells needs further corroboration.

In the context of study area it has to be asked whether there were or are cumulative environmental effects when several successive generations are exposed to rather persistent extreme diets. Replication in other settings would be valuable.

A distinct strength of our studies is that the population of the parish/study area was culturally and genetically homogeneous.

Acknowledgments

This overview is based on papers resulting from the work of myself, Lars-Olov Bygren, Sören Edvinsson, both of University of Umea, Micael Sjöström, Karolinska Institute, Stockholm, and Marcus Pembrey, University College London and Bristol University. They are, however, not responsible for the content of this paper.

References

1. Kuh, D., and Davey Smith, G. (1997) The life course and adult chronic disease: an historical perspective with particular reference to coronary heart disease, in *A Life Course Approach to Chronic Disease Epidemiology* (eds D. Kuh, and Y. Ben-Shlomo), Oxford University Press, Oxford, pp. 15–41.
2. Le Fanu, J. (1999) *The Rise and Fall of Modern Medicine*, Abacus, London.
3. Schwab, M., and Syme, S.L. (1997) On paradigms, community participation, and the future of public health. *Am. J. Public Health*, **87** (12), 2049–2051.
4. Churchman, C.W. (1979) *The Systems Approach and Its Enemies*, Basic Books, New York.
5. Forsdahl, A. (1977) Are poor living conditions during childhood an important risk factor for arteriosclerotic disease? *Br. J. Prev. Soc. Med.*, **31**, 91–95.
6. Barker, D.J.P., and Osmond, C. (1986) Infant mortality, childhood nutrition, and ischaemic heart disease in England and Wales. *Lancet*, **1**, 1077–1081.
7. Potischman, N., and Troisi, R. (1999) In-utero and early life exposures in relation to risk of breast cancer. *Cancer Causes Control*, **10**, 561–573.
8. Barker, D.J., Osmond, C., Thornburg, K.L., Kajantie, E., Forsen, T.J., and Eriksson, J.G. (2008) A possible link between the pubertal growth of girls and breast cancer in their daughters. *Am. J. Hum. Biol.*, **20** (2), 127–131.
9. Frankel, S., Gunnel, D.J., Peters, T.J., Maynard, M., and Smith, G.D. (1998) Childhood energy intake and adult mortality from cancer: the Boyd Orr Cohort Study. *BMJ*, **316**, 499–504.
10. Steffensen, F.H., Sorensen, H.T., Gillman, M.W. et al. (2000) Low birth-weight and preterm delivery as risk factors for asthma and atopic dermatitis in young adult males. *Epidemiology*, **11**, 185–188.
11. Richards, M., Hardy, R., Kuh, D., and Wadsworth, M.E.J. (2001) Birthweight and cognitive function in the British 1946 birth cohort: longitudinal population based study. *BMJ*, **322**, 199–203.
12. Barker, D.J.P. (1994) *Mothers, Babies, and Disease in Later Life*, BMJ Books, London.
13. Osmond, C., and Barker, D.J.P. (2000) Fetal, infant, and childhood growth are predictors of coronary heart disease, diabetes, and hypertension in adult men and women. *Environ. Health Perspect*, **108**, 545–553.
14. Olsen, J., Sörensen, H.T., Steffensen, F.H. et al. (2001) The association of indicators of fetal growth with visual acuity and hearing among conscripts. *Epidemiology*, **12**, 235–238.
15. Hultman, C.M., Sparen, P., Takei, N., Murray, R.M., and Cnattingus, S. (1999) Prenatal and perinatal risk factors for schizophrenia, affective psychosis, and reactive pscyhosis of early onset: case-control study. *BMJ*, **318**, 412–426.
16. Vickersk, M.H., Breier, B.H., McCarthy, D., and Gluckman, P.D. (2003) Sedentary behaviour during postnatal life is determined by the prenatal environment and exacerbated by postnatal hypercaloric nutrition. *Am. J. Physiol.*, **285**: R271–R273.
17. Barker, D.J. (2007) The origins of the developmental origins theory. *J. Intern. Med.*, **261** (5), 412–417.
18. Bateson, P., Barker, D., Clutton-Brock, T., Debal, D., D'udine, B., Foley, R.A., Gluckman, P., Godfrey, P., Kirkwood, T., Mirazón Lahr, M., McNamara, J., Metcalfe, M.B., Monaghan, P, Spencer, H.G., and Sultan, S.E. (2004) Developmental plasticity and human health. *Nature*, **430**, 419–421.
19. Gluckman, P.D., and Hanson, M.A. (2004) Living with the past: evolution, development, and patterns of disease. *Science*, **305**, 1733–1736.
20. Gluckman, P.D., and Hanson, M.A. (2007) Developmental plasticity and human disease: research directions. *J. Intern. Med.*, **261** (5), 461–471.
21. Anway, M.D., Cupp, A.S., Uzumcu, M., and Skinner, M.K. (2005) Epigenetic transgenerational actions of endocrine disruptors and male fertility. *Science*, **308** (5727), 1466–1469.

22 Boucher, B.J., Ewen, S.W., and Stowers, J.M. (1994) Betel nut (Areca catechu) consumption and the indication of glucose intolerance in adult CD1 mice and in their F1 and F2 offspring. *Diabetologia*, **37**, 49–55.
23 Campbell, J.H., and Perkins, P. (1998) Transgenerational effects of drug and hormonal treatments in mammals; a review of observations and ideas. *Prog. Brain Res.*, **73**, 535–553.
24 Bird, A. (2002) DNA methylation patterns and epigenetic memory. *Genes Dev.*, **16**, 6–21.
25 Chong, S., and Whitelaw, E. (2004) Epigenetic germline inheritance. *Curr. Opin. Genet. Dev.*, **14**, 692–696.
26 Dennis, C. (2003) Altered states. *Nature*, **42** (1), 686–688.
27 Jablonka, E. (2004) Epigenetic epidemiology. *Int. J. Epidemiol.*, **33**, 929–935.
28 Petronis, A. (2006) Epigenetics and twins: three variations on the theme. *Trends Genet.*, **22** (7), 347–350.
29 Reik, W., and Walter, J. (2001) Genomic imprinting: parental influence on the genome. *Nat. Rev.*, **2**, 21–32.
30 Strohman, R. (2002) Maneuvering in the complex path from genotype to phenotype. *Science*, **296**, 701–703.
31 Waterland, R.A., and Jirtle, R.L. (2003) Transposable elements: targets for early nutritional effects on epigenetic gene regulation. *Mol. Cell Biol.*, **23**, 5293–5300.
32 Johansson, E. (1983) *Kyrkböckerna Berättar*, Liber, Stockholm.
33 Kock, K. (1959) *Statistiska Centralbyrån 100 år. Minnesskrift med anledning av Tabellkommissionens omvandling år 1858 till ett allmänt statistiskt ämbetsverk*, Statistiska Centralbyrån, Stockholm.
34 Bygren, L.O., Kaati, G., and Edvinsson, S. (2001) Longevity determined by ancestors' overnutrition during their slow growth period. *Acta. Biotheoretica*, **49**, 53–59.
35 Kaati, G., Bygren, L.O., and Edvinsson, S. (2002) Cardiovascular and diabetes mortality determined by nutrition during parents' and grandparents' slow growth period. *Eur. J. Hum. Gen.*, **10**, 682–628.
36 Jörberg, L. (1972) *A History of Prices in Sweden 1732–1914*, CWK Gleerup, Lund.
37 Hellstenius, J. (1871) *Skördarna i Sverige och deras verkningar. (Harvests in Sweden and their repercussions)*, Statistisk Tidskrift, Stockholm, pp. 77–119.
38 Prader, A., Largo, R.H., Molinari, L., and Issler, C. (1988) Physical growth of Swiss children from birth to 20 years of age. First Zurich Longitudinal Study of Growth and Development. *Helvetica Paediatrica Acta.*, **43** (Suppl. 52).
39 Tanner, J.M. (1981) *A History of the Study of Human Growth*, Cambridge University Press, Cambridge.
40 Folkräkningen 1900 (2003) (The Swedish Census 1900). The Research Archives in Umeå, Umeå.
41 Folkmängd 1810–1990 (2005) Befolkningsutvecklingen på församlingsnivå vart tionde år under perioden 1810–1990 (Population changes at parish level at every tenth year 1810–1990). Demografiska databasen, Umeå.
42 Tabellverket 1849–1859 (2005) Demographic Data Base. Umeå.
43 McGill H.C. Jr. (1998) Nutrition in early life and cardiovascular disease. *Curr. Opin. Lipidol.*, **9**, 23–27.
44 Martyn, C.N., Barker, D.J., and Osmond, C. (1996) Mothers' pelvic size, fetal growth, and death from stroke and coronary heart disease in men in the UK. *Lancet*, **348**, 1264–1268.
45 Signorelli, L.B., and Trichopoulos, D. (1998) Perinatal determinants of adult cardiovascular disease and cancer. *Scand. J. Soc. Med.*, **26**, 161–165.
46 Kato, Y., Rideout, W.M., Hilton, K., Burton, S.C., Tsunoda, Y., and Surani, M.A. (1999) Developmental potential of mouse primordial germ cells. *Development*, **126**, 1823–1832.
47 Picton, H., Briggs, D., and Gosden, R. (1998) The molecular basis of oocyte growth and development. *Mol. Cell. Endocrinol.*, **145**, 27–37.
48 Pembrey, M. (1996) Imprinting and transgenerational modulation of gene expression; human growth as a model. *Acta. Genet. Med. Genellol.*, **45**, 111–125.
49 Gosden, R., Krapez, J., and Briggs, D. (1997) Growth and development of the mammalian oocyte. *Bioessays*, **9**, 875–882.

50 Pembrey, M. (2002) Time to take epigenetic inheritance seriously. *Eur. J. Hum. Genet.*, **10**, 669–171.

51 Bennett, S.T., Wilson, A.J., Esposito, L., Bouzekri, N., Undlien, D.E., Cucca, F., Nisticò, L., Buzzetti, R., Bosi, E., Pociot, F. et al. (1997) Insulin VNTR allelle-specific effect in type 1 diabetes depends on the identity untransmitted paternal allele. The IMDIAB Group. *Nat. Genet.*, **17**, 262–263.

52 Ong, K.K., Philips, D.I., Fall, C., Poulton, J., Bennet, S.T., Golding, J., Todd, J.A., and Dunger, D.B. (1999) The insulin gene VNTR, type 2 diabetes and birth weight. *Nat. Genet.*, **21** (3), 262–263.

53 Klenova, E.M., Morse, H.C. III, Ohlsson, R., and Lobanenkov, V.V. (2002) The novel BORIS+CTCF gene family is uniquely involved in the epigenetics of normal biology and cancer. *Semin. Cancer Biol.*, **12**, 399–414.

54 Loukinov, D.I., Pugacheva, E., Vatolin, S., Pack, S.D., Moon, H., Chernukhin, I.V., Mannan, P., Larsson, E., Kanduri, C., Vostrov, A.A. et al. (2002) BORIS, a novel male germ-line-specific protein associated with epigenetic reprogramming events, shares the same 11-zinc-finger domain with CTCF, the insulator protein involved in reading imprinting marks in the soma. *Proc. Natl. Acad. Sci. U. S. A.*, **99**, 6806–6811.

55 Nistal, M., and Paniagua, R. (1984) Occurrence of primary spermatocytes in the infant and child testis. *Andrologia*, **16**, 532–536.

56 Jaenisch, R., and Bird, A. (2003) Epigenetic regulation of gene expression: how the genome integrates intrinsic and environmental signals. *Nat. Genet.*, **33** (Suppl.), 245–254.

57 King, R.A., Rotter, J.I., and Motulsky, A.G. (2002) *The Genetic Basis of Common Diseases*, Oxford University Press, Oxford.

58 Petronis, A., Gottesman, I.I., Kan, P., Kennedy, J.L., Basile, V.S., Paterson, A.D., and Popendikyte, V. (2003) Monozygotic twins exhibit numerous epigenetic differences: clues to twin discordance? *Schizophr. Bull.*, **29**, 169–178.

59 Jablonka, E., and Lamb, M.J. (1995) *Epigenetic Inheritance and Evolution: The Lamarkian Dimension*, Oxford University Press, Oxford.

60 Petronis, A. (2001) Human morbid genetics revisited: relevance of epigenetics. *Trends Genet.*, **17** (3), 142–146.

61 Bjornsson, H.T., Fallin, M.D., and Feinberg, A.P. (2004) An integrated epigenetic and genetic approach to common human disease. *Trends Genet.*, **20**, 350–358.

62 Barker, D.J.P., and Martyn, C.N. (1992) The maternal and fetal origins of cardiovascular disease. *J. Epidemiol. Community Health*, **46**, 1–11.

63 Kuzawa, C.W. (2005) Fetal origins of developmental plasticity: are fetal clues reliable predictors of future nutritional environments? *Am. J. Hum. Biol.*, **17**, 5–21.

64 Oyama, S. (2000) *Evolution's Eye*, Duke University Press, Durham.

65 Oyama, S. (2000) *The Ontogeny of Information: Developmental Systems and Evolution*, Duke University Press, Durham.

66 Oyama, S. (1993) Constraints and development. *Neth. J. Zool.*, **3**, 6–16.

67 Oyama, S. (1994) Rethinking development, in *Handbook of Psychological Anthropology* (ed. P. Bock), Greenwood, Westport, pp. 185–196.

68 West-Eberhard, M.J. (2003) *Developmental Plasticity and Evolution*, Oxford University Press, New York.

69 Weiss, P.A. (1969) The living system: determinism stratified, in *Beyond Reductionism* (eds A. Koestler, and J.R. Smythies), MacMillan, New York, pp. 3–55.

70 Weiss, P.A. (1973) *The Science of Life: The Living System – A System for Living*, Futura Publishing Company, New York.

71 Weiss, P.A. (1978) Causality: linear or systematic? in *Psychology and Biology of Language and Thought: Essays in Honor of Eric Lenneberg* (eds G.A. Miller, and E. Lenneberg), Academic Press, New York, pp. 13–26.

72 Bruer, J.T. (1999) *The Myth of the First Three Years*, The Free Press, New York.

73 Strohman, R.C. (1993) Ancient genes, wise bodies, unhealthy people: the limits

of genetic thinking in biology and medicine. *Perspect Biol. Med.*, **37**, 112–144.
74 Lewontin, R. (1982) Organism and environment, in *Learning, Development and Culture* (ed. H.C. Plotkin), John Wiley & Sons, Inc., New York, pp. 151–170.
75 Lewontin, R. (1983) Gene, organism, and environment, in *Evolution: From Molecules to Men* (ed. D.S. Bendall), Cambridge University Press, Cambridge.
76 Broady, H. (1973) The systems view of man: implications for medicine, science, and ethics. *Perspect Biol. Med.*, **17**, 71–92.
77 Gottlieb, G. (2002) *Individual Development and Evolution: The Genesis of Novel Behavior*, Oxford University Press, New York.
78 Gottlieb, G. (1998) The significance of biology for human development: a developmental psychobiological systems perspective, in *Handbook of Child Psychology: Theoretical Models of Human Development*, vol. 1 (ed. R.M. Lerner), John Wiley & Sons, Inc., New York.
79 Wimsatt, W.C. (1999) Simple systems and phylogenetic diversity. *Philos. Sci.*, **65**, 267–314.
80 Vitzthum, V.J. (2003) A number no greater than the sum of its parts: the use and abuse of heritability. *Hum. Biol.*, **75**, 539–558.
81 Neuman-Held, E.M. (1999) The gene is dead – long live the gene: conceptualizing genes the constructionist way, in *Sociobiology and Bioeconomics. The Theory of Evolution in Biological and Economic Theory* (ed. P. Koslowski), Springer Verlag, Berlin, pp. 261–280.
82 Kellar, F.E. (2000) *The Century of the Gene*, Harvard University Press, Cambridge.
83 Barker, D.J.P., Bull, A.R., Osmond, C., and Simmonds, S.J. (1990) Fetal and placental size and risk of hypertension in adult life. *BMJ*, **301**, 259–262.
84 Barker, D.J.P. (2004) The developmental origins of adult disease. *J. Am. Coll. Nutr.*, **23** (Suppl. 6), S588–S595.
85 Bird, A. (2007) Perceptions of epigenetics. *Nature*, **447**, 396–398.
86 Hopkins, P.N., and Williams, R.R. (1981) A survey of 246 suggested coronary risk factors. *Atherosclerosis*, **40** (1), 1–5.
87 (a) Jablonka, E., and Lamb, M.J. (1995) *Evolution in Four Dimensions*, Oxford University Press, Oxford; (b) Jablonka, E., and Lamb, M.J. (2005) *Evolution in Four Dimensions*, The MIT Press, Cambridge.
88 Weissman, F., Lyko, F. (2003) Cooperative interaction between modifications and their function in the regulation of chromosome architecture. *BioEssays*, **25**, 792–797.
89 Pembrey, M.E., Bygren, L.O., Kaati, G., Edvinsson, S., Northstone, K., The Alspac Study Team, Sjöström, M. and Golding, J. (2006) *Eur. J. Hum. Genet.*, **14**, 159–166.
90 Judson, H., Hayward, B.E., Sheridan, E., and Bonthrone, D.T. (2002) A global disorder of imprinting in the female germ line. *Nature*, **416**, 539–542.
91 Young, L.E. (2001) Imprinting of genes and the Barker hypothesis. *Twin Research*, **4**, 307–314.
92 Muller, G.B., Newman, S.A. (eds.) (2003) *Origination of Organismal Form: Beyond Gene in Development and Evolutionary Biology* (Vienna Series in Theoretical Biology). The MIT press, Cambridge.
93 Kellar, E.F. (2002) *Making sense of life*. Harvard University Press, Cambridge.
94 Griffiths, P.E. and Gray, R.D. (1994) Developmental systems and evolutionary explanation. *Journal of Philosophy*, **XCI**, 277–304.
95 Whimsatt, W.C. (1994) The ontology of complex systems: levels, perspectives and causal thickets. *Canadian Journal of Philosophy*, **20**, S207–274.
96 Lewontin, R. (2000) *The triple helix: Genes, organism and the environment*. Harvard University Press, Cambridge.

Part III
Environmental and Toxicological Aspects

8
Genotoxic, Non-Genotoxic and Epigenetic Mechanisms in Chemical Hepatocarcinogenesis: Implications for Safety Evaluation

Wilfried Bursch

Abstract

According to the multistage concept of carcinogenesis, malignant cells develop stepwise via a sequence of intermediary cell populations whose phenotypes deviate increasingly from the original cell. Chemical compounds can modify the rate of carcinogenesis by affecting mutation frequency, growth rate, or expression of phenotypic deviations. Thus, genotoxic chemicals may cause DNA sequence mutations, giving rise to preneoplastic ("initiated") cells. It is generally assumed that – because this kind of genomic damage is irreversible and additive in the case of repeated exposure – no dose without a potential health effect may exist ("no-threshold"-concept). Subsequent clonal expansion of preneoplastic cells and development of frank neoplasia can be accelerated by so-called non-genotoxic carcinogens (tumor promoter), many of which disturb the growth regulatory network in tissues by interacting with receptors and downstream signaling, cytokines, and nutritional factors. Promoter doses below a certain threshold are not effective. Furthermore, tumor promotion, at least in the early stages, is considered to be reversible and thus in sharp contrast to the hazardous action of genotoxic agents.

Accumulating evidence suggests that chemicals may exert a carcinogenic action by epigenetic mechanisms, as exemplified by alterations of the DNA methylation pattern. Notably, both genotoxic and non-genotoxic carcinogens may cause similar and persistent epigenetic modifications of gene function without involving DNA sequence mutations. For safety evaluation, the potency of chemicals to induce epigenetic effects in target cells constitutes an important endpoint to elucidate the mode of action of chemicals.

8.1
Introduction

The multistage concept of carcinogenesis implies that malignant cells develop stepwise via a sequence of intermediary cell populations whose phenotypes deviate increasingly from the original cell [1–7]. Chemical compounds can modify the rate of carcinogenesis by affecting mutation frequency, growth rate, or expression of phenotypic deviations [1–7]. Current concepts of safety evaluation of chemicals emphasize making use of data on their cellular and molecular mechanisms [8–13]. The mechanisms underlying the adverse effects of carcinogens may be ascribed to two broad categories. The first mechanism is characterized by a direct interaction of a chemical or its reactive metabolites with the DNA of target cells, causing pre-cancerogenic DNA sequence mutations, which finally may lead to neoplastic transformation and development of neoplasia. For this mode of action it is considered that – according to current scientific knowledge – no dose without a potential health effect may exist. Therefore, safety evaluation has to pay special attention to CMR (carcinogenic, mutagenic, toxic to reproduction) properties of chemicals ([13], Directive 2005/90/EC of the European Parliament and of the Council of 18 January 2006). The second mechanism is non-genotoxic in nature and does not involve DNA sequence mutations. In this case, the chemical may affect target cells that either indirectly result in neoplastic transformation or promote the development of neoplasia from precancerous cells (tumor promotion). Tumor promotion often involves disturbance of the growth regulatory network in tissues due to the interaction of chemical (tumor promoter) with receptors and downstream signaling, cytokines, and nutritional factors. Promoter doses below a certain threshold are not effective [1, 2, 5, 6] and tumor promotion, at least in the early stages, is considered to be reversible [2, 5, 6, 14].

The classification of a chemical as either possessing or not possessing genotoxic (no-threshold phenomenon) or non-genotoxic (threshold phenomenon) properties has major consequences for the risk assessment approach to be taken and the advice to be given to risk managers [8–13]. For many years, elucidation of key events to characterize the mode of action of chemical carcinogens has been focussed on genotoxic vs. non-gentoxic properties. However, a number of more recent observations have revealed that non-mutational, epigenetic[1] events may be involved in all stages of cancer pathogenesis. Notably, epigenetic effects are heritable to daughter cells and in a number of cases, even may result in transgenerational adverse effects [7, 15–18]). Studies on hepatocarcinogenesis revealed that genotoxic and non-genotoxic chemicals may produce similar epigenetic changes [18]), thereby challenging the dichotomy between these modes of action. In this chapter, some implications for safety evaluation as related to genotoxic, non-genotoxic and epigenetic actions of chemicals – exemplified by hepatocarcinogenesis – are discussed.

---

1) The term "epigenetic" generally refers to heritable changes in DNA methylation, modification of histone or chromatin structure, all of which may stably alter gene transcription without involving DNA sequence mutations [7, 15].

8.2
Genotoxic and Non-Genotoxic Chemicals in Relation to the Multistage Model of Cancer Development

Mainly on the basis of experimental work with mouse skin, rat liver, and a few other tissues, tumor development can be broken down into sequential steps operationally defined in the context of each other, namely initiation, promotion and progression (Figure 8.1). In humans, approaches to elucidate molecular key events leading to cancer are often restricted to patients already diseased with frank neoplasia. Therefore, the scope of research is often limited to tumor biology rather than elucidation of the mechanisms of pathogenesis. In contrast, rodent liver currently offers one of the best model systems to study the stepwise development of cancer. Briefly, in this system putative prestages of cancer and their biological behavior have been well characterized; a variety of phenotypic markers are available that allow the identification of putative single initiated cells and of the succeeding clones (the latter as foci of altered hepatocytes) [1, 2, 4–6, 14, 19]. Different types of preneoplastic lesions can be distinguished according to phenotypic appearance and growth response to different tumor promoters suggesting that there may be more than one pathway to liver cancer [1, 2, 4–6, 14]. Adenomas are larger than foci and may show similar phenotypic alterations [14]. The underlying molecular mechanisms involve point mutations, small deletions, translocations of oncogenes and/or tumor suppressor genes which may accumulate during clonal expansion of initiated cells, eventually resulting in genomic instability and progression to hepatocellular carcinoma (HCC). Similar phenotypically altered lesions have been found in human liver [6, 7, 20].

8.2.1
Tumor Initiation

The first step or initiation can be induced by genotoxic chemicals such as Aflatoxin B1, polycyclic aromatic hydrocarbons, nitrosamines, and vinyl chloride [14]. Furthermore, initiation may be induced by viruses and radiation as well as reactive oxygen species (ROS, oxidative stress); exposure of cells to the latter predominantly results from chronic inflammation (Figure 8.1). Finally, initiation may also occur for unknown reasons ("spontaneously") as in most human cancers [14].

It is generally accepted that initiation occurs in single cells which gain a proliferation advantage over normal cells and gradually form larger clones (Figure 8.1, [1, 2, 5, 6, 19]). However, initiation alone does not lead, or only slowly leads, to tumor formation. Rather, the growth rate of "initiated" cell clones, and thereby cancer development, can be greatly accelerated by tumor promotion (see below). For toxicological risk assessment, important features of initiation are: (i) it may ensue upon single exposure to a genotoxic chemical; and (ii) the initiated genotype is transmitted to daughter cells (irreversibility). Thus, if an organism is exposed repeatedly even to low doses, the DNA sequence mutations caused by genotoxic substances are additive. As is to be expected, the incidence of tumors induced by

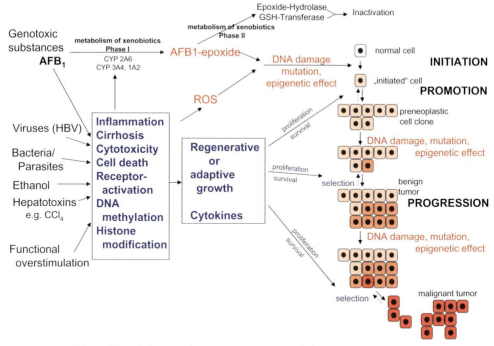

Figure 8.1 Risk factors in hepatocarcinogenesis and their synergistic action. AFB1, Aflatoxin B1; CYP, cytochrom P450; CCl_4, carbon tetrachloride; ROS, reactive oxygen species.

genotoxic carcinogens decreases with decreasing dose. However, because of the additivity of genomic damage, a safe dose below which a carcinogenic potential is unlikely to become manifest *cannot* be deduced ("no-threshold"-concept). For instance, for the purpose of food safety no "acceptable daily intake (ADI-value)" will be derived. In such cases, protection of human health is focussed on the ALARA-principle (as low as reasonably achievable); tools to quantify acceptable risk levels, such as one case per million, involve extrapolation by mathematical models or more recently, the margin of exposure (MOE)-concept to establish priority lists for risk management [12].

8.2.2
Tumor Promotion

Tumor promotion is operationally defined as a process accelerating and enhancing the formation of tumors after initiation. The essence of promotion is the selective multiplication of initiated cells, frequently associated with enzyme induction and growth or with tissue damage in the affected organs (Figure 8.1). Rodent non-genotoxic hepatocarcinogens comprise phenobarbital (PB), TCDD and related agents, DDT, α-, γ-hexachlorocyclohexane (HCH), ethinylestradiol, peroxisome proliferators (PP), ethanol, carbon tetrachloride (CCl_4), thioacetamide and many others [14]. These chemicals are extremely heterogenous with respect to chemical structure and general biological activity, but share the potency to induce adaptive or regenerative liver growth [2, 14, 21]. The initiated/preneoplastic cells – in rodent liver appearing as enzyme-altered foci – constitute the key target and tumor promotion is brought about by preferential increase of cell proliferation and concomitantly, inhibition of apoptosis in these cell populations [1, 2, 4–6, 14]. Notably, non-genotoxic carcinogens in most cases are tumor promoters, and it is frequently assumed that they are carcinogenic by promotion of spontaneously appearing initiated cells [2, 14].

In the context of safety evaluation, important features of promotion comprise: (i) A high degree of organ specificity, an important characteristic of many tumor-promoting agents, in contrast to initiation which may occur in several organs in parallel. Therefore, tumor promotion may determine the target organ of carcinogenic action. (ii) Promotion always requires a prolonged period of action, which may last weeks, months, or years. Tumor promotion, at least in the early stages, is considered to be reversible. (iii) Promoter doses below a certain threshold are not effective [1, 2, 5, 6]. In summary, these characteristics contrast sharply with the characteristics of initiation as outlined above. According to current concepts of risk assessment and management, based upon a given threshold dose, and applying uncertainty factors to account for species differences (human versus experimental animals), the amount and quality of toxicological data and so on, human exposure to a given chemical can be limited by an acceptable daily intake (ADI) value.

8.2.3
Tumor Progression

Individual cells in preneoplastic cell clones may undergo further mutational changes and new cell clones with higher proliferative potential may be selected (Figure 8.1; progression). In addition to many xenobiotic chemicals, caloric overnutrition and hormones (e.g., estrogens, androgens) have a tumor promoting potency in the liver [14]. Likewise, inflammation and cytokines released during liver regeneration as a cosequence of cytotoxic damage promote the development of neoplasia; chronic liver diseases, therefore, constitute a high risk for development of hepatocellular carcinoma [7, 21, 22].

8.2.4
Cellular and Molecular Mechanisms of Tumor Initiation and Promotion

As exemplified by the action of Aflatoxin B1 in the liver, the mycotoxin is metabolically activated to AFB1-epoxide (Figure 8.1; metabolism of xenobiotics, phase (I)). If not inactivated readily by epoxide-hydrolases or GSH-transferases (Figure 8.1; metabolism of xenobiotics, phase (II)) the reactive metabolite may covalently bind to DNA and, finally, may cause mutations (Figure 8.1; DNA damage, mutation).

AFB1-induced DNA damage has been shown in numerous *in vitro* and *in vivo* assays in bacteria, yeast and mammalian (including human) cells [23]. In geographical regions with high prevalence of Aflatoxin, approximately 30–50% of human hepatocellular carcinoma exhibited a mutation of codon 249 of the tumor suppressor gene p53 [3, 7]. The p53 protein ("guardian of the genome") is involved (i) in cell cycle arrest, necessary for DNA repair and, (ii) in the induction of apoptosis, eliminating cells with irreparable DNA damage; both processes protect from cancer development [3]. Therefore, a "loss of function" mutation of the p53-protein increases the probability of progression of malignancy. This example illustrates that mutagenic events frequently affect genes coding for key players of signal transduction pathways steering cell cycle, apoptosis or differentiation [3, 6, 7, 20, 24, 25].

The molecular mechanism underlying tumor promotion comprises activation of growth stimulating signal cascades, interacting with the altered genotype of initiated cells (e.g., synergistic action of hepatitis B-infection and Aflatoxin, Figure 8.1). A number of nuclear receptors have been found to be targeted by non-genotoxic carcinogens. For instance, 2,3,7,8-TCDD activates the aryl-hydrocarbon-hydroxylase(AH)-receptor), phenobarbital the CAR-receptor, peroxisomal proliferators (hypolidemic drugs, phthalates) bind to the "peroxisome proliferator activated receptor (PPAR)", all of which have in common the ability to mediate induction of liver growth in rodents (Figure 8.1) [2, 5, 8, 26].

The scientific discussion on the relevance of animal data for assessment of human health risks resulted in the development of a "Human Relevance Framework Concept" aiming at identifying a "mode of action (MOA)" [8, 11]. In principle, this concept involves a decision tree for toxicological evaluation of chemicals, addressing three key questions: (i) Are the experimental data valid ("weight of evidence"), to explain a given hazard in animals by one or more "mode(s) of action"? (ii) Are the key mechanism(s) as elucidated in animal experiments, as a matter of principle, active in humans as well? This question primarily addresses *qualitative* aspects. (iii) Considering toxicokinetics and -dynamics, is the MOA relevant for humans? This question primarily addresses *quantitative* aspects. Recently, based upon the huge amount of toxicological data available for phthalates, Klaunig *et al.* [8] applied the MOA-concept and succeeded in carving out cell proliferation and apoptosis as the essential, rate-limiting processes for the tumor promoting action of this class of compounds. Notably, peroxisome proliferation – the biological effect naming phthalates as "peroxisomal proliferators" – turned out not to be a relevant mode of action for human hepatocarcinogenesis. Consequently, risk assessment was revised and, for example, recently the ADI values for phthalates were adopted according to the improved state of knowledge [27–31].

Furthermore, the response of hepatocytes to the exogenous agents may be modulated by cytokines and nutrition [21, 32]. For instance, insulin-like growth factor (IGF I, II), transforming growth factor (TGF) alpha and beta, fibroblast growth factor (FGF) and hepatocyte growth factor (HGF) play significant roles in hepatocarcinogenesis of experimental animals and humans [21, 24, 25]. One of our prominent *in vivo* findings was that the cytokine transforming growth factor β1 (TGF-β1) constitutes a major regulator of apoptosis in rat liver, acting in concert with liver tumor promoter as well as nutritional factors to maintain liver cell number homeostasis. As outlined above, liver tumor promoters (e.g., phenobarbital) shift the ratio between cell birth and cell death in favor of cell birth, in particular in preneoplastic lesions of rat liver; thereby, cancer development is accelerated. The opposite applies to feed restriction, resulting in cancer prevention [33]. In addition, treatment of rats with TGF-β1 *in vivo* induced apoptosis in preneoplastic lesions of the liver, resulting in their rapid regression [34]. These observations prompted us to tackle the underlying mode(s) of action (MOA), particularly in view of potential interactions between liver tumor promoter, TGF-β1 and nutrition (glucose, amino acid supply). In a first approach, along with the general need for alternative test models in toxicology, we searched for a liver cell culture system yielding a high concordance with our *in vivo* findings. Recently, Sagmeister *et al.* [35] succeeded in establishing a new human hepatoma cell line denoted HCC-1.2 meeting this requirement. HCC-1.2 cells revealed to be sensitive to the pro-apoptotic action of TGF-β1 [32]. Apoptosis of HCC-1.2 cells was found to ensue via the intrinsic pathway, as demonstrated by caspase analysis (Western blot analysis revealed cleavage of caspase 3 and 9, but not of caspase 8; [32]). Furthermore, TGF-β1-induced apoptosis was inhibited by the liver tumor promoter phenobarbital [32]. As for nutritional factors, glucose deprivation exerted a pro-apoptotic effect, additive to TGF-β1 [32]. The pro-apoptotic action of glucose deprivation was antagonized by 2-deoxyglucose (2-DG; [32]), possibly by stabilizing the mitochondrial membrane involving the action of hexokinase II [36–39]). As the mitochondrial membrane stabilizing action of 2-DG has been reported to depend on its phosphorylation by hexokinase, we also tested the effect of the 5-thioglucose (5-TG). This substrate analog is an even stronger inhibitor of glycolysis than 2-DG, but 5-TG is practically non-phosphorylatable by hexokinase (hexokinase-K_m 2-DG: 0.027 mM, 5-TG: 4 mM; [40, 41]. Indeed, results of preliminary experiments suggest that 5-TG (0.5 mM) did not reduce apoptotic activity under glucose deprivation [32]. Taken together, these observations are in line with the concept suggested by Robey and Hay [39]. In addition to the potential role of mitochondria, evidence suggesting the involvement of the mTOR/S6K1 pathway in induction of cell death upon glucose withdrawal has been reported [42]. Finally, deprivation of branched-chaine amino acids (BCAA), but also of other amino acids AA, exerted a pro-apoptotic activity on HCC-1.2 cells; AA-deprivation and TGF-β1 appear to act additively [32].

Taken together, our recent data provide an example of how cytokines, nutritional signals and exogenous chemicals may interact at the level of signaling downstream of certain receptors along with sensors/key regulators of ana-/catabolism and, thereby, determine the balance between cell proliferation and cell death.

8.2.5
Epigenetic Effects of Genotoxic and Non-Genotoxic Hepatocarcinogens

A number of observations have revealed that non-mutational, epigenetic events may be involved in all stages of tumor development (Figure 8.1; DNA damage, epigenetic effect) [7, 15, 18, 43–46]. The term "epigenetic" generally refers to heritable changes in DNA methylation, modification of histone or chromatin structure, all of which may stably alter gene transcription *without* involving DNA sequence mutations [7, 15]. Cancer cells often simultaneously exhibit global DNA hypomethylation and gene-specific DNA hypermethylation, the simultaneous occurrence of which in the same cell may be explained by differential action of DNA methylating and DNA demethylating enzymes [7, 15]

For instance, for decades it has been known that omission of methyl-group sources from the diet induces liver tumors in rodents, suggesting a profound role of C1-metabolism in hepatocarcinogenesis [47–51]; the reader is referred to a recent review by Pogribny and colleagues [18]. Briefly, some of the important aspects highlighted by Pogribny *et al.* comprise (i) the sequence of pathological and molecular events in rodent liver tumor development resulting from methyl-deficiency is remarkably similar to the development of human HCC associated with viral hepatitis B and C infections, alcohol exposure, and metabolic liver diseases; (ii) feeding a methyl-deficient diet was found to cause a sustained global hypomethylation of liver DNA, suggested to be a promoting factor for clonal expansion of initiated hepatocytes; (iii) the changes in DNA methylation pattern were found to be specific to liver tissue and not to occur in other organs.

Furthermore, the significance of aberrant DNA-methylation in hepatocarcinogenesis was supported by a study of Grasl-Kraupp and colleagues [52]. Briefly, in human hepatocellular carcinoma (HCC) the ras-proto-oncogene is rarely mutated. Therefore, the conceivable inactivation of the putative tumor-suppressors and ras-associating proteins, NORE1A, NORE1B, and RASSF1A were studied, specifically to get insight into the role of mutational versus epigenetic gene silencing for HCC pathogenesis. The study revealed a considerable CpG-methylation of the NORE1B promoter in the majority of HCCs, along with frequent aberrant methylation of RASSF1A. The authors concluded that down-regulation of the Ras-interfering genes, NORE1B and RASSF1A, may be a strategy of malignant hepatocytes to escape from growth control in the presence of an unaltered ras [52]. Other examples highlighting the significance of changes in the DNA methylation pattern in hepatocarcinogenesis are p16/INK4A, E-cadherin, glutathione-S-transferase, suppressor of cytokine signaling (SOSC-1) soluble frizzeld-related protein (SFRP1), phosphate and tensin homolog (PTEN) [7].

As for the chemicals causing epigenetic effects, *genotoxic* hepatocarcinogens found to affect the DNA methylation in the liver–in addition to their genotoxic action–are N-nitrosomorpholine, 1,2-dimethylhydrazine, N-nitrosodiethylamine and Aflatoxin B1 [18, 53, 54]. Furthermore, studies by Tryndyak and colleagues [55] on tamoxifen-induced rat hepatocarcinogenesis revealed that tamoxifen resulted in an early and sustained change in DNA and histone methylation. As for *non-genotoxic* chemicals, peroxisome proliferators were found to cause global hypomethylation in mice together with regional hypermethylation [18]. Furthermore, in comparative studies with mouse strains highly susceptible (B6C3F1) and resistant (C3H/He) to hepatocarcinogenesis, phenobarbital was found to alter the DNA-methylation pattern, being most pronounced in the sensitive mouse strain (B6C3F1), suggesting that maintenance of normal DNA methylation patterns may be inversely correlated with tumor susceptibility [56–59]. Studies with primary mouse hepatocytes revealed that approximately 70% of the changes in the DNA methylation pattern induced by phenobarbital, diethanolamine and cholin-deficiency were identical [57]. In general, affected genes identified so far in these experimental set-ups are involved in invasion and metastasis, angiogenesis, tumor cell proliferation and survival [59]. These observations are in line with studies by ourselves revealing that the cancer susceptibility of C3H/He>B6C3F1>C57/Bl mice correlates positively with the degree of Phenobarbital-induced cell proliferation in these mouse strains [4].

On the other hand, targeting the DNA-methylation pattern may provide a means for chemoprevention. For instance, the cellular levels of the methyl donor S-adenosylmethionine (SAM) can be modulated by dietary supply of methyl groups, folic acid, vitamin B12 and alcohol [60]. Interestingly, SAM is anti-apoptotic in normal hepatocytes, but pro-apoptotic in liver cancer cells [61].

In summary, genotoxic and non-genotoxic hepatocarcinogens may cause similar epigenetic alterations, the adverse consequences of which eventually may be indistinguishable from those resulting from DNA sequence mutations.

8.2.6
How Carcinogens Alter the Microenvironment – Crucial Roles of Inflammation

Furthermore, the interaction of genotoxic and non-genotoxic chemicals in the liver is not restricted to the level of hepatocytes but – as summarized in Table 8.1 – involves interaction with non-parenchymal liver cells. The observations strongly suggest that the microenvironment in the liver (i.e., Kupffer and/or other non-parenchymal cells) has a profound effect in the complex processes leading to initiation, promotion and progression of hepatocarcinogenesis, in humans and also in rodents. To elucidate more fully the mechanism of tumor formation by liver carcinogens, the role(s) of the hepatic microenvironment should be investigated.

8.3
Concluding Remarks

It is generally accepted that genotoxic carcinogens may directly induce DNA lesions, resulting in DNA sequence mutations and giving rise to preneoplastic ("initiated") cells. For safety evaluation, a highly important biological feature is the irreversibility (additivity) of this kind of DNA damage. Thus, current scientific

Table 8.1 Some non-genotoxic and genotoxic carcinogens reported to activate Kupffer cells.

Substance	Endpoint	References
Peroxisome proliferators	Hepatocyte stimulating activity TNFα⇑; NF-κB ⇑	[62, 63]
Phenobarbital	Oxygen burst; phagocytosis ⇑	[64, 65]
Ethanol	TNFα⇑; IL-6 ⇑ (in humans), COX2 ⇑	[66, 67]
Carbon tetrachloride	HGF ⇑; TNFα, TGFβ, IL-1, COX2, CD14, IκB ⇑	[68, 69]
Arsenic	Oxygen burst (myeloid leukemia cells)	[70]
Nitrosamines	Oxygen burst	[71]

knowledge does not allow one to derive a safe exposure below which a given genotoxic agent may be considered to be without health risk ("no-threshold-concept"). In addition, mounting evidence suggests that epigenetic mechanisms such as heritable alterations of the DNA-methylation pattern – notably, without DNA sequence mutations – are involved in transforming a normal into a precancerous ("initiated") cell. Consequently, in order to elucidate the mode of action of chemicals and toxicological risk assessment, "initiation" requires better definition in molecular terms in order to understand more fully the mechanisms of (chemical) carcinogenesis. Furthermore, epigenetic mechanisms are not restricted to early steps of (hepato)carcinogenesis but are also involved in clonal expansion of precancerous cells. Thus, epigenetically re-programmed preneoplastic cells may gain a high potential to progress to frank neoplasia.

The safety concerns based on possible epigenetic programming of cells by chemicals have been strengthened by observations on endocrine disruptors. Endocrine disruptors are chemicals that may interfere with hormone signaling. *For instance,* o,p'-dichlorodiphenyltrichlorethane (o,p'-DDT), polychlorinated biphenyls (PCB), polychlorinated dibenzodioxins and dibenzofuranes (PCDD/PCDF) exhibit estrogenic activity; in total about 190 substances showing evidence for disturbances of estrogen, androgen, progestin, hypothalamic, pituitary or thyroid hormone homeostasis have been identified to date [72]. As for epigenetics, Waterland and Jirtle [73] reported that *in utero* or neonatal exposure of agouti mice to Bisphenol A stably altered the offspring epigenome (decreased cytosine-guanine dinucleotide methylation of Avy locus, visible as altered coat color distribution). The most prominent example, diethylstilbestrol (DES), was found to modify methylation of uterine genes which are permanently dysregulated upon developmental exposure of rodents [16]. Furthermore, transient exposure of gestating rats at the time of embryonic sex determination to vinclozilin was revealed to promote adult onset of spermatogenic defects and male subfertility [74]. Notably, 90% of all male progeny for four generations (F1–F4) developed these disease states after the direct exposure of the F0 gestating rat [74]. Finally, the etiology of transgenerational epigentic effects is not restricted to chemicals, as exemplified by Zambrano and coworkers [75], showing that embryonic caloric restriction may promote an F2 generation diabetes phenotype.

Taken together, genotoxic and non-genotoxic chemicals can cause persistent chemical modification of the genome without necessarily involving DNA sequence mutations. Stable, life-long and, in certain cases, even transgenerational epigenetic modulation of gene function creates a challenge for safety evaluation of chemicals. The potency of chemicals to induce epigenetic effects in target cells constitutes an important endpoint in elucidating the mode of action of chemicals.

References

1 Schulte-Hermann, R., Timmermann-Trosiener, I., Barthel, G., and Bursch, W. (1990) DNA synthesis, apoptosis, and phenotypic expression as determinants of growth of altered foci in rat liver during phenobarbital promotion. *Cancer Res.*, **50**, 5127–5135.

2 Schulte-Hermann, R., Bursch, W., and Grasl-Kraupp, B. (1995) Active cell death (apoptosis) in liver biology and disease. *Prog. Liver Dis.*, **13**, 1–35.

3 Olivier, M., Hussain, S.P., Caron de Fromentel, C., Hainaut, P., and Harris, C.C. (2004) TP53 mutation spectra and load: a tool for generating hypotheses on the etiology of cancer. *IARC Sci. Publ.*, **157**, 247–270.

4 Bursch, W., Chabicovsky, M., Wastl, U., Grasl-Kraupp, B., Bukowska, K., Taper, H., and Schulte-Hermann, R. (2005) Apoptosis in stages of mouse hepatocarcinogenesis: failure to counterbalance cell proliferation and to account for strain differences in tumor susceptibility. *Toxicol. Sci.*, **85**, 515–529.

5 Schwarz, M., and Appel, K.E. (2005) Carcinogenic risks of dioxin: mechanistic considerations. *Regul. Toxicol. Pharmacol.*, **43**, 19–34.

6 Pitot, H.C. (2007) Adventures in hepatocarcinogenesis. *Annu. Rev. Pathol.*, **2**, 1–29.

7 Wong, C.M., and Ng, I.O. (2008) Molecular pathogenesis of hepatocellular carcinoma. *Liver Int.*, **28**, 160–174.

8 Klaunig, J.E., Babich, M.A., Baetcke, K.P., Cook, J.C., Corton, J.C., David, R.M., DeLuca, J.G., Lai, D.Y., McKee, R.H., Peters, J.M., Roberts, R.A., and Fenner-Crisp, P.A. (2003) PPARalpha agonist-induced rodent tumors: modes of action and human relevance. *Crit. Rev. Toxicol.*, **33**, 655–780.

9 Fairhurst, S. (2003) Hazard and risk assessment of industrial chemicals in the occupational context in Europe: some current issues. *Food Chem. Toxicol.*, **41**, 1453–1462.

10 Nordberg, A., Ruden, C., and Hansson, S.O. (2008) Towards more efficient testing strategies – analyzing the efficiency of toxicity data requirements in relation to the criteria for classification and labelling. *Regul. Toxicol. Pharmacol.*, **50**, 412–419.

11 Holsapple, M.P., Pitot, H.C., Cohen, S.M., Boobis, A.R., Klaunig, J.E., Pastoor, T., Dellarco, V.L., and Dragan, Y.P. (2006) Mode of action in relevance of rodent liver tumors to human cancer risk. *Toxicol. Sci.*, **89**, 51–56.

12 Barlow, S., Renwick, A.G., Kleiner, J., Bridges, J.W., Busk, L., Dybing, E., Edler, L., Eisenbrand, G., Fink-Gremmels, J., Knaap, A., Kroes, R., Liem, D., Muller, D.J., Page, S., Rolland, V., Schlatter, J., Tritscher, A., Tueting, W., and Wurtzen, G. (2006) Risk assessment of substances that are both genotoxic and carcinogenic report of an International Conference organized by EFSA and WHO with support of ILSI Europe. *Food Chem. Toxicol.*, **44**, 1636–1650.

13 REACH (2007) Regulation (EC) No 1907/2006 of the European Parliament and of the Council of 18 December 2006 concerning the Registration, Evaluation, Authorisation and Restriction of Chemicals (REACH), establishing a European Chemicals Agency, amending Directive 1999/45/EC and repealing Council Regulation (EEC) No 793/93 and Commission Regulation (EC) No 1488/94

as well as Council Directive 76/769/EEC and Commission Directives 91/155/EEC, 93/67/EEC, 93/105/EC and 2000/21/EC.

14 Parzefall, W., and Schulte-Hermann, R. (2004) Mehrstufenkonzept der Kanzerogense und chenmischen kanzerogenese, in *Die Onkologie* (eds W. Hiddemann, H. Huber, R. Bartram and R. Claus), Springer Verlag, pp. 194–240.

15 Szyf, M. (2007) The dynamic epigenome and its implications in toxicology. *Toxicol. Sci.*, **100**, 7–23.

16 Newbold, R.R., Padilla-Banks, E., and Jefferson, W.N. (2006) Adverse effects of the model environmental estrogen diethylstilbestrol are transmitted to subsequent generations. *Endocrinology*, **147**, S11–S17.

17 Skinner, M.K., and Anway, M.D. (2007) Epigenetic transgenerational actions of vinclozolin on the development of disease and cancer. *Crit. Rev. Oncog.*, **13**, 75–82.

18 Pogribny, I.P., Rusyn, I., and Beland, F.A. (2008) Epigenetic aspects of genotoxic and non-genotoxic hepatocarcinogenesis: studies in rodents. *Environ Mol. Mutagen*, **49**, 9–15.

19 Low-Baselli, A., Hufnagl, K., Parzefall, W., Schulte-Hermann, R., and Grasl-Kraupp, B. (2000) Initiated rat hepatocytes in primary culture: a novel tool to study alterations in growth control during the first stage of carcinogenesis. *Carcinogenesis*, **21**, 79–86.

20 Teufel, A., Staib, F., Kanzler, S., Weinmann, A., Schulze-Bergkamen, H., and Galle, P.R. (2007) Genetics of hepatocellular carcinoma. *World J. Gastroenterol.*, **13**, 2271–2282.

21 Rodgarkia-Dara, C., Vejda, S., Erlach, N., Losert, A., Bursch, W., Berger, W., Schulte-Hermann, R., and Grusch, M. (2006) The activin axis in liver biology and disease. *Mutat. Res.*, **613**, 123–137.

22 Herzer, K., Sprinzl, M.F., and Galle, P.R. (2007) Hepatitis viruses: live and let die. *Liver Int.*, **27**, 293–301.

23 IARC (2002) International Agency for Research on Cancer (IARC) – summaries & evaluations. *AFLATOXINS*, **82**, 171.

24 Breuhahn, K., Longerich, T., and Schirmacher, P. (2006) Dysregulation of growth factor signaling in human hepatocellular carcinoma. *Oncogene*, **25**, 3787–3800.

25 Calvisi, D.F., Pascale, R.M., and Feo, F. (2007) Dissection of signal transduction pathways as a tool for the development of targeted therapies of hepatocellular carcinoma. *Rev. Recent Clin. Trials*, **2**, 217–236.

26 Konno, Y., Negishi, M., and Kodama, S. (2008) The roles of nuclear receptors CAR and PXR in hepatic energy metabolism. *Drug Metab. Pharmacokinet.*, **23**, 8–13.

27 EFSA (2005) Opinion of the scientific panel on food additives, flavourings, processing aids and materials in contact with food (AFC) on a request from the commission related to butylbenzylphthalate (BBP) for use in food contact materials. Adopted on 23 June 2005 by written procedure. *EFSA J.*, **241**, 1–14.

28 EFSA (2005) Opinion of the scientific panel on food additives, flavourings, processing aids and materials in contact with food (AFC) on a request from the commission related to di-isodecylphthalate (DIDP) for use in food contact materials. Adopted on 30 July 2005. *EFSA J.*, **245**, 1–14.

29 EFSA (2005) Opinion of the scientific committee on a request from EFSA related to a harmonised approach for risk assessment of substances which are both genotoxic and carcinogenic. *EFSA J.*, **282**, 1–31.

30 EFSA (2005) Opinion of the scientific panel on food additives, flavourings, processing aids and materials in contact with food (AFC) on a request from the commission related to di-isononylphthalate (DINP) for use in food contact materials. Adopted on 30 July 2005. *EFSA J.*, **244**, 1–18.

31 EFSA (2005) Statement of the scientific panel on food additives, flavourings, Processing Aids and Materials in Contact with Food on a request from the Commission on the possibility of allocating a group-TDI for butylbenzylphthalate (BBP), di-butylphthalate (DBP), bis(2-ethylhexyl) phthalate (DEHP), di-isononylphthalate (DINP) and di-isodecylphthalate (DIDP). http://www.efsa.europa.eu/EFSA/efsa_locale-1178620753812_1178620768397.htm.

32 Bursch, W., Karwan, A., Mayer, M., Dornetshuber, J., Frohwein, U., Schulte-Hermann, R., Fazi, B., Di Sano, F., Piredda, L., Piacentini, M., Petrovski, G., Fesus, L., and Gerner, C. (2008) Cell death and autophagy: cytokines, drugs, and nutritional factors. *Toxicology*, **254**, 147–157.

33 Grasl-Kraupp, B., Bursch, W., Ruttkay-Nedecky, B., Wagner, A., Lauer, B., and Schulte-Hermann, R. (1994) Food restriction eliminates preneoplastic cells through apoptosis and antagonizes carcinogenesis in rat liver. *Proc. Natl. Acad. Sci. U. S. A.*, **91**, 9995–9999.

34 Mullauer, L., Grasl-Kraupp, B., Bursch, W., and Schulte-Hermann, R. (1996) Transforming growth factor beta 1-induced cell death in preneoplastic foci of rat liver and sensitization by the antiestrogen tamoxifen. *Hepatology*, **23**, 840–847.

35 Sagmeister, S., Eisenbauer, M., Pirker, C., Mohr, T., Holzmann, K., Zwickl, H., Bichler, C., Kandioler, D., Wrba, F., Mikulits, W., Gerner, C., Shehata, M., Majdic, O., Streubel, B., Berger, W., Micksche, M., Zatloukal, K., Schulte-Hermann, R., and Grasl-Kraupp, B. (2008) New cellular tools reveal complex epithelial-mesenchymal interactions in hepatocarcinogenesis. *Br. J. Cancer*, **99**, 151–159.

36 Majewski, N., Nogueira, V., Bhaskar, P., Coy, P.E., Skeen, J.E., Gottlob, K., Chandel, N.S., Thompson, C.B., Robey, R.B., and Hay, N. (2004) Hexokinase–mitochondria interaction mediated by Akt is required to inhibit apoptosis in the presence or absence of Bax and Bak. *Mol. Cell*, **16**, 819–830.

37 Pastorino, J.G., Hoek, J.B., and Shulga, N. (2005) Activation of glycogen synthase kinase 3beta disrupts the binding of hexokinase II to mitochondria by phosphorylating voltage-dependent anion channel and potentiates chemotherapy-induced cytotoxicity. *Cancer Res.*, **65**, 10545–10554.

38 Byfield, M.P., Murray, J.T., and Backer, J.M. (2005) hVps34 is a nutrient-regulated lipid kinase required for activation of p70 S6 kinase. *J. Biol. Chem.*, **280**, 33076–33082.

39 Robey, R.B., and Hay, N. (2006) Mitochondrial hexokinases, novel mediators of the antiapoptotic effects of growth factors and Akt. *Oncogene*, **25**, 4683–4696.

40 Dills, W.L. Jr., Bell, L.S., and Onuma, E.K. (1981) Inhibitory effects of substrate analogs on lactate production from fructose and glucose in bovine spermatozoa. *Biol. Reprod*, **25**, 458–465.

41 Gottlob, K., Majewski, N., Kennedy, S., Kandel, E., Robey, R.B., and Hay, N. (2001) Inhibition of early apoptotic events by Akt/PKB is dependent on the first committed step of glycolysis and mitochondrial hexokinase. *Genes Dev.*, **15**, 1406–1418.

42 Hwang, S.O., and Lee, G.M. (2008) Nutrient deprivation induces autophagy as well as apoptosis in Chinese hamster ovary cell culture. *Biotechnol. Bioeng.*, **99**, 678–685.

43 Preston, R.J. (2007) Epigenetic processes and cancer risk assessment. *Mutat. Res.*, **616**, 7–10.

44 Ushijima, T. (2007) Epigenetic field for cancerization. *J. Biochem. Mol. Biol.*, **40**, 142–150.

45 Gronbaek, K., Hother, C., and Jones, P.A. (2007) Epigenetic changes in cancer. *APMIS*, **115**, 1039–1059.

46 Tischoff, I., and Tannapfe, A. (2008) DNA methylation in hepatocellular carcinoma. *World J. Gastroenterol.*, **14**, 1741–1748.

47 Ghoshal, A.K., and Farber, E. (1984) The induction of liver cancer by dietary deficiency of choline and methionine without added carcinogens. *Carcinogenesis*, **5**, 1367–1370.

48 Rao, P.M., Antony, A., Rajalakshmi, S., and Sarma, D.S. (1989) Studies on hypomethylation of liver DNA during early stages of chemical carcinogenesis in rat liver. *Carcinogenesis*, **10**, 933–937.

49 Lombardi, B., Chandar, N., and Locker, J. (1991) Nutritional model of hepatocarcinogenesis. Rats fed choline-devoid diet. *Dig. Dis. Sci.*, **36**, 979–984.

50 Feo, F., Pascale, R.M., Simile, M.M., De Miglio, M.R., Muroni, M.R., and Calvisi, D. (2000) Genetic alterations in liver carcinogenesis: implications for new

preventive and therapeutic strategies. *Crit. Rev. Oncog.*, **11**, 19–62.

51 Pascale, R.M., Simile, M.M., De Miglio, M.R., and Feo, F. (2002) Chemoprevention of hepatocarcinogenesis: S-adenosyl-L-methionine. *Alcohol*, **27**, 193–198.

52 Macheiner, D., Heller, G., Kappel, S., Bichler, C., Stattner, S., Ziegler, B., Kandioler, D., Wrba, F., Schulte-Hermann, R., Zochbauer-Muller, S., and Grasl-Kraupp, B. (2006) NORE1B, a candidate tumor suppressor, is epigenetically silenced in human hepatocellular carcinoma. *J. Hepatol.*, **45**, 81–89.

53 Zhang, Y.J., Chen, Y., Ahsan, H., Lunn, R.M., Lee, P.H., Chen, C.J., and Santella, R.M. (2003) Inactivation of the DNA repair gene O6-methylguanine-DNA methyltransferase by promoter hypermethylation and its relationship to aflatoxin B1-DNA adducts and p53 mutation in hepatocellular carcinoma. *Int. J. Cancer*, **103**, 440–444.

54 Zhang, Y.J., Chen, Y., Ahsan, H., Lunn, R.M., Chen, S.Y., Lee, P.H., Chen, C.J., and Santella, R.M. (2005) Silencing of glutathione S-transferase P1 by promoter hypermethylation and its relationship to environmental chemical carcinogens in hepatocellular carcinoma. *Cancer Lett.*, **221**, 135–143.

55 Tryndyak, V.P., Kovalchuk, O., Muskhelishvili, L., Montgomery, B., Rodriguez-Juarez, R., Melnyk, S., Ross, S.A., Beland, F.A., and Pogribny, I.P. (2007) Epigenetic reprogramming of liver cells in tamoxifen-induced rat hepatocarcinogenesis. *Mol. Carcinog.*, **46**, 187–197.

56 Watson, R.E., and Goodman, J.I. (2002) Effects of phenobarbital on DNA methylation in GC-rich regions of hepatic DNA from mice that exhibit different levels of susceptibility to liver tumorigenesis. *Toxicol. Sci.*, **68**, 51–58.

57 Bachman, A.N., Kamendulis, L.M., and Goodman, J.I. (2006) Diethanolamine and phenobarbital produce an altered pattern of methylation in GC-rich regions of DNA in B6C3F1 mouse hepatocytes similar to that resulting from choline deficiency. *Toxicol. Sci.*, **90**, 317–325.

58 Bachman, A.N., Phillips, J.M., and Goodman, J.I. (2006) Phenobarbital induces progressive patterns of GC-rich and gene-specific altered DNA methylation in the liver of tumor-prone B6C3F1 mice. *Toxicol. Sci.*, **91**, 393–405.

59 Phillips, J.M., and Goodman, J.I. (2008) Identification of genes that may play critical roles in phenobarbital (PB)-induced liver tumorigenesis due to altered DNA methylation. *Toxicol. Sci.*, **104**, 86–99.

60 Boobis, A.R., Doe, J.E., Heinrich-Hirsch, B., Meek, M.E., Munn, S., Ruchirawat, M., Schlatter, J., Seed, J., and Vickers, C. (2008) IPCS framework for analyzing the relevance of a noncancer mode of action for humans. *Crit. Rev. Toxicol.*, **38**, 87–96.

61 Lu, S.C., and Mato, J.M. (2008) S-Adenosylmethionine in cell growth, apoptosis and liver cancer. *J. Gastroenterol. Hepatol.*, **23** (Suppl. 1), S73–S77.

62 Hasmall, S., James, N., Hedley, K., Olsen, K., and Roberts, R. (2001) Mouse hepatocyte response to peroxisome proliferators: dependency on hepatic nonparenchymal cells and peroxisome proliferator activated receptor alpha (PPARalpha). *Arch. Toxicol.*, **75**, 357–361.

63 Rose, M.L., Rusyn, I., Bojes, H.K., Belyea, J., Cattley, R.C., and Thurman, R.G. (2000) Role of Kupffer cells and oxidants in signaling peroxisome proliferator-induced hepatocyte proliferation. *Mutat. Res.*, **448**, 179–192.

64 Laskin, D.L., Robertson, F.M., Pilaro, A.M., and Laskin, J.D. (1988) Activation of liver macrophages following phenobarbital treatment of rats. *Hepatology*, **8**, 1051–1055.

65 Kroll, B., Kunz, S., Klein, T., and Schwarz, L.R. (1999) Effect of lindane and phenobarbital on cyclooxygenase-2 expression and prostanoid synthesis by Kupffer cells. *Carcinogenesis*, **20**, 1411–1416.

66 Urbaschek, R., McCuskey, R.S., Rudi, V., Becker, K.P., Stickel, F., Urbaschek, B., and Seitz, H.K. (2001) Endotoxin, endotoxin-neutralizing-capacity, sCD14, sICAM-1, and cytokines in patients with various degrees of alcoholic liver disease. *Alcohol. Clin. Exp. Res.*, **25**, 261–268.

67 Wheeler, M.D., Kono, H., Yin, M., Nakagami, M., Uesugi, T., Arteel, G.E.,

Gabele, E., Rusyn, I., Yamashina, S., Froh, M., Adachi, Y., Iimuro, Y., Bradford, B.U., Smutney, O.M., Connor, H.D., Mason, R.P., Goyert, S.M., Peters, J.M., Gonzalez, F.J., Samulski, R.J., and Thurman, R.G. (2001) The role of Kupffer cell oxidant production in early ethanol-induced liver disease. *Free Radical Biol. Med.*, **31**, 1544–1549.

68 Noji, S., Tashiro, K., Koyama, E., Nohno, T., Ohyama, K., Taniguchi, S., and Nakamura, T. (1990) Expression of hepatocyte growth factor gene in endothelial and Kupffer cells of damaged rat livers, as revealed by in situ hybridization. *Biochem. Biophys. Res. Commun.*, **173**, 42–47.

69 Luckey, S.W., and Petersen, D.R. (2001) Activation of Kupffer cells during the course of carbon tetrachloride-induced liver injury and fibrosis in rats. *Exp. Mol. Pathol.*, **71**, 226–240.

70 Chou, W.C., Jie, C., Kenedy, A.A., Jones, R.J., Trush, M.A., and Dang, C.V. (2004) Role of NADPH oxidase in arsenic-induced reactive oxygen species formation and cytotoxicity in myeloid leukemia cells. *Proc. Natl. Acad. Sci. U. S. A.*, **101**, 4578–4583.

71 Teufelhofer, O., Parzefall, W., Kainzbauer, E., Ferk, F., Freiler, C., Knasmuller, S., Elbling, L., Thurman, R., and Schulte-Hermann, R. (2005) Superoxide generation from Kupffer cells contributes to hepatocarcinogenesis: studies on NADPH oxidase knockout mice. *Carcinogenesis*, **26**, 319–329.

72 Petersen, G., Rasmussen, D., and Gustavson, K. (2007) Study on enhancing the endocrine disruptor priority list with focus on low production volume chemicals. ENV.D.4/ETU/2005/028.

73 Waterland, R.A., and Jirtle, R.L. (2003) Transposable elements: targets for early nutritional effects on epigenetic gene regulation. *Mol. Cell Biol.*, **23**, 5293–5300.

74 Anway, M.D., Leathers, C., and Skinner, M.K. (2006) Endocrine disruptor vinclozolin induced epigenetic transgenerational adult-onset disease. *Endocrinology*, **147**, 5515–5523.

75 Zambrano, E., Martinez-Samayoa, P.M., Bautista, C.J., Deas, M., Guillen, L., Rodriguez-Gonzalez, G.L., Guzman, C., Larrea, F., and Nathanielsz, P.W. (2005) Sex differences in transgenerational alterations of growth and metabolism in progeny (F2) of female offspring (F1) of rats fed a low protein diet during pregnancy and lactation. *J. Physiol.*, **566**, 225–236.

9
Carcinogens in Foods: Occurrence, Modes of Action and Modulation of Human Risks by Genetic Factors and Dietary Constituents

M. Mišík, A. Nersesyan, W. Parzefall, and S. Knasmüller

Abstract

It has been well documented that dietary factors have an impact on age-related disorders such as cancer and cardiovascular diseases. However, the focus on specific groups has changed over recent years and on the basis of the development of powerful techniques and innovative findings, new concepts have emerged. Epigenetic aspects are one of these new concepts. This chapter summarizes major groups of food carcinogens as a basis for epigenetic evaluation.

9.1
Introduction

Almost 30 years ago, Doll and Peto [1, 2] published a widely cited study on avoidable risks of cancer in the industrialized world in which they claimed that dietary factors play a key role in the etiology of cancer in humans. Figure 9.1 shows their estimates in more detail. It can be seen, that approximately 35% of all cancer deaths were attributed to nutritional factors. The assumption that diet plays a key role in the onset of cancer was subsequently confirmed in several large prospective trials (for details see e.g. [3, 4]).

Already in the late 1970s, major groups of food carcinogens, such as nitrosamines (NAs), polycyclic aromatic hydrocarbons (PAHs), heterocyclic aromatic amines (HAAs) [5] as well as specific mycotoxins, including aflatoxin B_1 (AFB_1), the most potent human carcinogen known so far [6, 7], had been identified and characterized. In the 1980s, Bruce Ames from the Berkeley University of California, published a series of articles in which he emphasized that also plant-derived foods may also contain potent carcinogens [8–10].

In parallel with the identification of risk factors, it was found that the human diet contains a plethora of compounds which protect us against food-borne and environmental carcinogens by multiple modes of action (for reviews see [9, 11]).

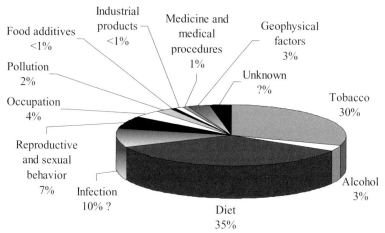

Figure 9.1 Proportions of cancer deaths attributed to different factors (according to Doll and Peto [2]).

The cancer risks caused by carcinogens contained in our foods are also affected not only by diet-related factors but also by our genetic background. A number of studies, for example investigation of twins in Northern Europe, demonstrated the large effect of heritability on cancer in specific sites such as the prostate and colon [12]. One of the reasons for the impact of genetic factors on cancer incidences in humans is the polymorphisms of genes encoding for enzymes which are involved in the metabolism of genotoxic carcinogens contained in our diet and also in DNA-specific genes which encode for repair enzymes [7].

This chapter gives a broad overview of the topics mentioned above. The first section describes major groups of carcinogens contained in the human diet, the second concerns the effects of putative cancer protective food components and the third describes the importance of enzyme polymorphism in regard to diet-related risks.

9.2
Genotoxic Carcinogens in Human Foods

The first carcinogens in human foods were discovered in the 1940s and 1950s and belonged to two classes, namely polycyclic aromatic hydrocarbons and nitrosamines [13, 14]; in the early 1970s, another important group, namely heterocyclic aromatic amines (HAAs), was identified by the Japanese scientist T. Sugimura (for reviews see [15–18]).

9.2.1
Polycyclic Aromatic Hydrocarbons

PAHs are ubiquitous environmental contaminants which are formed by incomplete combustion of organic compounds. Approximately 500 PAHs have been identified so far. Human exposure to PAHs by foods occurs via consumption of barbecued meats [14]. Another source of exposure is the consumption of fruits and vegetables contaminated with air particles containing PAHs that are formed by combustion [19]. Twenty five PAHs are commonly determined in analyses of foods [14]. In an analysis of PAHs in the UK diet, it was found that the major contributions come from cereals (about one-third), from oils and fats (about one-third) and from fruits, vegetables and sugars (about one-third) [20]. The total daily dietary load of PAHs in the UK was estimated to be 3.70 µg, 5.0–17.0 µg in the Netherlands, and 1.2 µg in New Zealand (for review see [14]). The carcinogenic potency of PAHs increases with the number of condensed benzene rings but not all structures convey carcinogenity. The target organs of their carcinogenic action in rodents are multiple, for example, lungs, stomach and breast. An ideal method to assess heath risks caused by these compounds in humans are by DNA-adduct measurements which can be linked directly to human cancer risks [21]. In this context it is notable that a biomonitoring study of Californian non-smoking firefighters showed that PAH adduct formation did not correlate with the extent of fire fighting but was associated with the frequency of consumption of charbroiled food [5].

The carcinogenic properties of PAHs are well documented and many of them have been classified as group 1 ("agents carcinogenic to humans"; benzo[a]pyrene), group 2A ("agents probably carcinogenic to humans"; dibenz[a,h]anthracene, dibenzo[a,l]pyrene) or as group 2B ("agents possibly carcinogenic to humans"; benz[a]anthracene, benzo[b]fluoranthene, dibenzo[a,h]pyrene, benzo[j]fluoranthene, benzo[k]fluoranthene, benzo[c]phenanthrene, dibenzo[a,i]pyrene) by the IARC[1)] [22].

The induction of tumors is assumed to be due to DNA-damage caused by reactive metabolites which are formed as a consequence of enzymatic conversion (i.e., after hydroxylation by cytochrome CYP1A1 and subsequent formation of DNA-reactive bay-region diols) [23]. One of the most important detoxification pathways is the enzymatic conjugation with glutathione which is catalyzed by a tissue and substrate specific glutathion-S-transferases (GSTs) [23, 24].

9.2.2
Nitrosamines

NAs are found in cured meats where they are formed as a consequence of reactions between amines contained in proteins with nitrosating agents. Another pathway is the endogenous nitrosation of amines under acidic conditions in the stomach as a

---

1) International Agency for Research on Cancer, Lyon, France.

consequence of the simultaneous presence of protein-rich foods and of vegetables containing nitrates and/or nitrite, or of consumption of water with high levels of nitrates [7]. Several hundred different NAs have been characterized [25] and the most important ones found in foods are N-nitrosodimethylamine (NDMA), N-nitrosodiethylamine (NDEA), N-nitrosopiperidine (NPIP) and N-nitrosopyrrolidine (NPYR). It is notable that tobacco smoke contains a number of specific nicotine-derived NAs which are considered to play a key role in the etiology of human lung cancer. The IARC placed NDMA and NDEA into the category of agents which are probably carcinogenic to humans (2A), and N-nitrosodibutylamine, NPIP and NPYR into the group considered as possibly carcinogenic (2B).

The activation of most NAs follows a common scheme, that is, hydroxylation of the α-C-atom by a specific cytochrome (CYP2E1) which leads to formation of unstable DNA-reactive metabolites [23]. The target tissues of tumor induction are multiple and major affected organs are liver, esophagus and brain [26–29].

In humans, the high incidence of nasal cancer in fishermen in the Southern province of China has been attributed to consumption of dry fish containing high levels of NAs (for details see [30]); also the increased incidence of esophageal cancer seen in areas with high nitrate and nitrite levels in drinking water was attributed to NAs formation [31].

9.2.3
Heterocyclic Aromatic Amines (HAAs) and Other Thermal Degradation Products

HAAs are formed as reaction products of glucose, amino acids and creatinine during frying of meats and were discovered by a Japanese scientist in fish [16, 17]. These compounds induced predominantly colon cancer in experiments with rodents and it was hypothesized that they may contribute to the high rates of gastro-intestinal cancer in individuals living in countries with high levels of meat consumption [32]. Three main factors affect their formation: (i) the type of meat (in particular, its creatinine level); (ii) the cooking method (temperature); and (iii) the preparation time. HAAs are found mainly in cooked muscle meats while other sources of dietary protein (milk, eggs, and organ meats such as liver) contain only trace amounts. The temperature is the most important factor in the formation of the amines, frying and barbecuing producing the largest amounts. It was shown that the HAA levels are increased threefold when the cooking temperature is increased from 200 °C to 250 °C [33].

More than 20 HAAs have been isolated so far, belonging to different chemical groups as derivates of γ-carboline, amino-α-carboline, dipyrido-imidazol, imidazoquinoline, imidazoquinoxaline, phenylimidazopyridine [17, 34, 35].

Scientists commonly use abbreviations for the different types of amines. The most important ones are PhIP (2-amino-1-methyl-6-phenylimidazo[4,5-b]pyridine) and AαC (2-amino-9H-pyrido[2,3-b]indole) which are most abundantly found in fried meat. Other compounds which were frequently used in model studies are IQ (2-amino-3-methylimidazo[4,5-f]quinoline) and Trp-P-1 (3-amino-1,4-dimethyl-5H-pyrido[4,3-b]indole).

The metabolism of HAAs is quite complex. Following hydroxylation of the N-atom by cytochrome P4501A2, the metabolites are acetylated or sulfated and glucoronated, predominantly in the liver. Following excretion of the latter metabolites into the gut via the bile they can be reactivated by β-glucuronidases of the intestinal microflora. The reactive metabolites may then damage DNA in colon mucosa cells (for a detailed description see [36]). Sulfotrasferases play a key role in the metabolic activation of amino imidazoles, and carbolines.

9.2.4
Mycotoxins

Aflatoxins, in particular AFB_1 account for the high prevalence of hepatocellular cancer in certain areas of the world (e.g., in Central Africa and China). These toxins are formed during storage of plant-derived foods by molds (by *Aspergillus flavus* and other fungi) which grow in tropical areas [37].

AFB_1 is enzymatically activated in the liver to a reactive epoxide, predominantly by the cytochrome P450 isoforms CYP1A2 and CYP3A4. The epoxide can be detoxified by conjugation with glutathione [38]. As AFB_1 causes a specific "fingerprint" mutation in a human suppressor gene (*p53*), it is possible to determine by sequencing of DNA samples, isolated from liver tumors, whether the mycotoxin is responsible for the disease [39].

Several other mycotoxins are also suspected to cause cancer in humans. For example, ochratoxin A which was found to be responsible for the outbreak of Balkan endemic nephropathy (BEN), a renal disorder in Eastern European countries in the 1980s. Several pieces of evidence have recently shifted opinion from ochratoxin A to aristolochic acid as the causative agent of BEN [40]. Although ochratoxin A might still be involved, (i) aristolactam–DNA adducts in renal tissues and urothelial cancers of affected patients have been found, (ii) "signature" mutations in the *p53* gene were identified in the upper urothelial cancers associated with this disease, and (iii) the renal pathophysiology and histopathology observed in endemic nephropathy most closely resemble those of aristolochic acid nephropathy.

Other potentially relevant carcinogenic mycotoxins are fumonisin B_1 and trichotecenes (nivalenol, deoxynivalenol). Contamination of foods with these compounds has been proposed to play a role in the etiology of esophageal cancers in Transkey (South Africa) and Asia [41] and experimental evidence of DNA-damage in mammalian cells is available [42].

9.2.5
Food Additives and Carcinogens in Plant-Derived Foods

The DNA-damaging and carcinogenic properties of food additives have been studied intensively and it can be assumed that the chemicals which are currently used by the industry can be regarded in general as safe.

"Classical" examples for hazardous additives which have been banned many decades ago are "butter yellow" which causes bladder cancer [43] and 2-(2-furyl)-

3-(5-nitro-2-furyl) acrylamide (furylfuramide), (AF-2) which was used in Japan as a preservative in the 1970s [44].

Besides the formation of HAAs thermal degradation and condensation of basic constituents occurs during food preparation. Acrylamide (AA) has been detected in baked foods containing starch and proteins. The levels were quite variable up to 3770 µg AA/kg in snacks (crisps and potato snacks [45]). AA was found to be a germ cell mutagen besides its neurotoxicity and toxicity for reproduction and development in rats and mice. AA is metabolically activated by CYP2E1 to the reactive and genotoxic metabolite glycidamide which is also formed in even smaller amounts directly during food preparation, as has been reported recently [46].

This finding illustrates that there may be many more genotoxic and carcinogenic compounds generated during the Maillard reaction of which we are still unaware.

The artificial sweetener sodium saccharin and the natural flavoring agent d-limonene were found to induce tumors in rodents, but it became clear in subsequent mechanistic studies, that the induction of bladder cancer by saccharin and formation of renal tumors by d-limonene are species specific [47–49]. In 2000, the National Toxicology Program of the National Institutes of Health (NIH) re-evaluated the evidence for human carcinogenicity of saccharin [50] and recommended in its "Report on Carcinogens, 9th Edition" that it should be removed from the list of potential carcinogens. This delisting took place in the 11th Edition [51], and the California Environmental Protection Agency (EPA) also removed saccharin from its Proposition 65 list of carcinogens.

d-Limonene produces renal cancer in male rats only. The mechanism is unique to the male rat since the syndrome is dependent on alpha$_2$µ-globulin in urine which is not present in the female rat and in humans, mice, dogs, and guinea-pigs. Therefore, no renal carcinomas will develop following exposure to hydrocarbons like d-limonene in either of these species. Indeed there is no evidence for genotoxicity or carcinogenicity in humans. The IARC classifies this compound at present under Class 3: *not classifiable as to its carcinogenicity to humans* [52]. On the contrary, limonene is now known as a significant chemopreventive agent [53] and is predicted to have potential value as a dietary cancer preventive compound in humans [54]. The FDA considers d-limonene to be GRAS (generally regarded as safe) as a food additive when used as a synthetic flavoring substance and adjuvant.

The first carcinogenic plant specific constituent which was identified is cycasin, a glycoside contained in cycad flour which was produced from palm trees and used for the production of sweets (cakes and biscuits) in tropical areas [55]. The compound is cleaved enzymatically by representatives of the intestinal flora and the aglycone is converted to a highly DNA-reactive derivative in the hepatic tissue and causes liver cancer [55].

In 1986, Ames published a provoking article in which he claimed that many plants which are used as foods contain carcinogens which can be regarded as "natural pesticides" [9]. Typical examples are capsaicin, the pungent principle of chilli and compounds like coumarin (in vanilla, and cassia), sesamol (in sesame seeds), safrol (in spices such as nutmeg, mace, cinnamon, anise and black pepper), estragol (tarragon, sweet fennel and anis), and agaritine (mushrooms). We evaluated recently the current state of knowledge on genotoxic and carcinogenic properties of these agents [56]. There is no doubt that some of them cause tumors in rodents. However, many of these compounds are contained in spices and therefore consumed only in minor quantities by men. Therefore, it can be assumed that their contribution to the overall human cancer risk is low or negligible. As already stressed by Ames and his coworkers [8, 10], our knowledge concerning the risks of synthetic compounds is by far higher than that on natural compounds which are commonly regarded as safe, and more experimental data should be available to assess the possible cancer risks of plant constituents in our daily diet more precisely.

9.2.6
Alcohol

One of the main diet-related cancer risks for men is alcohol. It has been shown that its consumption leads to a substantial increase in specific forms of cancer, for example in the esophagus, liver, large bowel and breast (for extensive reviews see [57–59]).

One of the key mechanisms which has been postulated to account for its carcinogenic effects is the induction of inflammation by the main metabolite acetaldehyde which was also shown to inhibit DNA-repair processes and to cause formation of stable DNA-adducts (N-methyldeoxyguanosine) [59]. In this context it is also notable, that the induction of CYP2E1, which catalyzes the activation of nitrosamines by chronic alcohol intake, may account for the strong synergistic effects between tobacco smoking and alcohol consumption in regard to cancer induction.

9.3
Contribution of Genotoxic Dietary Carcinogens to Human Cancer Risks

In 1992, Lutz and Schlatter published an interesting paper in which they assessed the contribution of different DNA-reactive compounds found in foods to the overall cancer risk in the Swiss population on the basis of exposure data and results from animal experiments [60]. They calculated, that according to the estimate of Doll and Peto [2] about 80 000 of the annual cancer cases per one million lives in Switzerland are due to dietary factors and assessed the contribution of individual groups of carcinogens. From these data, it became clear that only 25% of the overall dietary cancer risks could be explained by known chemical carcinogens contained in food. The "remaining" dietary cancer risk (about 75%) was postulated (at least partly) to be due to overweight [60].

Indeed, numerous investigations showed that excess body weight is associated with different forms of cancer. A recent paper of Renehan *et al.* [61] gives an overview on the current state of knowledge. The meta-analysis of studies published between 1966 and 2007 shows quite clear, that an excess BMI of 5 $kg\,m^{-2}$ in males correlates with an increased risk for esophageal carcinomas by 52%. In women, the highest risk rates were found in regard to endometrial cancer and gall bladder tumors (increase of the risks by ≥50%). Additionally, cancers in other organs are associated with overweight, for example the kidney (increase of the risk by 24% in ♂ and by 34% in ♀) and the colon (increase of the risk by 24% in ♂ and 9% in ♀) [61, 62].

9.4
Protective Effects of Dietary Components Towards DNA-Reactive Carcinogens

Apart from obesity, the overall consumption of diets containing specific protective compounds as well as individual susceptibility factors, may have a strong impact on diet-related human risks [63].

In the 1970s, the Japanese researcher Kada screened hundreds of plant extracts for antimutagenic effects in bacterial assays, in order to identify the most potent protective components [64, 65], while in the US, Wattenberg and coworkers investigated at the University of Minnesota investigated the protective properties of vegetables in animal cancer models and demonstrated that specific plant-derived diets prevent the induction of tumors by PAHs [66, 67].

To date, a large number of plant components and beverages have been identified which are protective towards dietary carcinogens [68]. Various examples of such compounds are listed in Table 9.1. It can be seen, that chemoprevention can be achieved by a broad variety of different mechanisms.

Probably one of the most important modes of action is the modulation of activities of drug metabolizing enzymes involved in the activation and/or detoxification of xenobiotics (including DNA-reactive carcinogens). The investigations of Kensler and his colleagues [83] contributed substantially to the current understanding of these processes. In brief, they showed, that the activation of a specific transcriptional factor (termed Nrf2) by dietary compounds (e.g., by sulforaphane found in broccoli and other constituents of cruciferous vegetables) leads, via interaction with the antioxidant responsive element (ARE), to the transcription of a broad variety of genes which encode for defense function, including detoxifying (phase II) enzymes which inactive carcinogens [84].

Table 9.1 also lists compounds which inactivate reactive oxygen species (ROS). It is know that human foods are a major cause of ROS-mediated damage [9, 82], which is a key event in both, cancer induction and promotion.

Another important principle of chemoprevention may be the induction of DNA-repair enzymes. However, it is quite unclear at present if effects detected with some dietary compounds, for example with spice ingredients such as vanillin and cinnamonaldehyde in bacterial test systems [85] can be extrapolated to the human situation and in general, only little is known about the impact of the diet on repair functions.

Table 9.1 Examples for dietary factors which protect against DNA-damaging tumor induction by food-related carcinogens.

Carcinogen	Protective factor	Mode of action	References
Aflatoxin B_1	Chlorophyll, lactic acid bacteria	Direct binding	[69, 70]
	Coffee and coffee-specific diterpenoids	Induction of glutathione S transferase – prevention of cancer	[71]
Benzo[a]pyren and other PAHs	Glucosinolates and cruciferous vegetables	Induction of glutathione S transferase which detoxified the metabolites	[72]
	Phenolic acids (e.g., ellagic acid)	Inactivation of DNA reactive metabolites (diol epoxide)	[73, 74]
Ochratoxin and patulin	Lactic acid bacteria in yogurt	Direct binding	[75]
Heterocyclic aromatic amines	Lactic acid bacteria, fiber, chlorophyll	Direct binding	[76]
	Brassica vegetables	Induction of glucouronosyl transferase	[77]
	Coffee diterpenoids	Inhibition of activation ($\downarrow$ CYP1A2 and $\downarrow$ of NAT2) and induction of detoxifying enzymes	[78]
Nitrosamines	Phenolics in plants	Prevention of formation in the stomach	[79, 80]
	Isothiocyanate (breakdown products of glucosinolate)	Inhibition of activation ($\downarrow$ CYP2E1)	[81]
Reactive enzyme species (ROS)	Antioxidant vitamins (C, A, E) and carotenoids	Direct scavenging	[82]
	Lactic acid bacteria, green tea, etc.	Induction of antioxidant enzymes	[82]

The development of new approaches which will enable a more detailed study of this topic, for example, the establishment of protocols for single cell gel electrophoresis assays may help to shed light on this exciting issue in the future (for details see [86]).

9.5
Gene Polymorphisms Affecting the Metabolism of Genotoxic Carcinogens

As described above, different groups of relevant dietary carcinogens are activated and/or detoxified by phase I and phase II enzymes, respectively. To our current knowledge, most of the genes encoding specific enzymes are polymorphic in humans; that is, they contain mutations which affect their functions. As a consequence, increased, but also decreased, levels of reactive metabolites of dietary contaminants may be generated, resulting in broad inter-individual variations in cancer susceptibility. For example genetic variants of CYP1A1, a phase I enzyme which metabolically activates PAHs, is known to exhibit higher inducibility and is associated with an increased lung cancer risk (see Table 9.2).

Genetic variants of phase II enzymes often lead to inefficient detoxification of reactive electrophilic metabolites, for example, (i) exposure to PAHs of individuals with the *GSTM1*0* genotype results in increased PAH–DNA adduct levels in peripheral blood lymphocytes (see Table 9.2); (ii) *GSTA1*B/*B* leads to decreased enzyme activity and colorectal cancer risks may increase in consumers of well-done red meats [91].

However, in general, adduct levels or cancer risks do not increase dramatically because parallel pathways through other enzyme isoforms can still inactivate the reactive metabolites. Table 9.2 provides further examples of enzyme polymorphisms/ variants in a non-exhaustive manner (for a detailed review see Reszka et al. [111]).

Polymorphisms occur also in genes encoding for DNA-repair factors and enzymes [112] and in this context it is notable that specific forms of cancer have been associated with such genes. For example *BRCA1* and *BRCA2* mutations lead to increased rates of breast cancer [104, 105] while *hMLH1* and *hMLH2* are involved in colorectal cancer in man [106]. Genes which play a role in base excision repair such as *XRCC1* and *XRCC2* as well as *ERCC4/XPF* appear to have an impact on human cancer etiology [113]. A "classical" example for an association of DNA-repair genes with skin tumors are Xeroderma pigmentosum patients which are defective in nucleotide excision repair (genes for seven possible factors: *XPA* to *XPG*) and thereby predisposed to increased risks [114, 115].

Single polymorphisms in *CYP*, *GST* or *NAT* genes or of DNA-repair genes show only moderate effects in terms of relative cancer risk, they are so-called low penetrance genes. However, genotype combinations appear to predispose to distinctly increased cancer risks. Some examples are shown in the lower part of Table 9.2, for example, the combined occurrence of the *GSTM1* null geno-/phenotype and the expression of a low activity variant of the epoxide hydrolase renders AFB_1

Table 9.2 Examples of the impact of enzyme polymorphisms on DNA-damaging and/or carcinogenic effects of dietary compounds.

Enzyme	Polymorphism geno-/phenotype	Effect on genotoxic dietary carcinogens	References
Cytochrome P450 (CYP) 1A1	CYP1A1*2 alleles (MsPI)	Increased lung cancer risk in homo- and heterozygotes increased inducibility, but secondary role in lung cancer risk	[87–89]
CYP1A2	CYP1A2*1K	Decreased activity, lower inducibility → probably less activation of aflatoxin B_1	[90]
CYP2A6	High activity phenotype	Dietary nitrosamine metabolism; colorectal cancer ↑	[91]
CYP2E1	Linked with a single or double 96 bp insertion in the regulatory region	Higher enzyme activity/inducibility; rectal cancer ↑ at intake of high dietary levels of nitrosamines and of salted/dried fish or oriental pickled vegetables	[92]
CYP2E1	CYP2E1 G1259C RsaI	Decreased CYP2E1 activity/inducibility; colorectal cancer ↑ in homozygous wild-type genotype (c1/c1) vs. the homozygous variant genotype (c2/c2)	[92]
Alcoholdehydrogenase	ADH2*1/2*1	Less-active form associated with increased risk of esophageal cancer in East Asians	[93]
Alcoholdehydrogenase	ADH3*1/2	Fast metabolizing form associated with increased risk of breast cancer in Caucasians	[94]
Aldehyde dehydrogenase	ALDH2*1/2*2 null allele	High risk of esophageal cancer in East Asians	[94]
Epoxide Hydrolase (EPHX1)	EPHX1 exon 3 (CT, Tyr113Arg) EPHX1 exon 4 (CA, His139Arg)	Low activity variant: decreased risk of lung cancer high activity variant: modestly increased risk of lung cancer	[95, 96]
Glutathion-S-Transferase (GST) A1	GSTA1*B/*B	Lower enzyme activity; colorectal cancer ↑ in consumers of well-done red meat	[91]
GST M1	GSTM1*0	PAH–DNA adduct levels ↑ in smokers, coke oven workers	[97, 98]
GST T1	GSTT1*0	Bulky PAH–DNA adduct levels slightly ↑	[99]
N-Acetyl Transferase (NAT) 1	NAT1*11	Increased activity. In postmenopausal women consuming high levels of red meat: Breast cancer risk ↑	[100]

Table 9.2 Continued.

Enzyme	Polymorphism geno-/phenotype	Effect on genotoxic dietary carcinogens	References
NAT 2	NAT2 rapid/ intermediate	Breast cancer ↑ (HAA from red meat)	[101]
DNA-repair factor Xeroderma Pigmentosum (XP)	XPD exon 10 AA (asp312Asn) XPD exon 23 CC (Lys751Gln)	Lung cancer risk ↑ in never-smokers and further increased lung cancer risk in current or recent smokers aromatic adduct levels ↑	[102]
XP DNA-repair factors	XPA-(A23G), XPC(poly-AT insert), XPD-Asp312Asn and	PAH DNA adduct levels ↑ in smokers, coke oven workers	[98]
Combined DNA-repair factors	XCCR1, XCCR3, XPD	With increased combination of variant alleles: bulky adduct levels (PAHs) ↑	[103]
DNA-repair factors	BRCA1, BRCA2 (high penetrance genes)	Loss of heterozygosity leads to impaired DNA repair of double strand breaks → breast cancer ↑	[104, 105]
DNA repair	MSH2, MLH1, PMS1, PMS2	Germline mutations in mismatch DNA repair genes → hereditary nonpolyposis colon cancer ↑	[106]
Combinations of genotypes			
Combined CYP1A1 GST M1	CYP1A1 Ile462Val GSTM1*0	Higher activity for activation of PAHs; increased risk of lung cancer in non-smokers (presumably dietary exposure)	[107]
Combined CYP1A2 NAT 2	CYP1A2 high activity NAT2 slow acetylators	Post-meal urinary mutagens (HAA from pan-fried hamburgers) ↑	[108]
Combined CYP1A2 NAT 2	CYP1A2 high activity NAT2 rapid acetylators	Preference for well-done red meat/ever smokers → colorectal cancer ↑	[109]
Combined GST M1 and EPXH1	GSTM1*0 EPHX1 exon 3 (CT, Tyr113Arg)	AFB_1 albumin adducts ↑, risk of $p53$ mutations ↑, and hepatocellular carcinoma ↑	[110]
Combined GST M1 and XPC or XPA	GSTM1*0 Xpc +/+ or XPA-A23A	BPDE DNA adducts ↑ in coke oven workers	[98]

exposed individuals at clearly increased risk for hepatocellular cancer. Other scenarios may be envisioned. For example, Tempfer et al. [116] reported on data collected in the frame of the prospective Nurses' Health Study which established seven SNPs[2] (*hPRB +331 G/A, AR CAG repeat, CYP19 TTTA(10), CYP1A1 Msp I, VDR FOK1, XRCC1 Arg194Trp,* and *XRCC2 Arg188His*) as small, but significant risk factors for spontaneous, non-hereditary breast cancer to which low-penetrance genetic risk factors of sporadic breast cancer may add, namely the SNPs *TGFBR1*6A, HRAS1, GSTP Ile105Val,* and *GSTM1* SNPs.

These examples demonstrate that multiple factors of susceptibility may modulate the health impact of dietary genotoxins on an individual level.

The complex interplay between food-derived carcinogens and metabolic, genetic, and protective factors is summarized in Figure 9.2.

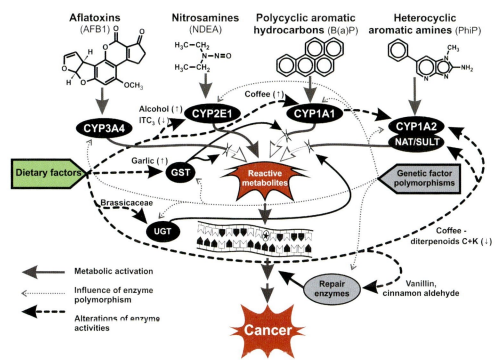

Figure 9.2 Schematic illustration of activation and detoxification pathways of genotoxic carcinogens contained in the human diet. CST – gluthation-S-transferase, UGT – glucuronosyl-transferase, NAT – N-acetyltransferase, SULT – sulfotransferase, ITC$_3$ – isothiocyanates, C + K – kahweol/cafestol.

---

2) Abbreviations: SNP, single nucleotide polymorphism; hPRB, human retinoblastoma geneproduct; AR, androgen receptor; VDR, vitamin D receptor; TGFBR1, TGFß1 receptor; hRAS1, highly polymorphic region of variable number of tandem repeats, 1 kb downstream of the human proto-oncogene *H-ras1*.

9.6
Concluding Remarks, Epigenetics and Outlook

Over the last 50 years, strong efforts have been made to identify hazardous and chemopreventive compounds in the diet in order to protect humans against cancer caused by nutrition. As a consequence, dietary strategies and recommendations have been developed and functional foods and supplements are produced in increasing amounts (e.g., see [115, 117]). However, studies indicate that the incidence of cancer is still increasing since the 1950s and it has been stressed that these effects cannot be explained solely by the higher life expectancy of the population (for details see [118]).

Furthermore, hazardous compounds in the diet must also be seen in the light of new epigenetic concepts such as the association of ancestral food supply with longevity and with cardiovascular and diabetic mortality, as shown by Pembrey *et al.* [119]. Will we get more evidence that environmental toxins cause epigenetic changes to the DNA in the developing germ line, and that these changes are maintained and carried along with the sperm to the next generation.

One possible solution to counteract this tendency is the further elucidation of preventive mechanisms and the identification of potent chemopreventive dietary factors. In this context it is notable, that recent findings made it questionable if generalized nutritional recommendations are indeed effective; for example it was shown in meta-analyses that intake of antioxidant supplements has adverse effects in terms of cancer risks in humans [114]; also, in the case of folate it was emphasized that excess uptake may lead to an increased risk of colon cancer [120]. Animal studies, in conjunction with clinical observations, suggest that folate elicits dual modulator effects on carcinogenesis, depending on the timing and dose of folate intervention and on the polymorphism of genes involved in its metabolism.

Probably one of the most promising concepts is the design of individual diets which are based on the determination of key parameters of cancer risks such as DNA-stability. The "Health Genome Concept" which has been developed by Fenech in recent years [121], provides exemplary details.

References

1 Doll, R. (1992) The lessons of life: keynote address to the nutrition and cancer conference. *Cancer Res.*, **52**, 2024s–2029s.

2 Doll, R., and Peto, R. (1981) The causes of cancer: quantitative estimates of avoidable risks of cancer in the United States today. *J. Natl. Cancer Inst.*, **66**, 1191–1308.

3 Boeing, H., Dietrich, T., Hoffmann, K., Pischon, T., Ferrari, P., Lahmann, P.H. *et al.* (2006) Intake of fruits and vegetables and risk of cancer of the upper aero-digestive tract: the prospective EPIC-study. *Cancer Causes Control*, **17**, 957–969.

4 World Cancer Research Fund, American Institute for Cancer Research (2007) *Food, Nutrition, Physical Activity, and the Prevention of Cancer: A Global Perspective*, World Cancer Research Fund, American Institute for Cancer Research, Washington, DC.

5 Rothman, N., Correa-Villasenor, A., Ford, D.P., Poirier, M.C., Haas, R., Hansen, J.A. *et al.* (1993) Contribution of occupation and diet to white blood cell polycyclic aromatic hydrocarbon-DNA adducts in wildland firefighters. *Cancer Epidemiol. Biomarkers Prev.*, **2**, 341–347.

6 Ferguson, L.R. (2002) Natural and human-made mutagens and carcinogens in the human diet. *Toxicology*, **181–182**, 79–82.

7 Goldman, R., and Shields, P.G. (2003) Food mutagens. *J. Nutr.*, **133** (Suppl. 3), S965–S973.

8 Ames, B.N. (1983) Dietary carcinogens and anticarcinogens. Oxygen radicals and degenerative diseases. *Science*, **221**, 1256–1264.

9 Ames, B.N. (1986) Food constituents as a source of mutagens, carcinogens, and anticarcinogens. *Prog. Clin. Biol. Res.*, **206**, 3–32.

10 Ames, B.N., and Gold, L.S. (1990) Dietary carcinogens, environmental pollution, and cancer: some misconceptions. *Med. Oncol. Tumor Pharmacother.*, **7**, 69–85.

11 Khan, N., Afaq, F., and Mukhtar, H. (2008) Cancer chemoprevention through dietary antioxidants: progress and promise. *Antioxid. Redox. Signal*, **10**, 475–510.

12 Lichtenstein, P., Holm, N.V., Verkasalo, P.K., Iliadou, A., Kaprio, J., Koskenvuo, M. *et al.* (2000) Environmental and heritable factors in the causation of cancer – analyses of cohorts of twins from Sweden, Denmark, and Finland. *N. Engl. J. Med.*, **343**, 78–85.

13 Lowenfels, A.B., and Anderson, M.E. (1977) Diet and cancer. *Cancer*, **39**, 1809–1814.

14 Phillips, D.H. (1999) Polycyclic aromatic hydrocarbons in the diet. *Mutat. Res.*, **443**, 139–147.

15 Sugimura, T. (1978) Let's be scientific about the problem of mutagens in cooked food. *Mutat. Res.*, **55**, 149–152.

16 Sugimura, T. (1982) Mutagens in cooked food. *Basic Life Sci.*, **21**, 243–269.

17 Sugimura, T. (1997) Overview of carcinogenic heterocyclic amines. *Mutat. Res.*, **376**, 211–219.

18 Sugimura, T., and Nagao, M. (1981) Carcinogenic, mutagenic, and comutagenic aromatic amines in human foods. *Natl. Cancer Inst. Monogr.*, 27–33.

19 Gilbert, J. (1994) The fate of environmental contaminants in the food chain. *Sci. Total Environ*, **143**, 103–111.

20 Dennis, M.J., Massey, R.C., McWeeny, D.J., Knowles, M.E., and Watson, D. (1983) Analysis of polycyclic aromatic hydrocarbons in UK total diets. *Food Chem. Toxicol.*, **21**, 569–574.

21 Gyorffy, E., Anna, L., Kovacs, K., Rudnai, P., and Schoket, B. (2008) Correlation between biomarkers of human exposure to genotoxins with focus on carcinogen-DNA adducts. *Mutagenesis*, **23**, 1–18.

22 International Agency for Research on Cancer (1987) *Overall Evaluations of Carcinogenicity to Humans*, Suppl. 7, IARC, Lyon, France.

23 Sheweita, S.A. (2000) Drug-metabolizing enzymes: mechanisms and functions. *Curr. Drug Metab.*, **1**, 107–132.

24 Eaton, D.L., Gallagher, E.P., Bammler, T.K., and Kunze, K.L. (1995) Role of cytochrome P4501A2 in chemical carcinogenesis: implications for human variability in expression and enzyme activity. *Pharmacogenetics*, **5**, 259–274.

25 Lijinsky, W. (1999) *N*-Nitroso compounds in the diet. *Mutat. Res.*, **443**, 129–138.

26 Peto, R., Gray, R., Brantom, P., and Grasso, P. (1991) Dose and time relationships for tumor induction in the liver and esophagus of 4080 inbred rats by chronic ingestion of *N*-nitrosodiethylamine or *N*-nitrosodimethylamine. *Cancer Res.*, **51**, 6452–6469.

27 Jakszyn, P., and Gonzalez, C.A. (2006) Nitrosamine and related food intake and gastric and oesophageal cancer risk: a

systematic review of the epidemiological evidence. *World J. Gastroenterol.*, **12**, 4296–4303.

28 Eichholzer, M., and Gutzwiller, F. (1998) Dietary nitrates, nitrites, and *N*-nitroso compounds and cancer risk: a review of the epidemiologic evidence. *Nutr. Rev.*, **56**, 95–105.

29 Dietrich, M., Block, G., Pogoda, J.M., Buffler, P., Hecht, S., and Preston-Martin, S. (2005) A review: dietary and endogenously formed *N*-nitroso compounds and risk of childhood brain tumors. *Cancer Causes Control*, **16**, 619–635.

30 Yu, M.C., Ho, J.H., Lai, S.H., and Henderson, B.E. (1986) Cantonese-style salted fish as a cause of nasopharyngeal carcinoma: report of a case-control study in Hong Kong. *Cancer Res.*, **46**, 956–961.

31 Zhang, X.L., Bing, Z., Xing, Z., Chen, Z.F., Zhang, J.Z., Liang, S.Y. *et al.* (2003) Research and control of well water pollution in high esophageal cancer areas. *World J. Gastroenterol.*, **9**, 1187–1190.

32 Alaejos, M.S., Gonzalez, V., and Afonso, A.M. (2008) Exposure to heterocyclic aromatic amines from the consumption of cooked red meat and its effect on human cancer risk: a review. *Food Addit. Contam.*, **25**, 2–24.

33 Pariza, M.W., Ashoor, S.H., Chu, F.S., and Lund, D.B. (1979) Effects of temperature and time on mutagen formation in pan-fried hamburger. *Cancer Lett.*, **7**, 63–69.

34 Turesky, R.J. (2002) Heterocyclic aromatic amine metabolism, DNA adduct formation, mutagenesis, and carcinogenesis. *Drug Metab. Rev.*, **34**, 625–650.

35 Nagao, M., and Sugimura, T. (eds) (2000) *Food Borne Carcinogens, Heterocyclic Amines*, John Wiley & Sons, Ltd, Chichester, New York, Weinheim, Brisbane, Singapure, Toronto.

36 Turesky, R.J. (2005) Interspecies metabolism of heterocyclic aromatic amines and the uncertainties in extrapolation of animal toxicity data for human risk assessment. *Mol. Nutr. Food Res.*, **49**, 101–117.

37 Chen, C.J., Yu, M.W., and Liaw, Y.F. (1997) Epidemiological characteristics and risk factors of hepatocellular carcinoma. *J. Gastroenterol. Hepatol.*, **12**, S294–308.

38 McLean, M., and Dutton, M.F. (1995) Cellular interactions and metabolism of aflatoxin: an update. *Pharmacol. Ther.*, **65**, 163–192.

39 Le Roux, E., Gormally, E., and Hainaut, P. (2005) Somatic mutations in human cancer: applications in molecular epidemiology. *Rev. Epidemiol. Sante. Publique.*, **53**, 257–266.

40 Grollman, A.P., Shibutani, S., Moriya, M., Miller, F., Wu, L., Moll, U. *et al.* (2007) Aristolochic acid and the etiology of endemic (Balkan) nephropathy. *Proc. Natl. Acad. Sci. U. S. A.*, **104**, 12129–12134.

41 Dutton, M.F. (1996) Fumonisins, mycotoxins of increasing importance: their nature and their effects. *Pharmacol. Ther.*, **70**, 137–161.

42 Knasmuller, S., Bresgen, N., Kassie, F., Mersch-Sundermann, V., Gelderblom, W., Zohrer, E. *et al.* (1997) Genotoxic effects of three Fusarium mycotoxins, fumonisin B1, moniliformin and vomitoxin in bacteria and in primary cultures of rat hepatocytes. *Mutat. Res.*, **391**, 39–48.

43 Weisburger, J.H., and Weisburger, E.K. (1968) Food additives and chemical carcinogens: on the concept of zero tolerance. *Food Cosmet. Toxicol.*, **6**, 235–242.

44 Tokiwa, H., Nakagawa, R., Horikawa, K., and Ohkubo, A. (1987) The nature of the mutagenicity and carcinogenicity of nitrated, aromatic compounds in the environment. *Environ Health Perspect.*, **73**, 191–199.

45 Boon, P.E., de Mul, A., van der Voet, H., van Donkersgoed, G., Brette, M., and van Klaveren, J.D. (2005) Calculations of dietary exposure to acrylamide. *Mutat. Res.*, **580**, 143–155.

46 Granvogl, M., Koehler, P., Latzer, L., and Schieberle, P. (2008) Development of a stable isotope dilution assay for the quantitation of glycidamide and its application to foods and model systems. *J. Agric. Food Chem.*, **56**, 6087–6092.

47 Williams, G.M., and Whysner, J. (1996) Epigenetic carcinogens: evaluation and risk assessment. *Exp. Toxicol. Pathol.*, **48**, 189–195.

48 Ellwein, L.B., and Cohen, S.M. (1990) The health risks of saccharin revisited. *Crit. Rev. Toxicol.*, **20**, 311–326.

49 Capen, C., Dybing, E., Rice, J., and Wilbourn, J. (1999) *Species Differences in Thyroid, Kidney and Urinary Bladder Carcinogenesis*, IARC, Lyon.

50 National Toxicology Program (1999) NTP Report on Carcinogens Background Document for Saccharin NTP.

51 National Toxicology Program (2005) *Report on Carcinogens*, 11th edn, U.S. Department of Health and Human Services, Public Health Service, National Toxicology Program. Appendix B.

52 International Agency for Research on Cancer (1999) *IARC Monographs on the Evaluation of Carcinogenic Risks to Humans*, vol. **73**, IARC, Lyon, France, pp. 307–327.

53 Crowell, P.L. (1999) Prevention and therapy of cancer by dietary monoterpenes. *J. Nutr.*, **129**, 775S–778S.

54 Tsuda, H., Ohshima, Y., Nomoto, H., Fujita, K., Matsuda, E., Iigo, M. *et al.* (2004) Cancer prevention by natural compounds. *Drug Metab. Pharmacokinet.*, **19**, 245–263.

55 Goldin, B.R. (1990) Intestinal microflora: metabolism of drugs and carcinogens. *Ann. Med.*, **22**, 43–48.

56 Ehrlich, V.A., Nersesyan, A., Hölzl, C., Ferk, F., Bichler, J., and Knasmüller, S. (2006) Kanzerogene und genotoxische Pflanzeninhaltsstoffe, in *Handbuch der Lebensmitteltoxikologie – Belastungen, Wirkungen, Lebensmittelsicherheit, Hygiene* (eds D. Dunkelberg, T. Gebel and A. Hartwig), Wiley-VCH Verlag GmbH, Weinheim, Germany, pp. 1915–1963.

57 Poschl, G., and Seitz, H.K. (2004) Alcohol and cancer. *Alcohol*, **39**, 155–165.

58 Bagnardi, V., Blangiardo, M., La Vecchia, C., and Corrao, G. (2001) Alcohol consumption and the risk of cancer: a meta-analysis. *Alcohol Res. Health*, **25**, 263–270.

59 Thomas, D.B. (1995) Alcohol as a cause of cancer. *Environ Health Perspect.*, **103** (Suppl. 8), 153–160.

60 Lutz, W.K., and Schlatter, J. (1992) Chemical carcinogens and overnutrition in diet-related cancer. *Carcinogenesis*, **13**, 2211–2216.

61 Renehan, A.G., Tyson, M., Egger, M., Heller, R.F., and Zwahlen, M. (2008) Body-mass index and incidence of cancer: a systematic review and meta-analysis of prospective observational studies. *Lancet*, **371**, 569–578.

62 Lane, G. (2008) Obesity and gynaecological cancer. *Menopause Int.*, **14**, 33–37.

63 Chesson, A., and Collins, A. (1997) Assessment of the role of diet in cancer prevention. *Cancer Lett.*, **114**, 237–245.

64 Kada, T., Inoue, T., Morita, K., and Namiki, M. (1986) Dietary desmutagens. *Prog. Clin. Biol. Res.*, **20**, 6245–6253.

65 Kada, T., Inoue, T., and Namiki, M. (1982) Environmental desmutagens and antimutagens, in *Environmental Mutagenesis and Plant Biology* (ed. E.J. Klekowski), Praeger Scientific, New York, pp. 133–152.

66 Wattenberg, L.W. (1977) Inhibition of carcinogenic effects of polycyclic hydrocarbons by benzyl isothiocyanate and related compounds. *J. Natl. Cancer Inst.*, **58**, 395–398.

67 Wattenberg, L.W. (1983) Inhibition of neoplasia by minor dietary constituents. *Cancer Res.*, **43**, 2448s–2453s.

68 Ribeiro, L.R., and Salvadori, D.M. (2003) Dietary components may prevent mutation-related diseases in humans. *Mutat. Res.*, **544**, 195–201.

69 El-Nezami, H.S., Polychronaki, N.N., Ma, J., Zhu, H.L., Ling, W.H., Salminen, E.K. *et al.* (2006) Probiotic supplementation reduces a biomarker for increased risk of liver cancer in young men from Southern China. *Am. J. Clin. Nutr.*, **83**, 1199–1203.

70 Simonich, M.T., Egner, P.A., Roebuck, B.D., Orner, G.A., Jubert, C., Pereira, C. *et al.* (2007) Natural chlorophyll inhibits aflatoxin B1-induced multi-organ carcinogenesis in the rat. *Carcinogenesis*, **28**, 1294–1302.

71 Cavin, C., Bezencon, C., Guignard, G., and Schilter, B. (2003) Coffee diterpenes prevent benzo[a]pyrene genotoxicity in

72 Wattenberg, L.W. (1990) Inhibition of carcinogenesis by naturally-occurring and synthetic compounds. *Basic Life Sci.*, **52**, 155–166.

73 Wood, A.W., Huang, M.T., Chang, R.L., Newmark, H.L., Lehr, R.E., Yagi, H. et al. (1982) Inhibition of the mutagenicity of bay-region diol epoxides of polycyclic aromatic hydrocarbons by naturally occurring plant phenols: exceptional activity of ellagic acid. *Proc. Natl. Acad. Sci. U. S. A.*, **79**, 5513–5517.

74 Huetz, P., Mavaddat, N., and Mavri, J. (2005) Reaction between ellagic acid and an ultimate carcinogen. *J. Chem. Inf. Model*, **45**, 1564–1570.

75 Fuchs, S., Sontag, G., Stidl, R., Ehrlich, V., Kundi, M., and Knasmuller, S. (2008) Detoxification of patulin and ochratoxin A, two abundant mycotoxins, by lactic acid bacteria. *Food Chem. Toxicol.*, **46**, 1398–1407.

76 Schwab, C.E., Huber, W.W., Parzefall, W., Hietsch, G., Kassie, F., Schulte-Hermann, R. et al. (2000) Search for compounds that inhibit the genotoxic and carcinogenic effects of heterocyclic aromatic amines. *Crit. Rev. Toxicol.*, **30**, 1–69.

77 Kassie, F., Rabot, S., Uhl, M., Huber, W., Qin, H.M., Helma, C. et al. (2002) Chemoprotective effects of garden cress (Lepidium sativum) and its constituents towards 2-amino-3-methyl-imidazo[4,5-f] quinoline (IQ)-induced genotoxic effects and colonic preneoplastic lesions. *Carcinogenesis*, **23**, 1155–1161.

78 Huber, W.W., Rossmanith, W., Grusch, M., Haslinger, E., Prustomersky, S., Peter-Vorosmarty, B. et al. (2008) Effects of coffee and its chemopreventive components kahweol and cafestol on cytochrome P450 and sulfotransferase in rat liver. *Food Chem. Toxicol.*, **46**, 1230–1238.

79 Bartsch, H., Ohshima, H., and Pignatelli, B. (1988) inhibitors of endogenous nitrosation – mechanisms and implications in human cancer prevention. *Mutat. Res.*, **202**, 307–324.

80 Bartsch, H., Nair, J., and Owen, R.W. (1999) Dietary polyunsaturated fatty acids and cancers of the breast and colorectum: emerging evidence for their role as risk modifiers. *Carcinogenesis*, **20**, 2209–2218.

81 Hecht, S.S. (2000) Inhibition of carcinogenesis by isothiocyanates. *Drug Metab. Rev.*, **32**, 395–411.

82 Knasmuller, S., Nersesyan, A., Misik, M., Gerner, C., Mikulits, W., Ehrlich, V. et al. (2008) Use of conventional and -omics based methods for health claims of dietary antioxidants: a critical overview. *Br. J. Nutr.*, **99E** (Suppl. 1), ES3–ES52.

83 Thimmulappa, R.K., Mai, K.H., Srisuma, S., Kensler, T.W., Yamamoto, M., and Biswal, S. (2002) Identification of Nrf2-regulated genes induced by the chemopreventive agent sulforaphane by oligonucleotide microarray. *Cancer Res.*, **62**, 5196–5203.

84 Yu, X., and Kensler, T. (2005) Nrf2 as a target for cancer chemoprevention. *Mutat. Res.*, **591**, 93–102.

85 Ohta, T. (1993) Modification of genotoxicity by naturally occurring flavorings and their derivatives. *Crit. Rev. Toxicol.*, **23**, 127–146.

86 Hoelzl, C., Lorenz, O., Haudek, V., Gundacker, N., Knasmüller, S., and Gerner, C. (2008) Proteome alterations induced in human white blood cells by consumption of Brussels sprouts: results of a pilot intervention study. *Proteomics Clin. Appl.*, **2**, 108–117.

87 Vineis, P., Veglia, F., Benhamou, S., Butkiewicz, D., Cascorbi, I., Clapper, M.L. et al. (2003) CYP1A1 T3801 C polymorphism and lung cancer: a pooled analysis of 2451 cases and 3358 controls. *Int. J. Cancer*, **104**, 650–657.

88 Taioli, E., Gaspari, L., Benhamou, S., Boffetta, P., Brockmoller, J., Butkiewicz, D. et al. (2003) Polymorphisms in CYP1A1, GSTM1, GSTT1 and lung cancer below the age of 45 years. *Int. J. Epidemiol.*, **32**, 60–63.

89 Agundez, J.A. (2004) Cytochrome P450 gene polymorphism and cancer. *Curr. Drug Metab.*, **5**, 211–224.

90 Aklillu, E., Carrillo, J.A., Makonnen, E., Hellman, K., Pitarque, M., Bertilsson, L. et al. (2003) Genetic polymorphism of CYP1A2 in Ethiopians affecting

induction and expression: characterization of novel haplotypes with single-nucleotide polymorphisms in intron 1. *Mol. Pharmacol.*, **64**, 659–669.

91 Sweeney, C., Coles, B.F., Nowell, S., Lang, N.P., and Kadlubar, F.F. (2002) Novel markers of susceptibility to carcinogens in diet: associations with colorectal cancer. *Toxicology*, **181–182**, 83–87.

92 Le Marchand, L., Donlon, T., Seifried, A., and Wilkens, L.R. (2002) Red meat intake, CYP2E1 genetic polymorphisms, and colorectal cancer risk. *Cancer Epidemiol. Biomarkers Prev.*, **11**, 1019–1024.

93 Terry, M.B., Gammon, M.D., Zhang, F.F., Knight, J.A., Wang, Q., Britton, J.A. et al. (2006) ADH3 genotype, alcohol intake and breast cancer risk. *Carcinogenesis*, **27**, 840–847.

94 Yokoyama, A., and Omori, T. (2003) Genetic polymorphisms of alcohol and aldehyde dehydrogenases and risk for esophageal and head and neck cancers. *Jpn. J. Clin. Oncol.*, **33**, 111–121.

95 Gsur, A., Zidek, T., Schnattinger, K., Feik, E., Haidinger, G., Hollaus, P. et al. (2003) Association of microsomal epoxide hydrolase polymorphisms and lung cancer risk. *Br. J. Cancer*, **89**, 702–706.

96 Kiyohara, C., Yoshimasu, K., Takayama, K., and Nakanishi, Y. (2006) EPHX1 polymorphisms and the risk of lung cancer – a HuGE review. *Epidemiology*, **17**, 89–99.

97 Chen, Y., Bai, Y., Yuan, J., Chen, W., Sun, J., Wang, H. et al. (2006) Association of polymorphisms in AhR, CYP1A1, GSTM1, and GSTT1 genes with levels of DNA damage in peripheral blood lymphocytes among coke-oven workers. *Cancer Epidemiol. Biomarkers Prev.*, **15**, 1703–1707.

98 Pavanello, S., Pulliero, A., Siwinska, E., Mielzynska, D., and Clonfero, E. (2005) Reduced nucleotide excision repair and GSTM1-null genotypes influence anti-B[a]PDE-DNA adduct levels in mononuclear white blood cells of highly PAH-exposed coke oven workers. *Carcinogenesis*, **26**, 169–175.

99 Palli, D., Masala, G., Peluso, M., Gaspari, L., Krogh, V., Munnia, A. et al. (2004) The effects of diet on DNA bulky adduct levels are strongly modified by GSTM1 genotype: a study on 634 subjects. *Carcinogenesis*, **25**, 577–584.

100 Zheng, W., Deitz, A.C., Campbell, D.R., Wen, W.Q., Cerhan, J.R., Sellers, T.A. et al. (1999) N-acetyltransferase 1 genetic polymorphism, cigarette smoking, well-done meat intake, and breast cancer risk. *Cancer Epidemiol. Biomarkers Prev.*, **8**, 233–239.

101 Deitz, A.C., Zheng, W., Leff, M.A., Gross, M., Wen, W.Q., Doll, M.A. et al. (2000) N-Acetyltransferase-2 genetic polymorphism, well-done meat intake, and breast cancer risk among postmenopausal women. *Cancer Epidemiol. Biomarkers Prev.*, **9**, 905–910.

102 Hou, S.M., Falt, S., Angelini, S., Yang, K., Nyberg, F., Lambert, B. et al. (2002) The XPD variant alleles are associated with increased aromatic DNA adduct level and lung cancer risk. *Carcinogenesis*, **23**, 599–603.

103 Matullo, G., Peluso, M., Polidoro, S., Guarrera, S., Munnia, A., Krogh, V. et al. (2003) Combination of DNA repair gene single nucleotide polymorphisms and increased levels of DNA adducts in a population-based study. *Cancer Epidemiol. Biomarkers Prev.*, **12**, 674–677.

104 Kinzler, K.W., and Vogelstein, B. (1997) Cancer-susceptibility genes. Gatekeepers and caretakers. *Nature*, **386**, 761–763.

105 Margolin, S., and Lindblom, A. (2006) Familial breast cancer, underlying genes, and clinical implications: a review. *Crit. Rev. Oncog.*, **12**, 75–113.

106 Peltomaki, P. (2001) Deficient DNA mismatch repair: a common etiologic factor for colon cancer. *Hum. Mol. Genet.*, **10**, 735–740.

107 Hung, R.J., Boffetta, P., Brockmoller, J., Butkiewicz, D., Cascorbi, I., Clapper, M.L. et al. (2003) CYP1A1 and GSTM1 genetic polymorphisms and lung cancer risk in Caucasian non-smokers: a pooled analysis. *Carcinogenesis*, **24**, 875–882.

108 Pavanello, S., Simioli, P., Mastrangelo, G., Lupi, S., Gabbani, G., Gregorio, P. et al. (2002) Role of metabolic

polymorphisms NAT2 and CYP1A2 on urinary mutagenicity after a pan-fried hamburger meal. *Food Chem. Toxicol.*, **40**, 1139–1144.

109 Le Marchand, L., Hankin, J.H., Wilkens, L.R., Pierce, L.M., Franke, A., Kolonel, L.N. et al. (2001) Combined effects of well-done red meat, smoking, and rapid N-acetyltransferase 2 and CYP1A2 phenotypes in increasing colorectal cancer risk. *Cancer Epidemiol. Biomarkers Prev.*, **10**, 1259–1266.

110 McGlynn, K.A., Rosvold, E.A., Lustbader, E.D., Hu, Y., Clapper, M.L., Zhou, T.L. et al. (1995) Susceptibility to hepatocellular-carcinoma is associated with genetic-variation in the enzymatic detoxification of aflatoxin B-1. *Proc. Natl. Acad. Sci. U. S. A.*, **92**, 2384–2387.

111 Reszka, E., Wasowicz, W., and Gromadzinska, J. (2006) Genetic polymorphism of xenobiotic metabolising enzymes, diet and cancer susceptibility. *Br. J. Nutr.*, **96**, 609–619.

112 Hoeijmakers, J.H.J. (2001) Genome maintenance mechanisms for preventing cancer. *Nature*, **411**, 366–374.

113 Zienolddiny, S., Campa, D., Lind, H., Ryberg, D., Skaug, V., Stangeland, L. et al. (2006) Polymorphisms of DNA repair genes and risk of non-small cell lung cancer. *Carcinogenesis*, **27**, 560–567.

114 Bjelakovic, G., Nikolova, D., Simonetti, R.G., and Gluud, C. (2004) Antioxidant supplements for prevention of gastrointestinal cancers: a systematic review and meta-analysis. *Lancet*, **364**, 1219–1228.

115 Kassie, F., and Knasmüller, S. (2004) Glucosinolates and the prevention of cancer, in *Functional Foods, Ageing and Degenerative Disease* (eds C. Remacle and B. Reusens), CRC Press, Woodhead Publishing Limited, Cambridge England, Boca Raton, Boston, New York, Washington, DC, pp. 615–623.

116 Tempfer, C.B., Hefler, L.A., Schneeberger, C., and Huber, J.C. (2006) How valid is single nucleotide polymorphism (SNP) diagnosis for the individual risk assessment of breast cancer? *Gynecol. Endocrinol.*, **22**, 155–159.

117 AICR (2007) *Food, Nutrition, Physical Activity and the Prevention of Cancer: a Global Perspectiv*, World Cancer Research Fund/American Institute for Cancer Research, Washington, DC, p. 517.

118 Belpomme, D., Irigaray, P., Sasco, A.J., Newby, J.A., Howard, V., Clapp, R. et al. (2007) The growing incidence of cancer: role of lifestyle and screening detection (Review). *Int. J. Oncol.*, **30**, 1037–1049.

119 Pembrey, M.E., Bygren, L.O., Kaati, G., Edvinsson, S., Northstone, K., Sjöström, M., Golding, J., and The ALSPAC Study Team (2006) Sex-specific, male-line transgenerational responses in humans. *Eur. J. Hum. Genet.*, **14**, 159–166.

120 Luebeck, E.G., Moolgavkar, S.H., Liu, A.Y., Boynton, A., and Ulrich, C.M. (2008) Does folic acid supplementation prevent or promote colorectal cancer? Results from model-based predictions. *Cancer Epidemiol. Biomarkers Prev.*, **17**, 1360–1367.

121 Fenech, M. (2005) The Genome Health Clinic and Genome Health Nutrigenomics concepts: diagnosis and nutritional treatment of genome and epigenome damage on an individual basis. *Mutagenesis*, **20**, 255–269.

Part IV
Nutritional Aspects

10
From Molecular Nutrition to Nutritional Systems Biology

Guy Vergères

Abstract

The human genome sequencing project headed by the Human Genome Organization (HUGO) which started in the early 1990s, has triggered dramatic technological and conceptual developments in the life sciences. In particular, HUGO led to the rise of the "-omic" technologies that, together with bioinformatics, now allow biological systems to be investigated holistically. Medical and pharmaceutical sciences have long realized that, in most cases, simple molecular mechanisms cannot account for deregulation in human metabolism. Consequently, these sciences now clearly make use of the tools derived from HUGO to take a systemic (global) approach to multifactorial diseases. Nutrition research is now learning from medical and pharmaceutical research by replacing the pharmacogenomics concept "patient/drug/treatment/omics" by the nutrigenomic concept "healthy consumer/food/prevention/omics". The technologies used in nutrigenomics, along the chain of molecular information leading from DNA to metabolites in the cell (genomics, transcriptomics, proteomics, metabolomics), as well as in nutrigenetics and nutri-epigenetics are described. Examples of these research strategies are also presented.

10.1
Impact of Life Sciences on Molecular Nutrition Research

The University of Harvard has created an artistically remarkable video entitled "The Inner Life of The Cell", which spectacularly illustrates the advances in life sciences over the past two decades [1]. This progress has been mainly driven by the development of new technologies and their concurrent application to the fields of medicine and pharmacology. With the support of these technologies molecular nutrition can now move closer to physiology to transform nutrition research into a translational science. The aim of this article is to present elements of the mutation that nutrition research is currently experiencing.

Epigenetics and Human Health
Edited by Alexander G. Haslberger, Co-edited by Sabine Gressler
Copyright © 2010 WILEY-VCH Verlag GmbH & Co. KGaA, Weinheim
ISBN: 978-3-527-32427-9

The chain of molecular information in a cell, which leads from DNA to metabolites via RNA and proteins, allows us to realize how life sciences have evolved. In the early 1980s a PhD student used to spend his entire thesis on the study of a single gene or protein. Advances in miniaturization and automation have profoundly changed molecular biology by opening the door to the -omic sciences, which now enable the comprehensive analysis of entire sets of molecules of the same kind along the chain of molecular information (genomics for DNA, transcriptomics for RNA, proteomics for proteins, metabolomics for metabolites). These technological improvements have dramatically increased the volume of data and have been, accordingly, accompanied by appropriate developments in the fields of statistics and bioinformatics, which now assist the "life scientist" in his interpretation of the data resulting from his experiments. Along that path of changes, the 21st century now witnesses the rise of systems biology, a holistic science, which integrates all -omic levels to extract quantitative information from laboratory and *in silico* experiments, allowing a phenotypic understanding of biological systems. Systems biology liberates the scientist from molecular details and allows him to address the function of cells, organs, and organisms, leading to the revival of physiology and, ultimately, to the realization of the true potential of this science [2].

The exhibition "Cradle to Grave by Pharmacopoeia" at the British Museum reveals that, on average, each British subject is prescribed 14 000 pills in his lifetime [3]. This impressive number explains why the molecular effects of drugs on organisms have been so intensively studied. In particular this has led to the emergence of the science of pharmacogenomics, which applies -omic technologies to investigate these interactions. Another way to look at drugs is by calculating our lifetime consumption, which can be estimated to ~3 kg, assuming an average weight of 0.2 g per pill. This quantity is evidently orders of magnitude below that of our lifetime consumption of food, which can be estimated at 60 000 kg. This simple comparison needs to be retained in relation to the potent pharmacological activity of drugs but, nonetheless, suggests that the molecular analytical strategies used in pharmacology could potentially be applicable to comprehensively study the molecular and physiological impact of nutrients on organisms. Intuitively, the expected differences between nutrition and pharmacology are twofold: firstly, compared to drugs, food has a complex chemical composition; secondly, food is not primarily expected to exert pharmacological activity. Thus, when compared to the action of a drug, the molecular response of an organism to a diet is expected to be broader in its molecular complexity and lower in its intensity.

That food could be pharmacologically active was nonetheless demonstrated, for example, in a nutritional clinical study aimed at improving the lipoprotein profile of hyperlipidemic patients [4]. The biochemistry of cholesterol and lipoproteins is well described and consists of several steps, among which are cholesterol biosynthesis and absorption, bile acid sequestration, fatty acid synthesis, lipoprotein lipase activation, inflammation, and lipid oxidation. The pharmaceutical industry has consequently developed drugs, such as statins, that interact with these biochemical pathways to improve the cholesterol-lipoprotein profiles of patients at risk of developing cardiovascular diseases. Parallel to these developments, molecular nutrition has identified components in food, such as red yeast rice, soy proteins, plant sterols, omega-3 fatty acids, vitamin C and vitamin E, which also interact with these pathways. Based on this knowledge, a dietary portfolio of cholesterol-lowering food was prepared, which was able to improve the lipoprotein profile of these patients to an extent similar to that achieved by statins.

This example is an excellent illustration of the pharmacological potential of a targeted nutrition that is based on existing molecular and physiological knowledge. It suggests that nutrition research can learn from pharmacological research by applying scientific and analytical strategies similar to pharmacogenomics to holistically investigate the molecular and physiological effects of nutrients on organisms. The challenge of nutritional systems biology for the 21st century is to achieve this goal [5].

Modern molecular nutrition research, which is evolving toward nutritional systems biology, is covered by the three major research fields of nutrigenomics, nutrigenetics and nutri-epigenetics. In the scientific literature, the term "nutrigenomics" often includes nutrigenetics and nutri-epigenetics. For clarity, the author considers these three research areas separately and uses the term "nutritional systems biology" to cover all of the three.

10.2
Nutrigenomics

The -omic technologies are particularly appropriate to investigate complex interactions between food, which has a broad chemical composition, and living organisms [6]. Nutrigenomics can thus be defined as the application of modern life sciences, in particular the -omic technologies, to the study of the interactions between food and living organisms, especially humans. The following sections will give a short technical introduction to the various -omic methods and present selected examples that illustrate their potential application to nutrition research and food science.

10.2.1
Genomics and Nutrition Research

Fred Sanger delivered pioneer work in the development of DNA sequencing, an achievement that was acknowledged with a Nobel Prize in 1980. A breakthrough in genomics was subsequently achieved with the human sequencing project, headed by the Human Genome Organization (HUGO), which started in 1990 and culminated in 2001 with the publication of the DNA sequence of the human genome, which is composed of 3.2 billion DNA bases [7]. Importantly, this work was accompanied by a dramatic increase in efficiency, and, ultimately, a decrease in the cost of DNA sequencing over the years. As a consequence, sequencing the entire genome of a single individual is no longer prohibitive from an economical point of view, as highlighted by the recent publication of the sequence of the genome of James D. Watson [8] and by plans of private companies to soon provide services for sequencing the genome of individuals at less than 1000 dollars [9].

A promising example of the application of genomics to nutrition research is the characterization of human microflora. The affordability of DNA sequencing now allows high throughput genomic characterization of the microbial population in the human gastrointestinal tract, the so-called human microbiome [10]. This work reveals that our intestinal microflora is not only complex in its composition but also dynamic in its interactions within the members of the microbial community, with the ingested food, as well as with cells of the intestinal epithelium and the immune system. The number of intestinal microbes also exceeds the human cells by one order of magnitude. Human beings can thus be considered as superorganisms composed of populations of dynamically interacting human and microbial cells. Furthermore, the genomic and metabolic characterization of this ecological system in murine models reveals that changes in the composition of the microflora lead to differentiated metabolism of the ingested nutrients and, consequently, to changes in the quality and the quantity of nutritive energy that can be extracted from food. That this conclusion may be relevant to humans is strengthened by a correlation between weight loss in obese persons undergoing calorie-restricted diets and the composition of their gut microflora [11]. Taken together, these studies suggest that the specific nature of the microflora differentially influences the metabolism of the food we ingest. It should, consequently, be possible to regulate metabolic disorders by modifying the composition of this microflora.

10.2.2
Transcriptomics and Nutrition Research

The human genome contains 20 000–25 000 genes that, depending on the nature of the environmental biological stimuli, are differentially expressed in the form of RNA. Microarray technologies now routinely allow the parallel measurement of the transcriptome, that is, the entire set of RNA molecules expressed in a given tissue, organ, or organism at a specific time and under specific conditions. Transcriptome analysis is not restricted to human material, since progress in sequencing technologies has led to the commercial availability of a large spectrum of microarrays that covers the animal, vegetal and microbial kingdoms. Transcriptomic strategies typically take a differential approach in that the relative expression of genes is compared either in different biological tissues (cell, organ, organism ...) exposed to the same conditions (biochemical, physiological, pathological ...) or in the same biological tissue exposed to different conditions. The genes that are differentially expressed are identified and the biological significance of these findings is discussed [12].

The ingestion of macronutrients and micronutrients modulates cellular gene expression [13]. A transcriptomic strategy can thus be taken to investigate the nutritional and physiological properties of specific foods in humans [14]. In particular, the blood cell transcriptome [15] is potentially a source of nutritional biomarkers for several reasons: (i) Human blood cells are easily accessible, compared to other tissues where biopsies are necessary. (ii) Blood cells are sentinels of the body that reflect the physiological status of the human organism. This property has long been recognized in medicine, as seen by the impact of clinical chemistry in medical diagnostics. (iii) Blood is an important component of the immune system that is particularly suited to studying the immune-modulating properties of bioactive components in food.

In line with the potential of blood analytics in nutrition research, we have launched a project aimed at characterizing the blood cell transcriptome of humans following the ingestion of dairy products. The analysis of the expressed transcriptome has revealed changes in the expression of genes coding for ribosomal proteins which are involved in metabolic processes, an observation that reflects the macronutrient properties of the ingested dairy products. In addition, changes in the expression of immunomodulatory genes suggest more specific properties for dairy products. For example, a decrease in the expression of the gene coding for toll-like receptor 2 (*TLR2*) is measured six hours post-ingestion (Sagaya et al., manuscript in preparation). This observation provides support at the molecular level for the already reported anti-inflammatory properties of dairy products in humans [16], a functionality that could be interesting in the context of the management of chronic inflammatory diseases. In this context, the blood cell transcriptome could be used to identify nutritional biomarkers in humans, which would be used to select lactic acid bacteria that ferment milk into dairy products with specific and/or enhanced nutritional and physiological properties. Keeping in mind the importance of the intestinal microflora in human physiology (see Section 10.2.1), as well as the evolutionary importance of dairy products in mammals, milk appears to be a strategic vector to deliver health-promoting bacteria to humans.

10.2.3
Proteomics and Nutrition Research

Moving along the chain of molecular information in the cell, the science of proteomics [17] emerged later than transcriptomics. There are biological and technological reasons for this delayed development. First, due to complex regulatory phenomena (alternative splicing, post-translational modifications ...), the total number of protein entities significantly exceeds the number of genes in a human organism and numbers above 100 000 have been advanced to characterize the size of the protein population. Secondly, the biochemistry of proteins is complex (a structure with 20 amino acids for proteins compared with 4 bases for DNA and RNA; numerous post-translational modifications; different physico-chemical properties separating water-soluble proteins from membrane-bound proteins; large dynamic range with, for example, the concentration of human serum proteins spanning over ten orders of magnitude). Finally, whereas genes can be easily and quantitatively copied and amplified using PCR technologies, laboratory technologies for copying proteins, such as *in vitro* translation systems, are significantly less robust. These hurdles explain why protein analytics has not yet reached the efficiency of other -omic technologies such as genomics and transcriptomics. Consequently, the simultaneous analysis of a thousand different proteins in a single experiment is already considered as an achievement by protein chemists. The technological developments in the field of proteomics are nonetheless impressive. The current strategies that are routinely used combine two-dimensional polyacrylamide gel electrophoresis (2D-PAGE), for separating the proteins in tissue extracts, with mass spectrometry (MS), for the identification of the differentially expressed proteins. Alternatively, proteins are separated by high pressure liquid chromatography (HPLC), or similar types of column-based chromatographic procedures, using columns that are directly coupled to the mass spectrometer. In any case, and for all -omic analytical strategies, the identification of differentially expressed proteins must be followed by a critical discussion of the physiological and/or clinical interpretation of the data.

In nutrition research, the application of -omic technologies is not restricted to the analysis of human tissues but can also be applied to the analysis of food. For example, a combination of proteomics and clinical immunology has been used to identify a new allergen in shrimp. In this work, proteins in shrimp extracts were first separated by 2D-PAGE and subsequently immuno-blotted for the presence of allergens using sera from allergic patients as a source of IgE antibodies. Identification by MS of the positive spots on the 2D-gels led to the identification of arginine kinase, a new allergen in shrimp [18]. This work could potentially open the door to food technology processes aimed at the removal of the identified allergens [19].

10.2.4
Metabolomics and Nutrition Research

Metabolomics is considered by many biologists as a highly relevant -omic science because metabolites, together with structural proteins, are at the end of the flow of molecular information in the cell. Metabolites thus ultimately translate the genetic information in our cells into biochemical work that supports the functioning of our organism. The number of metabolites identified in humans grows regularly: 6800 molecules were fully annotated in the Human Metabolome Database as of October 25th 2008 [20]. However, the size of the human metabolome is comparatively smaller than the size of the proteome which makes it more amenable to characterization. The advances in analytical methods such as nuclear magnetic resonance (NMR) and mass spectrometry coupled to gas chromatography (GC-MS) or to liquid chromatography (LC-MS), as well as the development of efficient bioinformatic tools, have also contributed significantly to the revival of metabolite analytics. Metabolomics is, consequently, experiencing a burst of interest in life sciences and the metabolome of body fluids such as serum and urine is now the subject of intensive investigation in medicine and pharmaceutical research. Human metabolism is in essence a physiological process that is, intimately coupled to nutrition. Metabolomics will, therefore, evolve into a strategic analytical tool for modern nutrition research [21].

The potential of metabolomics in nutrition research is illustrated by the analysis of the urine metabolite profile of human volunteers of various ethnic origins. Urinary metabolite excretion patterns based on NMR of East Asian and western population samples were significantly differentiated [22]. The authors of this study conclude that it is not genetic factors but environmental factors, in particular culturally-driven diets, that play a predominant role in differentiating the metabolic phenotypes of these populations. That the diet is a major discriminant of these urinary profiles is furthermore highlighted by the observation that subgroups of the same ethnic origin, but with different diets, such as vegetarians and meat consumers, could also be differentiated based on their urine NMR metabolite profiles. The molecules that participate most in this differentiation were identified and could potentially be used as nutritional markers.

10.3
Nutrigenetics

Single nucleotide polymorphisms (SNPs) are key modifications in the genome that can lead to functional changes in living organisms by altering the level of expression of genes coding for functional proteins and/or by altering the biochemical properties of the proteins these genes code for [23]. Although a total of 3 to 10 million SNPs have been estimated, which corresponds to circa 0.1% of the size of the human genome, the development of DNA sequencing technologies has rendered the comprehensive identification of these genetic variations economically feasible, even more since SNPs tend to be inherited in blocks, the so-called haplotype blocks, whose number has been estimated to be between 300 000 and 600 000. The existence of haplotypes thus enables geneticists to focus their resources on the identification of a manageable number of biomarkers, the tag SNPs, which unequivocally identify the corresponding haplotypes. One can foresee the commercialization of DNA chips, based on tag SNPs, that will soon allow the determination, in a single experiment, of the complete genotype of a single individual. Alternately, sequencing of the complete genome of an individual may become so economically competitive (see Section 10.2.1) that a genome-wide sequencing strategy may even be applied to fully characterize human genomes on an individual basis.

Nutrigenetics moves one step further than nutrigenomics in that it considers the impact of genetic diversity on the interactions of humans with nutrients [24]. Nutrigenetics may thus be defined as the application of modern life sciences to the study of the interaction between food and populations and/or individuals with different genetic backgrounds.

The association between coffee consumption, polymorphism of the *CYP1A2* gene, and the risk of myocardial infarction, is an excellent illustration of the potential and issues associated with nutrigenetics. We all know that exaggerated consumption of coffee may not be beneficial to our cardiovascular system. On the other hand, a meta-analysis has recently concluded that the association between coffee drinking and the risk of coronary heart disease remains controversial. A nutrigenetic study published in 2006 [25] has contributed to the debate by adding a genetic component to this topic. This study observed, in a case-control study, a statistically significant association between non-lethal myocardial infarction, elevated coffee consumption, and a *CYP1A2* genotype that codes for a cytochrome P450 enzyme with an impaired ability to metabolize, and thus detoxify, caffeine. In particular, the subjects under the age of 50, who carried at least one copy of the impaired genetic variant of *CYP1A2*, and who drank at least four cups of coffee daily had fourfold higher rates of non-lethal myocardial infarcts, than subjects with the same polymorphism but drinking no more than one cup of coffee daily. This effect was not observed in the population carrying both copies of the gene coding for the rapidly metabolizing cytochrome P450 enzyme. This population was apparently even protected from developing non-lethal myocardial infarcts when drinking moderate doses of coffee (1–3 cups of coffee).

The most obvious benefit of nutrigenetics is to unmask physiological effects of the diet that would otherwise not be observable if the subjects of nutritional trials were not analyzed for their genotypes. At the same time, each additional genotype investigated decreases the number of the subjects belonging to each of the genetic subgroups studied, which further accentuates a major problem in human trials, namely the availability of resources needed to carry studies with a statistically significant number of subjects. Before offering nutritional advice and services to the public, original findings, such as the data reported on coffee consumption in the previous paragraph, must be consolidated by other research groups in additional populations and using large numbers of subjects. The key question in nutrigenetics is the quantification of the impact of the identified genotype(s) on the association between food and physiological endpoints and, consequently, on our well-being and health. For monogenetic diseases or pathologies, such as lactose intolerance or phenylketonuria, the impact of genetic variations on health is relatively easy to estimate. Consequently, nutritional strategies aimed at mitigating the negative consequences of these interactions, such as the removal of lactose or phenylalanine from the diet, are evident and can be effectively implemented. In that regard, caffeine metabolism can be considered, from a biochemical point of view, as a relatively simple case since one single enzyme, cytochrome P450, appears to play a major role in the metabolism of this compound. Biology, however, is rarely simple, and additional factors, such as polymorphism of *N*-acetyltransferase, an enzyme also involved in the detoxication of caffeine, may also contribute to the individual susceptibility of humans to the side effects of coffee. In fact, most diseases, in particular the majority of chronic diseases, are polygenic by nature, which significantly complicates a quantitative evaluation of the impact of the relevant polymorphisms. Finally, in addition to SNPs, which have so far focused most of the resources in genetic research, tandem repeat polymorphisms also play a significant role in modulating the interaction of genes with our environment.

In conclusion, the relatively large list of nutrigenetic studies published to date cannot be evaluated with definitive statements. Thus, for the time being, nutrigenetics should generally be viewed with the label "ongoing research".

10.4
Nutri-Epigenetics

Human genetic information is transferred during the life cycle of individuals and generations via somatic cells and germ cells, respectively. Despite the fact that each of these cells basically possesses the same genetic information, the expression program of their genome is differentially regulated and involves changes in the chromosomal structure other than differences in the DNA sequence, which are specific to each tissue or organ. Epigenetic phenomena can intuitively be realized as each of us considers the development of his/her own organism over the years, from embryogenesis to older age. Epigenetics is the science that investigates these dynamic mechanisms. The molecular base beyond epigenetics lies at the level of the chromosomal structure and its regulation by chemical modifications of DNA and histones. The accessibility of transcription factors to gene promoters on the chromosomes is epigenetically regulated, on the one hand by chemical modification of cytosines in DNA, and on the other hand by post-translational modifications of the histones that regulate chromosomal structure (acetylation, methylation, phosphorylation …). Although the consequences of these modifications are complex, acetylation of histones is generally associated with an accessible nucleosomal structure that allows activation of gene expression by transcription factors, whereas methylation of cytosine residues on DNA, particularly at the so-called CpG islands in promoter regions of genes, is associated with gene expression silencing. An estimated 4% of the cytosines in the human genome, that is, ~30 million bases, may be subject to methylation! High throughput technologies are, however, available to detect DNA methylation. These technologies are directly derived from the classical molecular biological methods developed to sequence the human genome. In that respect, an international research consortium is currently active in mapping the human epigenome. This task is, however, daunting, as each human organism is composed of a multitude of cellular epigenomes, each of which is unique to a particular physiological condition in which the type of tissue or organ as well as their developmental stage, are key variables [26].

Nutri-epigenetics can be defined as the application of modern life sciences to the study of the modifications, other than changes in DNA sequence, of our chromosomes by nutrients [27]. The molecular link between epigenetics and nutrition is the so-called one-carbon metabolic pathway, also called the folate-methionine cycle. This pathway extracts a carbon atom from the diet and transfers it, in the form of a methyl group, to DNA, RNA, proteins or lipids. Folic acids, methionine, choline, and betaine are the major sources of this carbon atom in the diet whereas s-adenosylmethionine (SAM) is the ultimate methyl donor, that is, used by the DNA methyltransferases to methylate their target molecules in the cell, in particular DNA. Other nutrients such as vitamin B2, vitamin B6, vitamin B12, and zinc, act as co-factors for the enzymes involved in this process. The one-carbon pathway is not only central to nutrition from a scientific point of view, as it is involved in the methylation of DNA by the diet, but is also interesting from a philosophical point of view as it basically provides a molecular mechanism by which the voluntary act of selective eating, that is, nutrition, potentially frees us from genetic determinism by acting on the structure and activity of our genome, for better or for worse!

Epigenetic phenomena are evident when one looks at the evolution of monozygotic twins. Even if the genomes of twin pairs are essentially identical, phenotypic differences between twins already start at embryogenesis and become more obvious with aging. Such differentiation can only be explained by epigenetic mechanisms that take place in response to either a differential exposure to the environment or to stochastic changes at the cellular level. -Omic technologies allow characterization of monozygotic twins at the molecular level. A recent study observed that the number of differentially methylated CpG islands is 2.5-fold higher in 50 year old twins than in 3 year old twins [28]. That this molecular divergence has functional consequences is supported at the gene expression level, since a transcriptomic analysis of the expressed genome revealed that the number of differentially expressed genes in the older twins (~5000 genes) is increased fivefold, compared to the younger twins. A comprehensive analysis of the physiological consequences of these molecular changes will give clues to the mechanisms of aging and development of chronic diseases, in particular cancer.

The *agouti viable yellow* (A^{vy}) murine model is the most notable *in vivo* example of the epigenetic impact of diet on development [29]. The A^{vy} mutation is caused by retrotransposition of an intracisternal A particle (IAP) upstream of *agouti*, which normally regulates the production of a yellow pigment in the hair follicles. A cryptic promoter in the IAP drives ectopic *agouti* expression in murine tissues. These mice harbor a yellowish fur that is used as a laboratory marker for the expression of the A^{vy} allele. More importantly, these animals have a reduced life span and develop chronic diseases such as cancer, obesity, and diabetes. The promoter of the A^{vy} gene is under the control of methylation. As a consequence of methylation-induced downregulation of the ectopic expression of the A^{vy} gene, pregnant heterozygotic A^{vy}/a mice given a diet rich in methyl donors, particularly folate and choline, give birth to pups with a healthy phenotype.

Despite erasing the epigenetic information carried by the sperm and egg genomes during pre-implantation development and embryonic development, epigenetic marks can partially be transmitted meiotically over generations. Whether diet-induced hypermethylation at A^{vy} may be inherited transgenerationally was investigated in *agouti* mice, but the issue remains controversial. Diet-induced transgenerational heritage of epigenetic markers is also debated in humans. In particular, there seem to be transgenerational responses in gene expression to variable food availability, in particular during pregnancy and during the slow prepubertal growth period (SGP). For example, the nutrition of the grandmother during pregnancy influences the grandchildren's birth weight. Also, subjects, whose paternal grandparents had experienced at least 1 year of good availability of food during their SGP, revealed more premature deaths among them compared with those subjects where paternal grandparents had not had any such good year during their SGP [30]. These studies suggest that transgenerational responses exist that capture nutritionally related information from the previous generations and, through the female- or male-lineage, affect health-outcome in subsequent generations.

10.5
Nutritional Systems Biology

One of the most recent developments in life sciences is undoubtedly the rise in systems biology. Systems biology integrates the various components of living organisms, for example, genes, proteins, metabolites, and phenotypes in order to obtain a holistic and quantitative understanding of how biological systems work and respond to environmental stimuli. From a technical point of view, systems biology uses mathematical modeling and biological information to integrate all -omic information levels in cells, organs, or organisms, to comprehensively characterize the physiology of living organisms. As nutrients act at the molecular level on all planes of the cellular information flow, for example, at the level of gene expression, regulation of protein function, and metabolite production, systems biology should prove in the future to be particularly suited to promoting key developments in nutrition research [31].

So far, a true systems biology approach has not yet been taken in the field of nutrition research. First steps in that direction have, however, been made. For example, the anti-carcinogenic properties of flavones have been studied in a comprehensive manner, using a combination of transcriptomics, proteomics, metabolite analytics and *in vitro* cell biology [32]. This analytical work led to the conclusion that specific inhibition of the metabolic turnover of transformed cells is a key mechanism of the anticarcinogenic properties of flavones. From a larger perspective, the research strategy developed by the authors of this study is inspiring in that it paves the road for future nutrition research, not only to investigate the impact of diverse diets and nutrients *in vitro* and in animal models but also in humans. These efforts will lead to a comprehensive and quantitative understanding of the impact of the diet on the human organism that will ultimately be translated into efficient nutrition politics and policies, including dietary recommendations and food development.

10.6
Ethics and Socio-Economics of Modern Nutrition Research

In order for nutrition research to be efficiently translated into nutrition politics and policies, issues related to the ethical and socio-economical impact of these developments must be addressed [33]. The impact of nutrigenomics and nutri-epigenetics on the development of functional foods that will be beneficial to large fractions of populations, eventually during specific stages of the life cycle, can be foreseen. On the other hand, the benefits of nutrigenetics should help individuals or groups of individuals, who will receive specific dietary advice based on their specific genetic profiles. Thus, the issues raised by nutrigenomics and nutri-epigenetics very much concern the food industry (e.g., marketing of functional food), whereas nutrigenetics raises medical issues, already known from geneticists, that involve individual, familial, and professional spheres. A debate on these issues is needed which includes all stakeholders, including the research community, the food industry, regulatory and health authorities, consumers, and patients.

To end this article on a sketchy note that should speak equally to both the scientist and the layman, the accompanying scene that takes place at the restaurant "*My Food*" intuitively reveals the major issues raised by modern molecular nutrition research. Will this science be able to bring the awaited benefits? In other words, is the situation presented in this sketch science or fiction? In fact, one simply has to remove the laboratory from the restaurant "*My Food*" and, literally, to place it close by, for example, on the other side of the street, to quit fiction and to realize that nutrigenomics, nutrigenetics, and nutri-epigenetics are already part of our reality. This statement can, for example, be illustrated by the fact that a genetic test aimed at assessing our ability to metabolize caffeine (see Section 10.3) is already sold by a private company. Thus, the most relevant questions that must be addressed relate to the extent to which each of these sciences will concretely impact our health and well being as well as the time frame during which these progresses will be realized.

References

1. Harvard University (2007) Inner Life of the Cell, http://multimedia.mcb.harvard.edu/media.html (accessed August 14, 2009).
2. Strange, K. (2005) The end of "naive reductionism": rise of systems biology or renaissance of physiology? *Am. J. Physiol. Cell. Physiol.*, **288**, C968–C974.
3. The British Museum (2003) Cradle to Grave by Pharmacopoeia, http://www.britishmuseum.org/explore/highlights/highlight_objects/aoa/c/cradle_to_grave.aspx (accessed August 14, 2009).
4. Jenkins, D.J., Kendall, C.W., Faulkner, D.A., Nguyen, T., Kemp, T., Marchie, A., Wong, J.M. et al. (2006) Assessment of the longer-term effects of a dietary portfolio of cholesterol-lowering foods in hypercholesterolemia. *Am. J. Clin. Nutr.*, **83**, 582–591.
5. van Ommen, B., Cavallieri, D., Roche, H.M., Klein, U.I., and Daniel, H. (2008) The challenges for molecular nutrition research 4: the "nutritional systems biology level". *Genes. Nutr.*, **3**, 107–113.
6. Ferguson, L.R. (2006) Nutrigenomics: integrating genomic approaches into nutrition research. *Mol. Diagn. Ther.*, **10**, 101–108.
7. Lander, E.S., Linton, L.M., Birren, B., Nusbaum, C., Zody, M.C., Baldwin, J., Devon, K. et al. (2001) Initial sequencing and analysis of the human genome. *Nature*, **409**, 860–921.
8. Wheeler, D.A., Srinivasan, M., Egholm, M., Shen, Y., Chen, L., McGuire, A., He, W. et al. (2008) The complete genome of an individual by massively parallel DNA sequencing. *Nature*, **452**, 872–876.
9. Wolinsky, H. (2007) The thousand-dollar genome. Genetic brinkmanship or personalized medicine? *EMBO Rep.*, **8**, 900–903.
10. Turnbaugh, P.J., Ley, R.E., Hamady, M., Fraser-Liggett, C.M., Knight, R., and Gordon, J.I. (2007) The human microbiome project. *Nature*, **449**, 804–810.
11. Ley, R.E., Turnbaugh, P.J., Klein, S., and Gordon, J.I. (2006) Microbial ecology: human gut microbes associated with obesity. *Nature*, **444**, 1022–1023.
12. Gomase, V.S., and Tagore, S. (2008) Transcriptomics. *Curr. Drug. Metab.*, **9**, 245–249.
13. Desvergne, B., Michalik, L., and Wahli, W. (2006) Transcriptional regulation of metabolism. *Physiol. Rev.*, **86**, 465–514.
14. Elliott, R.M. (2008) Transcriptomics and micronutrient research. *Br. J. Nutr.*, **99**, S59–S65.
15. van Erk, M.J., Blom, W.A., van Ommen, B., and Hendriks, H.F. (2006) High-protein and high-carbohydrate breakfasts differentially change the transcriptome of human blood cells. *Am. J. Clin. Nutr.*, **84**, 1233–1241.
16. Buescher, E.S. (2001) Anti-inflammatory characteristics of human milk: how, where, why. *Avd. Exp. Med. Biol.*, **501**, 207–222.
17. Thongboonkerd, V. (2007) Proteomics. *Forum. Nutr.*, **60**, 80–90.
18. Yu, C.C., Lin, Y.F., Chiang, B.L., and Chow, L.P. (2003) Proteomics and immunological analysis of a novel shrimp allergen, Pen m 2. *J. Immunol.*, **170**, 445–453.
19. Arai, S. (2000) Functional food science in Japan: state of the art. *Biofactors.*, **12**, 13–16.
20. Wishart, D.S., Knox, C., Guo, A.C., Eisner, R., Young, N., Gautam, B., Hau, D.D. et al. (2009) HMDB: a knowledgebase for the human metabolome. *Nucleic. Acids. Res.*, **37**, D603–D610.
21. Brennan, L. (2008) Metabolomic applications in nutritional research. *Proc. Nutr. Soc.*, **67**, 404–408.
22. Holmes, E., Loo, R.L., Stamler, J., Bictash, M., Chan, Q., Ebbels, T., De Iorio, M. et al. (2008) Human metabolic phenotype diversity and its association with diet and blood pressure. *Nature.*, **453**, 396–400.
23. Gray, I.C., Campbell, D.A., and Spurr, N.K. (2000) Single nucleotide polymorphisms as tools in human genetics. *Hum. Mol. Genet.*, **9**, 2403–2408.
24. El-Sohemy, A. (2007) Nutrigenetics. *Forum. Nutr.*, **60**, 25–30.

25 Cornelis, M.C., El-Sohemy, A., Kabagambe, E.K., and Campos, H. (2006) Coffee, CYP1A2 genotype, and risk of myocardial infarction. *JAMA*, **295**, 1135–1141.

26 Callinan, P.A., and Feiberg, A.P. (2006) The emerging science of epigenomics. *Hum. Mol. Genet.*, **15**, R95–R101.

27 Gallou-Kabani, C., Vigé, A., Gross, M.S., and Junien, C. (2007) Nutri-epigenomics: lifelong remodelling of our epigenomes by nutritional and metabolic factors and beyond. *Clin. Chem. Lab. Med.*, **45**, 321–327.

28 Fraga, M.F., Ballestar, E., Paz, M.F., Ropero, S., Setien, F., Ballestar, M.L., Heine-Suñer, D. et al. (2005) Epigenetic differences arise during the lifetime of monozygotic twins. *Proc. Natl. Acad. Sci. U. S. A.*, **102**, 10604–10609.

29 Waterland, R.A., Travisano, M., and Tahiliani, K.G. (2007) Diet-induced hypermethylation at agouti viable yellow is not inherited transgenerationally through the female. *FASEB J.*, **21**, 3380–3385.

30 Kaati, G., Bygren, L.O., Pembrey, M., and Sjöström, M. (2007) Transgenerational response to nutrition, early life circumstances and longevity. *Eur. J. Hum. Genet.*, **15**, 784–790.

31 Van Ommen, B., Cavallieri, D., Roche, H.M., Klein, U.I., and Daniel, H. (2008) The challenges for molecular nutrition research: the "nutritional systems biology level". *Genes. Nutr.*, **3**, 107–113.

32 Herzog, A., Kindermann, B., Döring, F., Daniel, H., and Wenzel, U. (2004) Pleiotropic molecular effects of the pro-apoptotic dietary constituent flavone in human colon cancer cells identified by protein and mRNA expression profiling. *Proteomics*, **4**, 2455–2456.

33 Bergmann, M.M., Görman, U., and Mathers, J.C. (2008) Bioethical considerations for human nutrigenomics. *Annu. Rev. Nutr.*, **28**, 447–467.

11
Effects of Dietary Natural Compounds on DNA Methylation Related to Cancer Chemoprevention and Anticancer Epigenetic Therapy

Barbara Maria Stefanska and Krystyna Fabianowska-Majewska

Abstract

Aberrations in function and expression of a wide variety of genes are the hallmarks of cancer cells. A growing body of evidence shows that these aberrations are affected by genetics and epigenetics which cooperate at all stages of cancer development. Epigenetic alterations (especially DNA hypermethylation) resulting in the silencing of key tumor suppressor genes have attracted significant attention. In contrast to genetic changes, epigenetic modifications are reversible and responsive to developmental, physiological, environmental, dietary, and pathological signals. A number of studies provide evidence that some natural bioactive compounds found in food and herbs can influence gene expression via modulation of the DNA methylation process. It has been demonstrated that some polyphenols (i.e., catechins, quercetin, myricetin, genistein, resveratrol) and vitamins (i.e., retinoic acid, vitamin D_3) exert a profound inhibitory effect on DNA methyltransferase enzyme activity and/or *DNA methyltransferase* gene expression and contribute to reactivation of methylation-silenced tumor suppressor genes in cancer cells leading to blocking of cancer development. This raises the possibility that the natural compounds can be a new approach to anticancer therapy. Moreover, their presence in widely available food products can bring about the high efficacy of the compounds in cancer chemoprevention. This chapter gives a short overview of the mechanisms of DNA methylation inhibition by the selected bioactive food components and their potential use in cancer prevention and/or epigenetic therapy.

11.1
Introduction

Cancer initiation and progression require concurrent changes in expression of multiple genes. While genetic alterations account for some of these changes, epigenetic modifications (especially aberrations in DNA methylation patterns) have attracted a significant amount of attention as a potential cause of a concerted change in regulation of expression of numerous genes during carcinogenesis [1–3]. DNA methylation is one of the mechanisms of epigenetic regulation of gene expression. Thus, any alterations in DNA methylation pattern may affect gene transcriptional activities. The hallmarks of cancer cells are hypermethylation of some DNA regions (particularly promoters of tumor suppressor genes) and genomic hypomethylation. Both modifications are connected with incorrect formation of active (i.e., euchromatic DNA, undermethylated, accessible for proteins of transcription complex) and inactive (i.e., heterochromatic DNA, hypermethylated) domains in DNA, which can contribute to activation of oncogene expression and transcriptional silencing of tumor suppressor genes, respectively [3, 4]. As aberrations in DNA methylation patterns are reversible and start at very early stages of cancer development, it is reasonable to consider them as targets for chemoprevention. Chemoprevention is a strategy of preventing, blocking or reversing the process of carcinogenesis through the use of natural or synthetic compounds which are capable of modulation of different actions leading to cancer. Since the normal DNA methylation patterns and the normal function of enzymes catalyzing DNA methylation reaction are affected, for example, by intracellular levels of folate, vitamin B_{12}, methionine, and choline [2], which are methyl donor precursors or enzyme cofactors contained in food, it seems that environmental and dietary elements can have a major impact on the epigenome and the health of individuals. Moreover, a growing body of literature demonstrates that some natural bioactive compounds, constituents of food and herbs, may have a great influence on DNA methylation patterns. Studies of the actions of a few natural compounds, for example, tea catechins (catechin, epicatechin, (–)-epigallocatechin-3-gallate (EGCG)), bioflavonoids (quercetin, fisetin, myricetin), genistein from soybean, and coffee polyphenols (caffeic acid, chlorogenic acid), indicated that they are able to prevent or reverse promoter hypermethylation-induced silencing of key tumor suppressor genes and inhibit cancer development [5–8]. For other natural compounds, that is, all-*trans* retinoic acid (ATRA), vitamin D_3, and resveratrol, it was also demonstrated that their presence in the culture medium can lead to inhibition of promoter methylation of the selected tumor suppressor genes and to induction of gene expression in cancer cells [9, 10]. This raises the possibility that the natural compounds, which participate in regulation of DNA methylation, may be an effective approach to cancer prevention. Therefore, investigation of the physiological, therapeutic, and chemopreventive properties of the natural compounds has become a stern challenge for nutritionists, pharmacologists, and medical researchers.

11.2
DNA Methylation Reaction

In mammalian cells, DNA methylation occurs mainly at the 5-position of the cytosine pyrimidine ring. 5-Methylcytosine accounts for 3–8% of cytosine residues located predominantly in CpG sequences [11, 12]. Approximately 30% of CpG sequences are found in CG-rich regions known as CpG islands which account for 3–4% of genomic DNA. In normal cells, CpG islands are mostly unmethylated and located in regulatory regions (within promoters and their proximity) of half of all genes such as housekeeping genes, tissue-specific genes and tumor suppressor genes [12]. Promoters of some oncogenes also contain CpG islands but they are mainly methylated in normal cells [3]. DNA methylation within promoters of genes can be involved in the formation of a condensed chromatin structure leading to repression of gene transcription [3]. In normal mammalian cells, DNA methylation plays a significant role in regulation of multiple functions such as oncogene repression, control of expression of genes crucial for cell proliferation, differentiation, and normal development, parental imprinting, X chromosome inactivation, and preservation of chromosomal integrity by silencing of transpozones [3, 12].

DNA methylation is catalyzed by enzymes called DNA methyltransferases which transfer a methyl group from *S*-adenosyl-L-methionine (SAM, a donor of a methyl group) onto the 5-position of the cytosine pyrimidine ring (Figure 11.1) [12]. In mammals, DNA methyltransferase 1 (DNMT1), DNA methyltransferase 2 (DNMT2), DNA methyltransferase 3A and 3B (DNMT3A and DNMT3B) and DNA methyltransferase 3L (DNMT3L) have been identified [12]. The main enzyme responsible for maintenance of the DNA methylation pattern is DNMT1. DNMT1 copies a methylation pattern from a parental strand template to a daughter strand after DNA replication. DNMT1 also possesses *de novo* methylation activity similar to DNMT3A and DNMT3B methyltransferases, which are referred to as *de novo* methyltransferases in charge of *de novo* methylation of specific DNA sites (for instance repetitive sequences) during embryonic development [2]. Interestingly, the recent results demonstrate that methyltransferases from the DNMT3 group can also participate in maintenance of DNA methylation patterns [1]. However, because DNMT1 is the most abundant in proliferating cells, changes in its activity are often associated with hypermethylation-induced silencing of transcription of tumor suppressor genes. In human cancer cells, promoter hypermethylation was

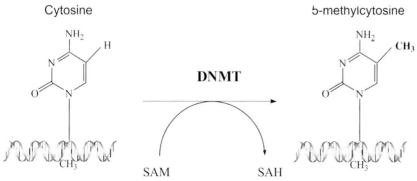

Figure 11.1 Chemistry of DNA methylation reaction.

mostly observed in the following genes: *GSTP1* (transferase S-glutathione class p) (e.g., in prostate, kidney, and breast cancer cells), $p16^{CDKN2A}$ (cyclin-dependent kinase inhibitor 2A) (e.g., in lung, kidney, bladder, breast cancer cells, and lymphoma), *MGMT* (O^6-methylguanine methyltransferase) (e.g., in lung cancer), *hMLH1* (human mutL homolog 1) (e.g., in endometrial cancer and lymphoma), *CDH1* (E-cadherin) (e.g., in lung, kidney, and endometrial cancer cells), *RARβ* (retinoic acid receptor β) (e.g., in prostate, lung, kidney, breast cancer cells, cervical neoplasia, and lymphoma), and *APC* (adenomatous polyposis coli) (e.g., in kidney, bladder, breast, and prostate cancer cells) [13]. In addition, the existing data imply that DNMT1 activity or *DNMT1* gene expression may be several-fold higher in cancer cells (e.g., in breast [14], endometrial [15], prostate cancers [16], and hepatocellular carcinomas [17]), as compared with normal cells.

11.3
Implication of the Selected Natural Compounds in DNA Methylation Regulation

11.3.1
ATRA, Vitamin D₃, Resveratrol, and Genistein

Vitamins (i.e., retinoic acid and vitamin D_3) and phytoestrogens (i.e., resveratrol and genistein) (Figure 11.2) are natural compounds which exert their effects mainly by binding to related members of the nuclear receptor families that, when bound to their ligands, modulate transcription through cognate response elements in the promoters of their target genes [18]. ATRA binds to the retinoic acid receptors (RARs) α, β and γ; and vitamin D_3 to the vitamin D receptor (VDR). RARs

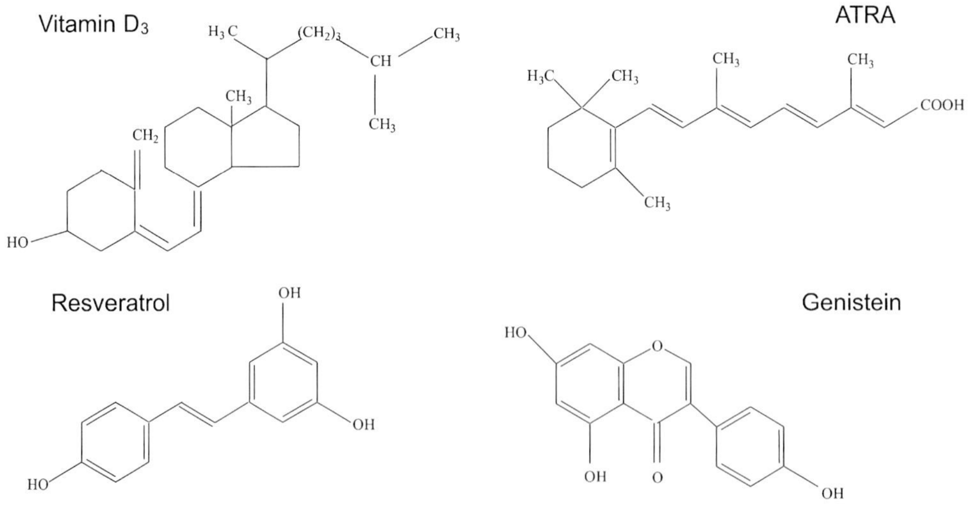

Figure 11.2 Structures of vitamins and phytoestrogens involved in regulation of DNA methylation.

heterodimerize with the retinoid X receptors (RXRs), and VDR can homodimerize or heterodimerize with RXR or RAR. As far as resveratrol and genistein are concerned, they can act as estrogen agonists or antagonists, depending on their concentrations, tissue type and tested genes as well as the presence of estradiol [19, 20].

ATRA is a vitamin A derivative and is generated from the β-carotene molecule (precursor of vitamin A) absorbed from fruit and vegetables. β-Carotene is composed of two retinyl groups and is broken down by β-carotene dioxygenase to two molecules of retinal. One of these molecules is converted into retinoic acid in an oxidation reaction [21]. Retinoic acid, bound to its nuclear receptors, regulates transcription of target genes important for cell proliferation, differentiation, and apoptosis through retinoic acid response elements. These effects seem to be crucial for the role of retinoic acid and its derivatives in chemoprevention and therapy of a variety of cancers, for example, leukemia or head and neck cancer [22], and also breast cancer [23].

Vitamin D_3 (cholecalciferol) is synthesized from the cholesterol derivative 7-dehydrocholesterol in the epidermis which requires ultraviolet radiation and is dependent on sun exposure. In addition, vitamin D_3 can be obtained from the diet (animal sources). The metabolic activation of vitamin D_3 is achieved through the following enzymatic steps: (i) hepatic hydroxylation of vitamin D_3 at the 25-position to generate 25-hydroxycholecalciferol, (ii) 1α-hydroxylation of 25-hydroxycholecalciferol in the kidney to generate 1α,25-dihydroxycholecalciferol (calcitriol) which is a biologically active metabolite that binds to VDR and regulates expression of target genes involved in cell proliferation, differentiation, and apoptosis. Calcitriol mediates the maintenance of skeletal health, calcium and bone homeostasis, enhancement of intestinal calcium transport, regulation of immune cells, modulation of hormone secretion, and maintenance of epidermal integrity [21, 24]. The active vitamin D_3 metabolite can be generated not only in the kidney but also in other tissues where vitamin D 1α-hydroxylase is active and highly expressed, for example, in skin, breast, prostate, and colon [21, 24]. A number of literature data support a role of calcitriol in growth regulation and differentiation of mammary cells, depending on the stage of breast tissue development and lactation [24]. Aberrations in vitamin D_3 metabolism and/or lack of VDR can increase the susceptibility of mammary cells to transformation, as some epidemiological studies have implied [24–26]. These studies provided convincing evidence that vitamin D and its receptor can be targets for breast cancer prevention and therapy.

Resveratrol (3,4′,5-trihydroxystilbene) is a dietary antioxidant polyphenol, found in a wide variety of plant species, including mulberries, peanuts, grapes, apricots, and pineapples [27]. Resveratrol can regulate estrogen-responsive genes through interaction with ERα and ERβ [28]. The polyphenol was shown to modulate the risk of cardiovascular disease (atherosclerosis) and mammary cancer, to inhibit chemical carcinogenesis in rodents, to inhibit growth promotion of preneoplastic lesions by effects on multiple signaling pathways, to inhibit cell proliferation, to block cell cycle progression in numerous types of human cancer cell lines, and to induce apoptosis [28–30].

Genistein (4′,5,7-trihydroxyisoflavone) is a major isoflavone from soy bean. The polyphenol was shown to prevent or slow down carcinogenesis in rats (mammary cancer) [30] and mice (leukemia) [31, 32], reduce mammary carcinogenesis and up-regulate mRNA expression of the *BRCA1* gene in rats exposed to carcinogen, that is, 7,12-dimethylbenz[a]anthracene (DMBA) [33], and inhibit carcinogenesis or metastasis in various types of cancers (e.g., colon and esophageal cancers) [34, 35]. Moreover, in *in vitro* studies (e.g., with human esophageal squamous cell carcinoma cell lines KYSE 510 and KYSE 150, or myeloid and lymphoid leukemic cell lines) genistein inhibited cancer cell proliferation and angiogenesis and induced apoptosis and cell cycle arrest (at the G2/M phase) [29, 31, 35]. It is noteworthy that the chemopreventive role of genistein was demonstrated in breast and prostate cancers as well as in head and neck cancer [22, 30].

A number of studies have demonstrated that ATRA, vitamin D_3, resveratrol, and genistein participate in both inhibition of cell proliferation and induction of differentiation or apoptosis in normal and transformed cells. The effects can be referred to regulation of a variety of genes, not only through direct binding to cognate response elements in gene promoters, but also through modulation of epigenetic events (especially DNA methylation patterns), as a few studies have revealed [9, 10]. Treatment of non-invasive MCF-7 breast cancer cells (representing an early stage of cancer development) with ATRA, vitamin D_3 or resveratrol for 72 h resulted in reduction of promoter methylation of *RARβ*, *PTEN*, and *APC* genes (by 20–50%). The most relevant effect, that is, up to 50% reduction of *PTEN* promoter methylation, was observed in the presence of resveratrol, whereas the insignificant inhibitory effect was exerted by vitamin D_3 on *RAR* promoter methylation (approximately 5% reduction). The changes in promoter methylation of the tested genes were accompanied by an increase in their expression levels in the presence of almost each natural compound (data unpublished). All tested compounds (i.e., ATRA, vitamin D_3, and resveratrol) caused approximately 30% increase in *PTEN* expression. A rise in *RARβ* and *APC* mRNA levels was observed only in cells treated with ATRA (36% and 16%, respectively) or vitamin D_3 (54% and 50%, respectively). In addition, in MCF-7 cells, vitamin D_3 and resveratrol caused a decrease in the *DNMT1* mRNA level (by 14% and 56%, respectively). In highly-invasive MDA-MB-231 cells, the compounds at very high concentrations led to reduction of *PTEN* promoter methylation (by 12–32%) [9, 10], although it was associated with an increase in the *PTEN* mRNA level (by 35%) only in the presence of vitamin D_3 (data unpublished). In the case of genistein, treatment of KYSE 510 cells with the phytoestrogen for a few days resulted in partial reversal of *p16*, *RARβ*, and *MGMT* promoter hypermethylation and in the gene reactivation [35]. Studies carried out on KYSE 150 cells and prostate cancer LNCaP and PC3 cells treated with genistein also demonstrated partial demethylation and *RAR* reactivation [5, 35]. Moreover, *in vitro* studies with nuclear extracts from human esophageal cancer KYSE 510 cells revealed that genistein considerably inhibits DNA methyltransferase activity in a dose-dependent manner and suppresses histone deacetylase (HDAC) activity, but to a limited extent [35]. In the light of the literature data it was assumed that demethylation and reactivation of methylation-silenced tumor suppressor genes by these natural chemicals (i.e., ATRA, vitamin D$_3$, resveratrol, and genistein) can result from their involvement in down-regulation of DNMT1 enzymatic activity and/or *DNMT1* gene expression. This indirect effect on DNMT1 activity can be attributed to the capability of the vitamins and phytoestrogens to stimulate *p21$^{WAF1/CIP1}$* and *PTEN* expression (by all selected chemicals), to activate retinoblastoma (Rb) protein and to inhibit E2F transcription factor activity (by ATRA and vitamin D$_3$), to inactivate ERα (by ATRA and vitamin D$_3$), and to inhibit AP-1 complex activity (by ATRA, resveratrol, and genistein).

11.3.1.1 Involvement of p21$^{WAF1/CIP1}$ and Rb/E2F Pathway in Regulation of DNMT1

An increase in expression of *p21$^{WAF1/CIP1}$* can influence DNMT1 activity owing to competition of p21$^{WAF1/CIP1}$ and DNMT1 proteins for the same binding site on proliferating cell nuclear antigen (PCNA) [36, 37]. PCNA is a ring-shaped protein interacting with a number of other proteins (e.g., with DNMT1) to increase their local concentration at replicated DNA sites. Numerous experimental data have revealed reversed dependence between p21$^{WAF1/CIP1}$ and DNMT1 protein levels both in normal and cancer cells [37, 38]. Further studies showed that the N-terminal region of DNMT1 has core PCNA binding activity and contains a typical PCNA binding motif shared with p21$^{WAF1/CIP1}$ protein. It was observed that the affinity of DNMT1 is much higher for DNA bound by PCNA than for free DNA. As a result, DNA bound by PCNA was methylated more efficiently by DNMT1 than free DNA [36]. The same PCNA loop-structures required for binding both p21$^{WAF1/CIP1}$ and DNMT1 suggested that p21$^{WAF1/CIP1}$ may be able to antagonize DNMT1 activity. It turned out that only a two- to three-fold molar excess of the full length p21$^{WAF1/CIP1}$ over PCNA is sufficient to completely block the DNMT1–PCNA interaction [36]. Moreover, p21$^{WAF1/CIP1}$-defective cancer cells exhibited a much higher level of DNA methylation than normal cells [36].

It is also remarkable that the regulatory region of the *DNMT1* gene (i.e., P1 and P3 promoters [11]) is regulated by the Rb/E2F pathway and contains E2F binding sites [11, 39]. In NIH 3T3 murine fibroblasts [39], human prostate cancer and human prostate normal cells [40], it was proved that E2F transcription factor binds directly to the E2F recognition sequence in the *DNMT1* promoter region and regulates transcription of *DNMT1* both positively and negatively in a cell cycle-dependent manner. E2F bound to *DNMT1* promoter activates *DNMT1* transcription at the late G1/S phase. At the G0/G1 phase, however, E2F represses *DNMT1* transcription by recruiting histone deacetylase (HDAC) and forming the E2F-Rb-HDAC complex. Active dephosphorylated Rb protein binds E2F and inhibits E2F target genes [40]. Importantly, the Rb-E2F complex can recruit HDAC and repress transcription by histone deacetylation and alteration in the chromatin structure [39]. It was also demonstrated that DNMT1 can directly interact with Rb protein via the N-terminal regulatory region [41]. Rb is able to disrupt the DNMT1-DNA binary complex and inhibit DNMT1 activity. Thus, Rb may play a role in hypomethylation of the cellular DNA. On the other hand, the interaction between Rb and DNMT1 can participate in down-regulation of E2F-responsive genes via generation of the DNMT1-Rb-E2F-HDAC complex [41]. Involvement of the Rb/E2F pathway in the regulation of DNMT1 activity and gene transcription is confirmed by studies implying that in breast cancer tissue samples [42] and prostate cancer cells [40] inactivation of the Rb/E2F pathway is associated with induction of DNA hypermethylation of CpG island-containing genes. In contrast, overexpression of *Rb* leads to hypomethylation of the cellular DNA [40].

Both ATRA and vitamin D_3 up-regulate $p21^{WAF1/CIP1}$ transcription and activate Rb protein leading to inhibition of cell cycle progression in the G1 phase, as shown in MCF-7 cells [43] and gastric cancer cells [44] treated with ATRA as well as in U937 and HL-60 leukemic cells and squamous carcinoma cell lines of the head and neck exposed to vitamin D_3 [18, 45]. $p21^{WAF1/CIP1}$ is an inhibitor of cyclin-dependent kinases (CDKs), for example, CDK2, which inactivates Rb protein through phosphorylation. Inhibition of CDK2 action leads to activation of Rb protein that binds to E2F transcription factor blocking increase in the expression of E2F-responsive genes, including *DNMT1* [11, 40] and many genes involved in DNA synthesis and S-phase progression. Furthermore, in MCF-7 cells treated with ATRA, a fall in expression of gene encoding E2F was detected [18]. Studies on determining the mechanism of the ATRA and vitamin D_3-dependent increase in $p21^{WAF1/CIP1}$ expression revealed the presence of retinoic acid and vitamin D response elements in $p21^{WAF1/CIP1}$ promoter [18]. Moreover, it was suggested that ATRA can up-regulate $p21^{WAF1/CIP1}$ expression indirectly via stimulation of proteins, that is, STAT1 (signal transducer and activator of transcription 1) and IRF-1 (interferon regulatory factor-1), involved in the regulation of $p21^{WAF1/CIP1}$ expression [46].

Resveratrol is also able to markedly induce $p21^{WAF1/CIP1}$ expression and inhibit cancer cell proliferation by cell cycle blockade at the S-phase, as demonstrated in MCF-7 cells [19]. Furthermore, in a wide variety of breast cancer cell lines the polyphenol increased *BRCA1* expression, accompanied by stimulation of *p53* and $p21^{WAF1/CIP1}$ gene expression [27]. Moreover, several reports showed that genistein inhibits human neuroblastoma SK-N-MC cell proliferation via cell cycle arrest (in the G2/M phase) and, similarly to resveratrol, up-regulation of $p21^{WAF1/CIP1}$ expression [47].

11.3.1.2 Involvement of the AP-1 Transcriptional Complex in Regulation of DNMT1

Several studies revealed the influence of c-Jun [11] and c-Fos [48], which are major components of the AP-1 complex, on up-regulation of *DNMT1* expression. In addition, multiple AP-1 recognition sites were identified within the *DNMT1* regulatory region (i.e., within P2, P3, and P4 promoters) [11]. This raised the possibility that the elevated level of PTEN protein, which negatively regulates the Shc/Ras/Raf/MAPK/AP-1 oncogenic signaling pathway [49, 50], can contribute to a decrease in *DNMT1* expression. In contrast, studies with human breast cancer cells indicated that estrogen receptors activated by estrogens stimulate activity of c-Ha-ras protein, which, in turn, leads to induction of the Shc/Ras/Raf/MAPk/AP-1 pathway [51]. Therefore, it was assumed that estrogen receptors cause an increase in

DNMT1 transcription via activation of the AP-1 complex. Thus, inactivation of the estrogen receptors can bring about down-regulation of *DNMT1* expression.

In myeloid leukemic cells (e.g., in the HL-60 cell line) treated with ATRA or vitamin D_3, up-regulation of *PTEN* expression was observed [52]. However, the mechanism of the effect has not yet been explained. Since PTEN is a negative regulator of the Shc/Ras/Raf/MAPK/AP-1 oncogenic signal transduction pathway [49], elevated PTEN level leads to inhibition of AP-1 activity and down-regulation of AP-1-responsive genes, including *DNMT1* [11]. Furthermore, ATRA can directly suppress AP-1 transcriptional activity as was observed in MCF-7 cells [18]. Several studies suggested that activated RARs can block AP-1 activity by inhibiting the action of Jun amino-terminal kinase (JNK) which phosphorylates and enhances c-Jun activity [53]. It was also proposed that RARs are able to antagonize AP-1 activity by competition for a limited amount of transcriptional coactivators, such as CREB-binding protein (CBP) and p300 protein [53]. Further studies with HeLa cells provided evidence that RAR can disrupt the ability of c-Jun to homodimerize with itself and heterodimerize with c-Fos [53]. The results showed that RARs possibly interact with the conserved bZIP regions (mediating dimerization) of c-Jun and c-Fos proteins.

It was demonstrated that ATRA and vitamin D_3 negatively regulate ERα activity. In MCF-7 cells, ATRA suppressed transcription of ER-responsive genes, and vitamin D_3 contributed to a decrease in the *ER* mRNA level [54]. Although the mechanism of the above effects remains to be elucidated, it is assumed that vitamin D_3 can act via a still unidentified response element in the *ER* gene promoter.

Similarly to ATRA and vitamin D_3, resveratrol and genistein also influence *PTEN* gene expression. Exposure of MCF-7 cells to the phytoestrogens resulted in a slight, but statistically significant, increase in *PTEN* mRNA and protein levels [55]. Simultaneously, other experimental data indicated that the phytoestrogens inhibit PI3K, MAPK, and Akt signaling pathways [20, 56, 57], which are cellular responses downstream of PTEN's lipid phosphatase activity [50]. It was proposed that resveratrol and genistein may affect *PTEN* expression through stimulation of proteins, that is, PPARγ, p53 and/or Egr-1, involved in up-regulation of *PTEN* transcription [50, 55]. More importantly, genistein was shown to influence PPARγ activation [55]. Because PTEN inactivates the Shc/Ras/Raf/MAPK/AP-1 pathway, an increase in PTEN level can contribute to a drop in expression of genes controlled by the AP-1 complex, including *DNMT1* [11].

11.3.2
Polyphenols with a Catechol Group

Certain polyphenols, that is, tea catechins (catechin, epicatechin, EGCG), bioflavonoids (quercetin, fisetin, myricetin), and coffee polyphenols (caffeic acid, chlorogenic acid) (Figure 11.3), have been demonstrated to exert a profound inhibitory effect on DNA methylation and *DNMT* activity. First, treatment of MCF-7 and MDA-MB-231 breast cancer cells with caffeic acid or chlorogenic acid led to a

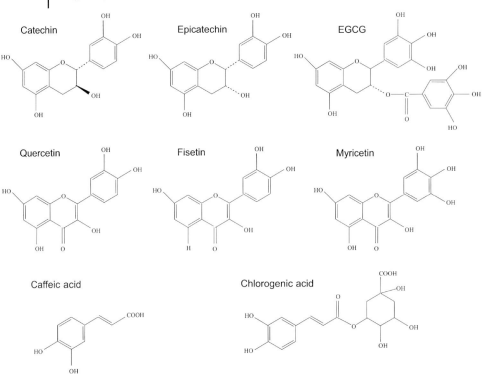

Figure 11.3 Structures of catechol-containing polyphenols.

decrease in *RARβ* promoter methylation [8]. Secondly, assays of enzymatic DNA methylation *in vitro* showed that the tea catechins and bioflavonoids inhibit the prokaryotic SssI DNMT-mediated DNA methylation as well as the human DNMT1-mediated DNA methylation in a concentration-dependent manner, but with varying potencies and efficacies [7]. EGCG, a major polyphenol from green tea, turned out to be more efficacious than other tested compounds in inhibition of DNA methylation catalyzed by SssI DNMT and human DNMT1. Moreover, EGCG caused demethylation and reactivation of *p16*, *RARβ*, *MGMT*, and *hMLH1* genes in human esophageal cancer KYSE 510 cells [58]. Studies carried out on human esophageal cancer KYSE 150 cells, colon cancer HT-29 cells, and prostate cancer PC3 cells treated with EGCG also demonstrated reactivation of some methylation-silenced genes (i.e., *p16* and *RARβ*) [5, 6]. Additionally, the presence of EGCG in a culture medium of MCF-7 and MDA-MB-231 cells contributed to partial reversal of *RARβ* promoter hypermethylation [7].

Considerable initial research efforts have already been made to find out the mechanism responsible for inhibition of DNA methylation by the above-mentioned tea catechins, bioflavonoids, and coffee polyphenols. The results showed that their structures share a catechol group which may play a key role in their action on DNA methylation. It is known that food compounds with a catechol group are excellent substrates for the methylation mediated by catechol-*O*-

methyltransferase (COMT). The COMT-mediated methylation reaction results in depletion of the methyl donor SAM and formation of SAH which is a very potent feedback inhibitor of DNA methylation [8]. It is reasonable to suggest that various catecholic dietary polyphenols may be an important cumulative modulator of the cellular DNA methylation process. It is noteworthy that the studies implied that EGCG can exert a dual inhibitory effect on DNMT-mediated DNA methylation. First, it can indirectly, like other catechol-containing polyphenols, inhibit DNMT activity through increased formation of SAH. Secondly, EGCG can interact directly with the catalytic site of the human DNMT1. Molecular modelling studies indicated that EGCG is well accommodated in a hydrophilic pocket of DNMT1 and effectively tethered within the DNMT1 binding site by at least four hydrogen bonds [6]. The experiments revealed that the high affinity of EGCG to bind DNMT1 is independent of COMT and dependent on and stabilized by Mg^{2+} ions [7]. It was assumed that the crucial role in binding of EGCG to DNMT1 is linked to the galloyl moiety since EGCG analogues without a gallic acid moiety were poor inhibitors of DNMT1 activity. Moreover, myricetin, which has a pyrogallic acid moiety, similar to the galloyl group of EGCG, was a stronger inhibitor of DNMT activity than other bioflavonoids (e.g., quercetin or fisetin) in the presence of Mg^{2+} ions [7].

The polyphenols with a catechol group were shown to possess antioxidant properties and to reduce the risk of chronic diseases, including cancer. They induce cell cycle arrest, differentiation and/or apoptosis of cancer cells, inhibit proliferation and invasiveness of cancer cells and regulate redox-mediating signaling (e.g., AP-1 and NF-κB) [29, 59–61]. These effects can partly arise from the ability of the dietary chemicals to inhibit DNA methylation (via an increase in SAH level) and activate methylation-silenced tumor suppressor genes. On the other hand, through exerting these effects the compounds may indirectly inhibit DNMT1 activity and/or *DNMT1* expression similarly to other polyphenols (i.e., resveratrol and genistein) and vitamins (i.e., ATRA and vitamin D_3).

11.4
Conclusions and Future Perspectives

Vitamins and polyphenolic compounds are potent bioactive molecules with anticarcinogenic properties since they can interfere with the initiation and progression of cancer by affecting cell proliferation, differentiation, apoptosis, angiogenesis, and metastasis. These effects can partly be connected to inhibition of DNA methylation and re-establishment of normal DNA methylation patterns. A number of tumor suppressor genes have been reported to be transcriptionally silenced by the promoter hypermethylation during cancer development. Therefore, inhibition of DNA methyltransferase activity may be a putative mechanism that contributes to repression of cancer growth and/or cancer initiation. It indicates the intriguing possibility of the use of certain dietary chemicals with activity of DNA methylation inhibitors in chemoprevention and/or anticancer epigenetic therapy. *In vitro* studies have provided encouraging results, proving the role of the bioactive food components in re-activation of methylation-silenced tumor suppressor genes and preventing, slowing down or reversing the process of carcinogenesis. These effects were not so pronounced in some experiments on animals. Bioavailability, metabolism and/or interactions with reactive molecules in animal and human organisms can contribute to a diminution of the effects of the dietary chemicals, which should be elucidated in further studies. Nevertheless, it should be emphasized that more than two-thirds of human cancers have been postulated to be preventable by modifying lifestyle. It was reported that approximately 35% of human cancer mortality is attributable to diet alone. According to a broad range of epidemiological studies, individuals who consume fruit and vegetables several times (at least five) a day have a markedly lower risk of developing cancer.

Since not only hypermethylation of DNA but also deacetylation of histones are known to be key epigenetic mechanisms of transcriptional silencing of many regulatory genes, it becomes interesting to determine the effect of the combination of dietary DNMT inhibitors with dietary histone deacetylase inhibitors (e.g., sulforaphane from broccoli) on gene activities and cancer development. As the initial experiments give promising results, the combined therapy may be an effective approach to cancer chemoprevention.

References

1 Jones, P.A., and Baylin, S.B. (2007) The epigenomics of cancer. *Cell*, **128**, 683–692.
2 Szyf, M. (2005) DNA methylation and demethylation as targets for anticancer therapy. *Biochem. (Moscow)*, **70**, 651–669.
3 Szyf, M., Pakneshan, P., and Rabbani, S.A. (2004) DNA methylation and breast cancer. *Biochem. Pharmacol.*, **68**, 1187–1197.
4 Balch, C., Montgomery, J.S., Paik, H.-I., Kim, S., Huang, T.H.-M., and Nephew, K.P. (2005) New anti-cancer strategies: epigenetic therapies and biomarkers. *Front. Biosci.*, **10**, 1897–1931.
5 Fang, M., Chen, D., and Yang, C.S. (2007) Dietary polyphenols may affect DNA methylation. *J. Nutr.*, **137**, 223S–228S.
6 Fang, M.Z., Wang, Y., Ai, N., Hou, Z., Sun, Y., Lu, H., Welsh, W., and Yang, C.S. (2003) Tea polyphenol (−)-epigallocatechin-3-gallate inhibits DNA methyltransferase and reactivates methylation-silenced genes in cancer cell lines. *Cancer. Res.*, **63**, 7563–7570.
7 Lee, W.J., Shim, J.-Y., and Zhu, B.T. (2005) Mechanisms for the inhibition of DNA methyltransferases by tea catechins and bioflavonoids. *Mol. Pharmacol.*, **68**, 1018–1030.
8 Lee, W.J., and Zhu, B.T. (2006) Inhibition of DNA methylation by caffeic acid and chlorogenic acid, two common catechol-containing coffee polyphenols. *Carcinogenesis*, **27**, 269–277.
9 Krawczyk, B., and Fabianowska-Majewska, K. (2006) Alteration of promoter methylation of selected tumor suppressor genes by adenosine analogs in combination with natural compounds in MCF-7 cells. *Proceedings of the Free Papers, UICC World Cancer Congress, Washington.* pp. 53–56.
10 Krawczyk, B., Rudnicka, K., and Fabianowska-Majewska, K. (2007) Effects of nucleoside analogues on promoter methylation status of selected tumour suppressor genes in MCF-7 and MDA-MB-231 breast cancer cell lines. *Nucleosides Nucleotides Nucleic Acids*, **26**, 1043–1046.

11 Bigey, P., Ramchandani, S., Theberge, J., Araujo, F.D., and Szyf, M. (2000) Transcriptional regulation of the human DNA methyltransferase (dnmt1) gene. *Gene.*, **242**, 407–418.

12 Hermann, A., Gowher, H., and Jeltsch, A. (2004) Biochemistry and biology of mammalian DNA methyltransferases. *Cell. Mol. Life. Sci.*, **61**, 2571–2587.

13 Paluszczak, J., and Baer-Dubowska, W. (2006) Epigenetic diagnostics of cancer – the application of DNA methylation markers. *J. Appl. Genet.*, **47**, 365–375.

14 Girault, I., Tozlu, S., Lidereau, R., and Bièche, I. (2003) Expression analysis of DNA methyltransferases 1, 3A, and 3B in sporadic breast carcinomas. *Clin. Cancer Res.*, **9**, 4415–4422.

15 Xiong, Y., Dowdy, S.C., Xue, A., Shujuan, J., Eberhardt, N.L., Pedratz, K.C., and Jiang, S.-W. (2005) Opposite alterations of DNA methyltransferase gene expression in endometrioid and serous endometrial cancers. *Gynecol. Oncol.*, **96**, 601–609.

16 Patra, S.K., Patra, A., Zhao, H., and Dahiya, R. (2002) DNA methyltransferase and demethylase in human prostate cancer. *Mol. Carcinog.*, **33**, 163–171.

17 Oh, B.K., Kim, H., Park, H.J., Shim, Y.H., Choi, J., Park, C., and Park, Y.N. (2007) DNA methyltransferase expression and DNA methylation in human hepatocellular carcinoma and their clinicopathological correlation. *Int. J. Mol. Med.*, **20**, 65–73.

18 Wang, Q., Lee, D., Sysounthone, V., Chandraratna, R.A.S., Christakos, S., Korah, R., and Wieder, R. (2001) 1,25-dihydroxyvitamin D_3 and retinoic acid analogues induce differentiation in breast cancer cells with function- and cell-specific additive effects. *Breast Cancer Res. Treat.*, **67**, 157–168.

19 Pozo-Guisado, E., Alvarez-Barrientos, A., Mulero-Navarro, S., Santiago-Josefat, B., and Fernandez-Salguero, P.M. (2002) The antiproliferative activity of resveratrol results in apoptosis in MCF-7 but not in MDA-MB-231 human breast cancer cells: cell-specific alteration of the cell cycle. *Biochem. Pharmacol.*, **64**, 1375–1386.

20 Pozo-Guisado, E., Lorenzo-Benayas, J., and Fernandez-Salguero, P.M. (2004) Resveratrol modulates the phosphoinositide 3-kinase pathway through an estrogen receptor α-dependent mechanism: relevance in cell proliferation. *Int. J. Cancer*, **109**, 167–173.

21 Murray, R.K., Granner, D.K., Mayes, P.A., and Rodwell, V.W. (eds) (1995) *Harper's Biochemistry*, PZWL, Warsaw, pp. 1–955.

22 Baer-Dubowska, W. (2003) Chemoprevetion – prophylaxis and adjuvant therapy of head and neck cancers. *Postepy w chirurgii glowy i szyi*, **2**, 3–14.

23 Fujii, T., Yokoyama, G., Takahashi, H., Namoto, R., Nakagawa, S., Toh, U., Kage, M., Shirouzu, K., and Kuwano, M. (2008) Preclinical studies of molecular-targeting diagnostic and therapeutic strategies against breast cancer. *Breast Cancer*, **15**, 73–78.

24 Welsh, J., Wietzke, J.A., Zinser, G.M., Byrne, B., Smith, K., and Narvaez, C.J. (2003) Vitamin D-3 receptor as a target for breast cancer prevention. *J. Nutr.*, **133**, 2425S–2433S.

25 Shin, M.J., Holmes, M.D., Hankinson, S.E., Wu, K., Colditz, G.A., and Willet, W.C. (2002) Intake of dairy products, calcium and vitamin D and risk of breast cancer. *J. Natl. Cancer Inst.*, **94**, 1301–1310.

26 Grant, W.B. (2002) An ecologic study of dietary and solar ultraviolet-B links to breast carcinoma mortality rates. *Cancer*, **94**, 272–281.

27 Le Corre, L., Fustier, P., Chalabi, N., Bignon, Y.J., and Bernard-Gallon, D. (2004) Effects of resveratrol on the expression of a panel of genes interacting with the BRCA1 oncosuppressor in human breast cell lines. *Clin. Chim. Acta.*, **344**, 115–121.

28 Bowers, J.L., Tyulmenkov, V.V., Jernigan, S.C., and Klinge, C.M. (2000) Resveratrol acts as a mixed agonist/antagonist for oestrogen receptors alpha and beta. *Endocrinology*, **141**, 3657–3667.

29 Cornwell, T., Cohick, W., and Raskin, I. (2004) Dietary phytoestrogens and health. *Phytochem.*, **65**, 995–1016.

30 Whitsett, T.G. Jr., and Lamartiniere, C.A. (2006) Genistein and resveratrol: mammary cancer chemoprevention and

mechanisms of action in the rat. *Expert. Rev. Anticancer Ther.*, **6**, 1699–1706.

31 Raynal, N.J.-M., Momparler, L., Charbonneau, M., and Momparler, R.L. (2008) Antileukemic activity of genistein, a major isoflavone present in soy products. *J. Nat. Prod.*, **71**, 3–7.

32 Shen, J., Tai, Y.C., Zhou, J., Stephen Wong, C.H., Cheang, P.T., Wong, W.S., Xie, Z., Khan, M., Han, J.H., and Chen, C.S. (2007) Synergistic antileukemia effect of genistein and chemotherapy in mouse xenograft model and potential mechanism through MAPK signalling. *Exp. Hematol.*, **35**, 75–83.

33 Cabanes, A., Wang, M., Olivo, S., DeAssis, S., Gustafsson, J.A., Khan, G., and Hilakivi-Clarke, L. (2004) Prepubertal estradiol and genistein exposures up-regulate BRCA1 mRNA and reduce mammary tumorigenesis. *Carcinogenesis*, **25**, 741–748.

34 Arai, N., Strom, A., Rafter, J.J., and Gustafsson, J.A. (2000) Estrogen receptor β mRNA in colon cancer cells: growth effects of estrogen and genistein. *Biochem. Biophys. Res. Commun.*, **270**, 425–431.

35 Fang, M.Z., Chen, D., Sun, Y., Jin, Z., Christman, J.K., and Yang, C.S. (2005) Reversal of hypermethylation and reactivation of $p16^{INK4a}$, *RAR*, and *MGMT* genes by genistein and other isoflavones from soy. *Clin. Cancer. Res.*, **11**, 7033–7041.

36 Iida, T., Suetake, I., Tajima, S., Morioka, H., Ohta, S., Obuse, C., and Tsurimoto, T. (2002) PCNA clamp facilitates action of DNA cytosine methyltransferase 1 on hemimethylated DNA. *Genes. Cells*, **7**, 997–1007.

37 Milutinovic, S., Knox, J.D., and Szyf, M. (2000) DNA methyltransferase inhibition induces the transcription of the tumor suppressor $p21^{WAF1/CIP1/sdi1}$. *J. Biol. Chem.*, **275**, 6353–6359.

38 Milutinovic, S., Brown, S.E., Zhuang, Q., and Szyf, M. (2004) DNA methyltransferase 1 knock down induces gene expression by a mechanism independent of DNA methylation and histone deacetylation. *J. Biol. Chem.*, **279**, 27915–27927.

39 Kimura, H., Nakamura, T., Ogawa, T., Tanaka, S., and Shiota, K. (2003) Transcription of mouse *DNA methyltransferase 1 (Dnmt1)* is regulated by both E2F-Rb-HDAC-dependent and -independent pathways. *Nucleic. Acids. Res.*, **31**, 3101–3113.

40 McCabe, M.T., Davis, J.N., and Day, M.L. (2005) Regulation of DNA methyltransferase 1 by the pRb/E2F1 pathway. *Cancer Res.*, **65**, 3624–3632.

41 Robertson, K.D., Ait-Si-Ali, S., Yokochi, T., Wade, P.A., Jones, P.L., and Wolffe, A.P. (2000) DNMT1 forms a complex with Rb, E2F1 and HDAC1 and represses transcription from E2F-responsive promoters. *Nat. Genet.*, **25**, 338–342.

42 Agoston, A.T., Argani, P., De Marzo, A.M., Hicks, J.L., and Nelson, W.G. (2007) Retinoblastoma pathway dysregulation causes DNA methyltransferase 1 overexpression in cancer via MAD2-mediated inhibition of the anaphase-promoting complex. *Am. J. Pathol.*, **170**, 1585–1593.

43 Pratt, M.A., Niu, M., and White, D. (2003) Differential regulation of protein expression, growth and apoptosis by natural and synthetic retinoids. *J. Cell. Biochem.*, **90**, 692–708.

44 Wu, Q., Chen, Z., and Su, W. (2001) Growth inhibition of gastric cancer cells by all-trans retinoic acid through arresting cell cycle progression. *Chin. Med. J. (Engl.)*, **114**, 958–961.

45 Hager, G., Kornfehl, B., Weigel, G., and Formanek, M. (2004) Molecular analysis of p21 promoter activity isolated from squamous carcinoma cell lines of the head and neck under the influence of 1,25(OH)2 vitamin D_3 and its analogs. *Acta. Otolaryngol.*, **124**, 90–96.

46 Arany, I., Whitehead, W.E., Ember, I.A., and Tybing, S.K. (2003) Dose-dependent activation of p21WAF1 transcription by all-trans-acid in cervical squamous carcinoma cells. *Anticancer Res*, **23**, 495–497.

47 Ismail, I.A., Kang, K.S., Lee, H.A., Kim, J.W., and Sohn, Y.K. (2007) Genistein-induced neuronal apoptosis and G2/M cell cycle arrest is associated with MDC1

up-regulation and PLK1 down-regulation. *Eur. J. Pharmacol.*, **575**, 12–20.

48 Bakin, A.V., and Curran, T. (1999) Role of DNA 5-methylcytosine transferase in cell transformation by fos. *Science.*, **283**, 387–390.

49 Chuang, J.H., Ostrowski, M.C., Romigh, T., Minaguchi, T., Waite, K.A., and Eng, C. (2006) The ERK1/2 pathway modulates nuclear PTEN-mediated cell cycle arrest by cyclin D1 transcriptional regulation. *Hum. Mol. Genet.*, **15**, 2553–2559.

50 Krawczyk, B., Rychlewski, P., and Fabianowska-Majewska, K. (2006) PTEN- tumour suppressor protein: regulation of protein activity and gene expression. *Postepy. Biologii. Komorki.*, **33**, 365–380.

51 Pethe, V., and Shekhar, P.V.M. (1999) Estrogen inducibility of c-Ha-*ras* transcription in breast cancer cells. Identification of functional estrogen-responsive transcriptional regulatory elements in exon 1/intron 1 of the c-Ha-ras gene. *J. Biol. Chem.*, **274**, 30969–30978.

52 Hisatake, J., O'Kelly, J., Uskokovic, M.R., Tomoyasu, S., and Koeffler, H.P. (2001) Novel vitamin D_3 analog, 21-(3-methyl-3-hydroxy-butyl)-19-nor D_3, that modulates cell growth, differentiation, apoptosis, cell cycle, and induction of PTEN in leukemic cells. *Blood*, **97**, 2427–2433.

53 Zhou, X.-F., Shen, X.-Q., and Shemshedini, L. (1999) Ligand-activated retinoic acid receptor inhibits AP-1 transactivation by disrupting c-Jun/c-Fos dimerization. *Mol. Endocrinol.*, **13**, 276–285.

54 Swami, S., Krishnan, A.V., and Feldman, D. (2000) 1α,25-dihydroxyvitamin D_3 down-regulates estrogen receptor abundance and suppresses estrogen action in MCF-7 human breast cancer cells. *Clin. Cancer Res.*, **6**, 3371–3379.

55 Waite, K.A., Sinden, M.R., and Eng, C. (2005) Phytoestrogen exposure elevates PTEN levels. *Hum. Mol. Genet.*, **14**, 1457–1463.

56 Gong, L., Li, Y., Nedeljkovic-Kurepa, A., and Sarkar, F.H. (2003) Inactivation of NF-kappaB by genistein is mediated via Akt signaling pathway in breast cancer cells. *Oncogene.*, **22**, 4702–4709.

57 Yu, R., Hebbar, V., Kim, D.W., Mandlekar, S., Pezzuto, J.M., and Kong, A.N. (2001) Resveratrol inhibits phorbol ester and UV-induced activator protein 1 activation by interfering with mitogen-activated protein kinase pathways. *Mol. Pharmacol.*, **60**, 217–224.

58 Lu, Q., Qiu, X., Hu, N., Wen, H., Su, Y., and Richardson, B.C. (2006) Epigentics, disease, and therapeutic intervention. *Aging. Res. Rev.*, **5**, 449–467.

59 Nair, S., Li, W., and Kong, A.-N.T. (2007) Natural dietary anti-cancer chemopreventive compounds: redox-mediated differential signaling mechanisms in cytoprotection of normal cells versus cytotoxicity in tumor cells. *Acta. Pharmacol. Sin.*, **28**, 459–472.

60 Ramos, S. (2007) Effects of dietary flavonoids on apoptotic pathways related to cancer chemoprevention. *J. Nutr. Biochem.*, **18**, 427–442.

61 Tapiero, H., Tew, K.D., Ba, G.N., and Mathé, G. (2002) Polyphenols: do they play a role in the prevention of human pathologies? *Biomed. Pharm.*, **56**, 200–207.

12
Health Determinants Throughout the Life Cycle
Petra Rust

Abstract

Individual health promotion becomes more important considering the rapidly growing older population. Identification of individual risk factors like genetic diversity of populations, differences in food availability and intake, as well as nutritional behavior and lifestyles poses big challenges for personal nutrition. Various studies have observed that the risk of many diseases is not just determined by risk factors in mid-adult life, but begins during fetal development, in childhood, and adolescence. Food diversity shows evidence of being most important for individual health.

12.1
Introduction

We eat many different foodstuffs throughout our life and are exposed to a complex mixture of nutrients and non-nutrients. On the one hand these nutrients are important for an adequate supply of energy and essential substances to enable us to grow and function; on the other hand there are food compounds which do not have an obvious nutritional role but can protect us from a variety of age- or diet-related diseases. Thus, optimal food choice is not just to avoid malnutrition but also to promote health.

Health is a basic human right and is essential for social and economic development. Pre-requisites for health are peace, shelter, education, social security, social relations, food, income, empowerment, a stable eco-system, sustainable resource use, social justice, respect for human rights and equity. Demographic trends such as urbanization, an increase in the number of older people and the prevalence of chronic diseases pose new problems all over the world. New and re-emerging infectious diseases and greater recognition of mental health problems require an urgent response [1].

Environmental factors including nutrition, lifestyle, and socioeconomic factors as well as individual factors such as age, sex and genetic disposition are important influencing factors on human health. Smoking, unhealthy diet, excessive alcohol consumption and poor physical activity are important determinates of disease and mortality [2]. Studies suggest that smoking is accountable for 4.1% of the global burden of disease, while alcohol, inactivity and poor nutrition contribute 4%, 1.3% and 1.8%, respectively [3].

Chronic noncommunicable diseases (NCDs) including cardiovascular diseases, cancer, chronic obstructive pulmonary disease and diabetes are the biggest causes of death in the world [4]. As populations become older the proportion of deaths due to noncommunicable diseases will rise significantly. Globally, deaths from cancer will increase from 7.4 million in 2004 to 11.8 million in 2030, and deaths from cardiovascular diseases will increase from 17.1 million to 23.4 million in the same period [5].

It is predicted that the four leading causes of death in the world in 2030 will be ischemic heart disease, cerebrovascular disease (stroke), chronic obstructive pulmonary disease (COPD) and lower respiratory infections (mainly pneumonia), obesity and diabetes are also showing worrying trends. Therefore, it is very important to mobilize resources to support prevention programs focusing on major modifiable risk factors, such as tobacco use, unhealthy diet and physical inactivity, at population and individual levels.

There are three interacting factors affecting the risk of developing age- or diet-related diseases: life-stage, lifestyle and genes. As people get older, their organisms are less effective at avoiding diseases and their immune systems are less able to distinguish them from pathogens. The resulting deficiencies lead to those diseases associated with older age: cancer, cardiovascular disease (CVD), type II diabetes, cataract, arthritis, and so on. Poor diet can speed up this process; 80% of case-controlled studies suggest that a diet rich in fruits and vegetables can reduce this risk. Most important things to reduce disease risk are a healthy body weight, moderate consumption of alcohol, non-smoking and regular physical activity. These factors will determine whether most of the population are at high or low risk of developing diseases. However, diet interacts with genes and individual genetic differences in response to diet are discussed. For example nutrigenetics found a relationship between folate and the gene for MTHFR – 5,10-methylenetetrahydrofolate reductase. There is a variant in the gene for MTHFR that produces a less efficient form of the enzyme. Individuals with the less efficient enzyme and low dietary folate accumulate homocysteine and have less methionine, which increases their risk of vascular disease and premature cognitive decline. Greater folate intake or folate supplementation of these individuals leads to a fast metabolism of the increased homocysteine levels. But, while a higher intake of folate may be beneficial for some individuals, it may be shown in the future that increased intake has unexpected risks for other individuals or sub-populations [6].

Epigenetics is just beginning to reveal its possible implications in nutrition. Nutrition can exert influence on a genome: DNA methylation appears to provide a format for long-term dietary (re-)programming of the genome, suggesting that nutritional supplementation may have unexpected adverse effects on gene regulation, and well-adapted diets applied pre- and post-natally may have a fundamental and long-lasting positive impact. Some evidence shows that chronic diseases in adulthood are due to persistent influences during early-life nutrition [7, 8].

It is now generally accepted that the risk of many diseases is not just determined by risk factors in mid-adult life, but begins in childhood or adolescence, and likely even earlier, i.e., during fetal development [9, 10].

Nutritional requirements depend on the stage in life. Pregnancy and lactation represent physiological states for which specific nutritional needs apply beyond classical recommendations. Early in life, growth and development are primary biological objectives. Teenage years and the change to reproductive fertility cause a significant change in hormonal status, with physiological and metabolic consequences, many of which change dietary needs and responses [11, 12]. With increasing numbers of elderly people, their physiological, metabolic and even microbial states as well as their special nutritional needs are being documented.

Early dietary exposure can also influence a person's response to later diet by influencing the individual's gut microbiome. Such dietary influences can be achieved directly by microorganisms present in foods or by manipulation of subsets of microorganisms by food components that can only be fermented/utilized by certain bacterial populations. Ley *et al.* [13] suggest that the microflora is a central factor in human metabolism, immunity, sensation, disease resistance, inflammation and comfort.

12.2
Pre- and Postnatal Determinants

The four relevant factors in fetal life are: intra-uterine growth retardation (IUGR); premature delivery of a normal growth-for-gestational-age fetus; overnutrition *in utero*; and intergenerational factors.

There is evidence, that intra-uterine growth retardation is associated with a greater risk of coronary heart disease (CHD), stroke, and diabetes. Some studies found high blood pressure in childhood to be associated with low birth weight; this association may depend on the type of growth retardation or impairment *in utero*. There is also good evidence for reduced fetal growth and increased risk of metabolic syndrome in middle life [14] Regarding the association of fetal growth and glucose tolerance or dyslipidaemia, results are conflicting [15]. The increased risk for the development of NCDs may not be the low birth weight per se but rather a pattern of growth, restricted fetal growth followed by rapid postnatal catch-up growth.

However, high birth weight can also be a risk factor; it is associated with increased risk of diabetes and cardiovascular diseases. Kuh and Ben-Shlomo [16] found also a relationship between higher birth weight and increased risk of breast and other cancers.

Imprinting during prenatal development has been shown to cause methylation differences across entire regions of a fetal genome and can result from different nutrient imbalances and deficiencies of the mother [17].

Animal studies propose that adipocyte hyperplasia early in development is one of the factors that could account for the high predisposition to adult obesity in children who are overweight [18]. These studies documented that specific dietary factors early in life stimulate or inhibit the proliferation of adipocytes. This effect could influence the response to diets later in life.

12.3
Determinants During Infancy and Adulthood

Postnatal feeding conditions also play a major role: although evidence on the effect of breastfeeding on dyslipidemias and blood pressure is conflicting, the risk for several chronic diseases of childhood and adolescence (type 1 diabetes, celiac disease, some childhood cancers, and inflammatory bowel disease) have been associated with infant formula feeding and short term breastfeeding [19].

A trend toward overweight and obesity in adolescence due to unhealthy eating and physical inactivity can be recognized worldwide. Some long-term cohort studies have shown that high blood pressure in adolescence or young adulthood is strongly related to later risk of stroke or CHD [20]. High serum cholesterol levels, both in middle age and in early life, are associated with an increased risk of disease (especially cardiovascular diseases) later on. More than 60% of overweight children have at least one additional risk factor for cardiovascular disease, such as raised blood pressure, hyperlipidemia, or hyperinsulinemia, and more than 20% have two or more risk factors [21].

High blood pressure, IGT and dyslipidemia in children and adolescents are associated with unhealthy lifestyles: diets containing excessive amounts of fat (especially saturated fats), refined carbohydrates, cholesterol and salt, and inadequate intake of fiber and potassium, and inactivity. Low exercise and smoking have been found as independent predictors of cardiovascular diseases and stroke in later life [15].

12.4
Determinants in Adults and Older People

In adults smoking, obesity, physical inactivity, high blood cholesterol, high blood pressure, and excessive alcohol consumption are the known risk factors for cardiovascular diseases and diabetes. Moreover, genetics and programming play an important role in disease development.

The dramatic changes in fertility and mortality rates during the 20th century have led to a rapidly ageing world in the 21st century. Therefore, prevention of disabilities and generally poor health in older years is very important. Being mobile for daily life activities, subjective feelings of well-being, and economic security contribute to the quality of life in older people. As the elderly population increases vastly throughout the world, specific interventions will have an impact on a great number of people [22].

12.5
Interactions Throughout the Lifecycle

Low birth weight, followed by adult obesity, has been shown to be a high risk factor for coronary heart diseases [23], as well as diabetes [24]. Several studies have shown an interaction between rapid catch-up growth in weight after intra-uterine growth retardation and increased risk of disease, for example, the highest risk of CHD or metabolic syndrome. The exact period in life in which increased growth may increase risk of later disease is currently unclear.

Obesity in childhood and adolescence is related to higher risk of later CHD or diabetes. If obesity in childhood persists into adulthood, the morbidity and mortality is greater than if the obesity developed in the adult [25].

12.6
Intergenerational Effects

Young girls who grow poorly become stunted women and more often give birth to low birth-weight babies who are then likely to continue the cycle by being stunted in adulthood and so on. Studies suggest intergenerational factors in obesity such as obesity of the parents, maternal gestational diabetes, and maternal birth weight [26]. Unhealthy lifestyles, in particular smoking, have a direct effect on the health of the next generation, for example, smoking during pregnancy and low birth weight, and the increased risk of respiratory disease [27].

Most important for chronic disease prevention are healthy diets, physical activity and non-smoking. Established risk factors for CHD, stroke and diabetes are hypertension, obesity and dyslipidemia. Globally, risk factors are rising, especially obesity. Risk factors in early life may have negative impact throughout the life cycle and can even affect the health of the next generation. Therefore, an adequate pre- and postnatal nutritional environment is very important for health in later life.

Knowledge of the relevance of essential nutrients led to a major public health breakthrough by ensuring essential nutrient supply. Although most of the population are adequately nourished, individuals within the population may profit through more specific recommendations. Moreover, in developed countries, caloric over-nutrition increasingly coincides with micronutrient deficiencies due to a one-sided diet and this contributes to the rapidly growing epidemic of obesity and diabetes.

Food diversity is proving to be most important to individual health, especially beyond the scope of essential nutrients. The association of bioactive food molecules and the risk and progression of chronic and degenerative diseases has been investigated in several studies. The response among individuals to those bioactive compounds may be important and may set the stage for future personalized diet and health [28, 29].

The real concern about these early manifestations of chronic disease, besides that they are occurring earlier and earlier, is that once developed they tend to last throughout life. However, there is evidence that they can be corrected.

Identification of individual risk factors considering the genetic diversity of populations, the complexity of foods, cultures and lifestyles, and the variety of metabolic processes poses huge challenges for personal dietary advice. Individual health promotion must be recognized as an essential element of health promotion. It is necessary to enable people to increase control over, and to improve, their health.

Human food choices have always been rooted in personal preferences and individual experiences, including sensory acuity, cultural habits and the personal economic situation. The nutrition community has recognized that different physiological events require significant adaptations to diet. For example, pregnant women, active athletes and elderly people have specific nutrient requirements, and these needs should guide dietary recommendations [30].

References

1 WHO (1997) The Jakarta Declaration on Health Promotion into the 21st Century, http://www.ldb.org/vl/top/jakdec.htm (accessed 24 August 2009).

2 WHO (2000) The World Health Report 2000, World Health Organization, Geneva.

3 Ezzati, M., VanderHoorn, S., Rodgers, A., Lopez, A., Mathers, C., and Murray, C. (2003) Estimates of global and regional health gains from reducing multiple major risk factors. *Lancet*, **362** (9380), 271–280.

4 Habib, S.H., and Saha, S. (2008) Burden of non-communicable disease: global overview. *Diabetes and Metabolic Syndrome: Clinical Research and Reviews*, available online 6 June 2008.

5 WHO (2008) World Health Statistics 2008. http://www.who.int/whosis/whostat/2008/en/index.html (accessed 24 August 2009).

6 Astley, S.B. (2007) An introduction to nutrigenomics developments and trends. *Genes Nutr.*, **2** (1), 11–13.

7 Dolinoy, D.C., Weidman, J.R., and Jirtle, R.L. (2007) Epigenetic gene regulation: linking early developmental environment to adult disease. *Reprod. Toxicol.*, **23**, 297–307.

8 Thaler, R., Karlic, H., Rust, P., and Haslberger, A.G. (2008) Epigenetic regulation of human buccal mucosa mitochondrial superoxide dismutase gene expression by diet. *Br. J. Nutr.*, **8**, 1–7.

9 WHO (2002) Diet, Physical Activity and Health, World Health Organization, Geneva. (documents A55/16 and A55/16 corr. 1).

10 WHO (2002) Programming of Chronic Disease by Impaired Fetal Nutrition: Evidence and Implications for Research and Intervention Strategies, World Health Organization, Geneva (documents WHO/NHD/02.3 and WHO/NPH/02.1).

11 Koletzko, B., Larque, E., and Demmelmair, H. (2007) Placental transfer of long-chain polyunsaturated fatty acids (LC-PUFA). *J. Perinat. Med.*, **35**, S5–S11.

12 Olsen, S.F., Halldorsson, T.I., Willett, W.C. et al. (2007) Milk consumption during pregnancy is associated with increased infant size at birth: prospective cohort study. *Am. J. Clin. Nutr.*, **86**, 1104–1110.

13 Ley, R.E., Turnbaugh, P.J., Klein, S., and Gordon, J.I. (2006) Microbial ecology: human gut microbes associated with obesity. *Nature*, **444**, 1022–1023.

14 Barker, D.J.P., Martyn, C.N., Osmond, C., Hales, C.N., and Fall, C.H.D. (1993) Growth in utero and serum cholesterol concentrations in adult life. *Br. Med. J.*, **307**, 1524–1527.

15 Aboderin, I., Kalache, A., Ben-Shlomo, Y., Lynch, J.W., Yajnik, C.S., Kuh, D., and Yach, D. (2002) Life Course Perspectives on Coronary Heart Disease, Stroke and Diabetes: Key Issues and Implications for Policy and Research, World Health Organization, Geneva (document WHO/NMH/NPH/02.1).

16 Kuh, D., and Ben-Shlomo, Y. (1997) *Preface in A Life Course Approach to Chronic Disease Epidemiology*, Oxford University Press, Oxford. Preface.

17 Cutfield, W.S., Hofman, P.L., Mitchell, M., and Morison, I.M. (2007) Could epigenetics play a role in the developmental origins of health and disease? *Pediatr. Res.*, **61**, 68–75.

18 Ailhaud, G., Massiera, F., Weill, P., Legrand, P., Alessandri, J.M., and Guesnet, P. (2006) Temporal changes in dietary fats: role of n-6 polyunsaturated fatty acids in excessive adipose tissue development and relationship to obesity. *Prog. Lipid. Res.*, **45**, 203–236.

19 Davis, M.K. (2001) Breastfeeding and chronic disease in childhood and adolescence. *Pediatr. Clin. North Am.*, **48** (1), 125–141.

20 McCarron, P., Smith, G.D., Okasha, M., and McEwen, J. (2000) Blood pressure in young adulthood and mortality from cardiovascular disease. *Lancet*, **355**, 1430–1431.

21 Dietz, W.H. (2001) The obesity epidemic in young children. *Br. Med. J.*, **322**, 313–314.

22 Darnton-Hill, I., Nishida, C., and James, W.P.T. (2004) A life course approach to diet, nutrition and the prevention of chronic diseases. *Public. Health Nutr.*, **7** (1A), 101–121.

23 Yajnik, C.S. (2002) The lifecycle effects of nutrition and body size on adult obesity, diabetes and cardiovascular disease. *Obes. Rev.*, **3**, 217–224.

24 Lithell, H.O., McKeigue, P.M., Berglund, L., Mohsen, R., Lithell, U.B., and Leon, D.A. (1996) Relation of size at birth to non-insulin dependent diabetes and insulin concentrations in men aged 50–60 years. *Br. Med. J.*, **312**, 406–410.

25 Freedman, D.S., Dietz, W.H., Srinivasan, S.R., and Berenson, G.S. (1999) The relation of overweight to cardiovascular risk factors among children and adolescents: the Bogalusa Heart Study. *Pediatrics*, **103**, 1175–1182.

26 Godfrey, K.M., and Barker, D.J.P. (2001) Fetal programming and adult health. *Public. Health Nutr.*, **4** (2B), 611–624.

27 Montgomery, S.M., and Ekbom, A. (2002) Smoking during pregnancy and diabetes mellitus in a British longitudinal birth cohort. *Br. Med. J.*, **324**, 26–27.

28 Kussmann, M., and Daniel, H. (2008) Editorial overview. *Curr. Opin. Biotechnol.*, **19**, 63–65.

29 Kussmann, M., Raymond, F., and Affolter, M. (2006) Omics-driven biomarker discovery in nutrition and health. *J. Biotechnol.*, **124**, 758–787.

30 Kussmann, M., and Fay, L.B. (2008) Nutrigenomics and personalized nutrition: science and concept. *Personalized Med.*, **5** (5), 447–455.

Part V
Case Studies

13
Viral Infections and Epigenetic Control Mechanisms

Klaus R. Huber

Abstract

Eukaryotic cells and higher organisms are challenged continuously by invading organisms. Among the many mechanisms of defense, accumulating evidence shows that the options implied in RNA silencing are utilized throughout the plant and animal kingdom to fight off virus attacks. Virus-derived siRNAs accumulate in plant and insect infected tissues, fungus strains that lacked Dicer activity were found to be highly susceptible to virus infections, and recent evidence indicates that this defense machinery controls – to a certain degree – HIV infections, and, in turn, HIV actively suppresses the expression of certain miRNA clusters of infected cells. It is no surprise that the acquisition of heritable and adaptive mechanisms of defense appears necessary to maintain the fitness of higher organisms. Even the outcomes of the adaptive immune responses are dependent on the coordinate acquisition of gene expression programs that favor one cell type over the other, providing the cells with the potential for differentiation into alternative lineages. All these processes are regulated at both the transcriptional and epigenetic level and utilized to ward off infections.

13.1
The Evolutionary Need for Control Mechanisms

Epigenetic interactions between infectious creatures and higher organisms have an enormous level of complexity. All involved parties have evolved measures and countermeasures to gain incremental advantages. As implicated by the word "evolved", these interactions have existed for hundreds of millions of years.

Various aspects of this interactive network need to be considered:

- Epigenetic regulation controls the cells' defense mechanisms after attack by a virus.
- Epigenetic regulation can be enforced upon the genetic machinery of eukaryotic cells by invading viruses.

- The fate of immune competent cells during their development is regulated by epigenetic factors.
- Previous infections have epigenetic consequences on future contagions.

Roughly 50% of our genome consists of various repetitive sequences, the majority of which are mobile elements [1]. They can be classified as class I DNA transposons and class II retrotransposons [2, 3]. Their mobility within the genome can alter gene expression (and even create new genes). In addition, the GC-rich nature of some mobile elements can introduce new GC islands through their insertion at new sites in the genome. Furthermore, over 500 000 recognizable copies of endogenous retroviruses are spread within the genome. Several of these invading elements already entered the genome before the hominoid–Old World Monkey split [4]. All these mobile elements are major components of genomic variation and a driving force of primate evolution. Whether or not retrotransposons are leftovers from – or precursors of – retroviruses is unresolved, but at least the endogenous retroviruses that invaded our genome millions of years ago are infectious agents that now interact with our gene regulation. This exemplifies the complex nature of epigenetic regulation during infections. More than just a momentary sedation of an infection must be considered, rather a continuum of individual interactions between billions of virions with millions of cells, and thousands of succeeding generations are involved. In general terms, after the emergence of self-replicating genomes during evolution, genomic parasites evolved as well. Therefore, the acquisition of heritable and adaptive mechanisms of defense appears necessary to maintain the fitness of higher organisms.

13.2
Control by RNA Silencing

RNA silencing (or interference – RNAi) is one of the mechanisms for gene regulation and a defense mechanism against viruses. The pathway for defense against both RNA and DNA viruses most probably involves microRNA (miRNA) [5] which is transcribed by RNA polymerase II from corresponding miRNA genes. MicroRNA duplexes are formed by an enzyme (that cuts RNA = RNAse: Dicer) and incorporated into a silencing complex (RISC) that is induced by invading RNA. In this complex, invading RNA is degraded or its translation is inhibited via epigenetic mechanisms. The degree of inhibition is guided by the perfect – or imperfect – sequence complementarity between miRNA and the hostile mRNA (Figure 13.1) [6].

The RNAi pathways all require the multidomain ribonuclease Dicer [7]. RNA interference is an ancient gene-silencing process inferred by the architecture of the Dicer proteins which are highly conserved, thus, all Dicer molecules evolved from a common ancestral enzyme. The finding that the Dicer from *Gardia intestinalis* can substitute for *Saccharomyces pombe* demonstrates that the earliest eukaryotic organisms were capable of RNA interference-like processes [8].

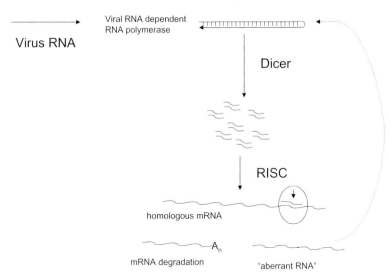

Figure 13.1 Proposed mechanism of RNA silencing [9]: Replicating RNA viruses produce dsRNA by a RNA dependent RNA polymerase. RNAse-III type enzymes (Dicer) cut dsRNA into small interfering RNAs (siRNA). The siRNAs guide endonucleolytic cleavage of homologous mRNA in association with the RNA-induced silencing complex (RISC).

Dicer processes dsRNA into small fragments called short interfering RNAs (siRNAs) or the microRNAs (miRNAs). The dsRNAs that trigger RNA interference and Dicer activity are made by the cell in the cytoplasm or the nucleus with genomic sequences by either transcription through inverted DNA repeats, simultaneous synthesis of sense and antisense RNAs, or by *viral replication* and the *activity of cellular or viral RNA-dependent RNA polymerases*. The resulting small RNAs can subsequently activate epigenetic gene silencing in the cytoplasm and at the genome level. This is achieved by posttranscriptional degradation of complementary mRNAs and (at least shown in plants) transcriptional gene silencing by methylation of homologous DNA sequences or chromatin remodeling [9].

13.3
Viral Infections and Epigenetic Control Mechanisms

13.3.1
RNA Silencing in Plants

The observation that virus-derived siRNAs accumulate in plant and insect infected tissues strongly suggests that RNA silencing has antiviral roles. Plant viruses are inducers, as well as targets, of RNA silencing-based antiviral defense. The replication intermediates or the folded viral RNAs activate RNA interference by the generation of small interfering RNAs.

In plants, the antiviral defense is augmented further by the production of dsRNA from viral templates through the action of RNA-dependent RNA polymerases [10]. In addition, signals to uninfected tissues are sent by mobile silencing molecules that infer antiviral immunity [11]. However, the whole issue is not yet resolved because it can be argued that Dicer-like activities suffice to control virus infections without the need for an RNA silencing complex. Even further complexity of this mechanism is implicated by the findings that the recovery from virus-induced symptoms is not necessarily accompanied by a commensurate reduction in viral RNA levels [12].

For counter protection, viruses obviously try to overcome this defense by producing suppressor proteins against these resistance complexes [13]. Many of the identified silencing-suppressor proteins bind long double-stranded RNA or siRNAs and thereby prevent assembly of the silencing effector complexes [14].

13.3.2
RNA Silencing in Fungi

The role of RNA silencing in fungi was tested by examining the effects of Dicer mutations on mycovirus infection of *Cryphonectria parasitica*. The mutant fungus strains that lacked Dicer activity were found to be highly susceptible to mycovirus infections, whereas the wild-type strain was not severely debilitated by the virus. These results provide direct evidence that a Dicer-like gene is involved in antiviral defense in fungi [15].

Equally, RNA interference might control virus infections in insects. Two evolutionary diverse viruses were recognized as pathogenic triggers by interaction of virus dsRNA and *Drosophila* Dicer and degraded by the RISC complex. Again, both viruses encode potent expression of suppressors of RNA interference as a counter defensive strategy [16].

13.3.3
RNA Silencing in Mammals

In mammals, the interferon (IFN) system is a central innate antiviral defense mechanism, while the involvement of RNA interference to control virus infection is uncertain. Recent evidence, however, indicates that this defense machinery controls – to a certain degree – HIV infections, and, in turn, HIV actively suppresses the expression of certain miRNA clusters of infected cells. This suppression yet again is required for efficient viral replication [17]. Thus, Dicer (and Drosha – another RNAse) play a central role in human cells in defense against HIV, and HIV has evolved mechanisms for counterattack. Also, *in vitro* experiments with cells that lack interferon alpha and beta genes show higher levels of influenza A virus when Dicer was removed from the cells by a knockdown mechanism [18]. Various promising therapeutic trials are underway utilizing synthetic small interfering RNAs in an attempt to inhibit viral replication.

13.4
Epigenetics and Adaptive Immune Responses

Another key element in antiviral defense in higher organisms is represented by the adaptive immune system. Lymphoid cells specialize during their development from their earliest stages to their later dedication as immune cell types. Various cell fate choices result in a diverse, but balanced, immune system. These are dependent on the coordinate acquisition of gene expression programs that favor one cell type over the other, providing the cells with the potential for differentiation into alternative lineages. These processes are called lineage priming, and this is regulated at both the transcriptional and epigenetic level [19]. It is controlled, among other factors, by specific changes in chromatin structure in the vicinity of lineage-specific genes by histone modifications. So far, however, the interactions of the molecular signals that generate and regulate flexible chromatin structures are not understood in detail. The nucleosome remodeling and deacetylase complex, together with associated DNA binding factors, are certainly implicated in early lymphocyte development. Fluctuating concentrations of nuclear factors with disparate activities during development are responsible for transcriptional flexibility by establishing a balanced equilibrium between activating and silencing components. In all, the hematopoeitic system will contribute to future models of epigenetic gene regulation because some of the epigenetic states of lineage priming have already been deduced and a substantial number of transcription regulators of cell fate identified [20]. In due course, the epigenetic control of this arm of flexible defense against infections will be understood in greater detail.

Many viruses stay dormant once they have infected their hosts and elicited an immune response (i.e., herpes viruses). The conferred immunity to further infections with the same or related virus strains and the resulting consequences to human health are not yet appreciated in detail. This latency can be regarded as a (short time) heritable change of the vulnerability toward viral infections – maybe not at an organismal level but certainly at a cellular level. Because this "heritable" change happens without involvement of cellular DNA, this process can be considered epigenetic.

In summary, epigenetic regulation controls the cells defense mechanisms after attack by a virus but can also be enforced upon the genetic machinery of eukaryotic cells by invading viruses. Furthermore, the fate of immune competent cells during their development is regulated by epigenetic factors and viral infections have epigenetic consequences on future contagions.

References

1 Lander, E.S. *et al.* (2001) *Nature,* **409,** 860.
2 Smit, A.F. (1996) *Curr. Opin. Genet. Dev.,* **6,** 743.
3 Deininger, P.L., and Batzer, M.A. (2002) *Genom. Res.,* **12,** 1455.
4 Han, K. *et al.* (2007) *Science,* **316,** 238.
5 Voinnet, O. (2005) *Nat. Rev. Genet.,* **6,** 206.
6 Saumet, A., and Lecellier, C.H. (2006) *Retrovirology,* **3,** 3.

7 Bernstein, E. et al. (2001) Nature, **409**, 363.
8 MacRae, I.J., Zhou, K., Li, F. et al. (2006) Science, **311**, 195.
9 Matzke, M., Matzke, A.J., and Kooter, J.M. (2001) Science, **293**, 1080.
10 Mourrain, P. et al. (2000) Cell, **101**, 533.
11 Schwach, F. et al. (2005) Plant Physiol., **138**, 1842.
12 Jovel, J., Walker, M., and Sanfacon, H. (2007) J. Virol., **81** (22), 12285.
13 Wang, X.H. et al. (2006) Science, **312**, 452.
14 Csorba, T. et al. (2007) J. Virol., **81** (21), 11768.
15 Segers, G.C. et al. (2007) Proc. Natl. Acad. Sci. U. S. A., **104** (31), 12902.
16 Wang, X.H. et al. (2006) Science, **312**, 452.
17 Triboulet, R. et al. (2007) Science, **315**, 1579.
18 Matskevich, A.A., and Moelling, K. (2007) J. Gen. Virol., **88** (Pt 10), 2627.
19 Bernstein, B.E. et al. (2006) Cell, **125**, 315.
20 Kioussis, D., and Georgopoulos, K. (2007) Science, **317**, 620.

14
Epigenetics Aspects in Gyneacology and Reproductive Medicine
Alexander Just and Johannes Huber

Abstract

Epigenetics regulates human genotypical activity throughout the entire lifecycle. Preexisting schemes in our genome's methylization and acetylization are modified through interaction with our environment and nutrition. Because the genome is not static, it is highly adaptable, allowing adjustment to the environment. Further, pregnancy, birth and early childhood determine our lives themselves. Scientific knowledge of epigenetics will influence different aspects in medicine care in the future. Our intention is to describe some special facts in this field. Epigenetics frees us of the theory of a destiny that is predefined genetically. We have much more responsibility for our own and our descendants' daily lives.

Pregnancy, birth and early childhood determine our lives. For a long time our body's destiny was assumed to be predetermined in genes. However, the idea of the genome as an innate constant, an immutable human blueprint, crumbles in the face of new molecular-biological observations.

The "Ubercode of DNA" does not cause changes in the underlying DNA sequence but in the transcription [1]. Epigenetics therefore regulates human genotypical activity throughout the entire lifecycle. Reversible methylation and acetylation processes may be the cause of many biochemical and molecular processes. Pre-existing schemes in our genome's methylization and acetylization are modified through interaction with our environment and nutrition. This may lead to destabilization of the genome and long lasting changes in cellular processes (Figure 14.1).

US-geneticist MH just compared the genome to a revolving door: "... genes are constantly arriving, others are leaving ..."

In any case, our genome is not static, it is highly adaptable allowing adjustment to the environment: sequences cut themselves out of the thread of life and jump to be inserted in other positions. Genes work and create proteins – depending on their packaging, which in turn is being perturbed by the environment (Figure 14.2).

Epigenetics and Human Health
Edited by Alexander G. Haslberger, Co-edited by Sabine Gressler
Copyright © 2010 WILEY-VCH Verlag GmbH & Co. KGaA, Weinheim
ISBN: 978-3-527-32427-9

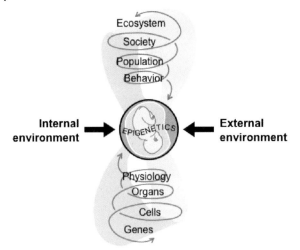

Figure 14.1 Changes in the external and internal environments, which interact to influence fetal development. Such environmental changes can occur at all levels of biological organization, from the molecular to the organism's behavior and place in society, and tend to be amplified in their consequences as they ascend through these levels, ultimately they may be epigenetic in nature. (David Crews, Endocrinology 147: S4–S10, 2006).

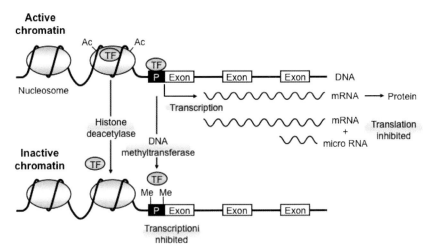

Figure 14.2 Epigenetic modification of histones or of DNA itself controls access of transcription factors (TFs) to the DNA sequence, thereby modulating the rate of transcription to messenger RNA (mRNA). (Peter D. Gluckman, N Engl J Med 2008; 359: 61–73).

Pregnancy and the first years of life are sensitive to these environmental perturbations, a period during which structures are being determined and the outcome of any alteration might not manifest itself for decades. We have to live with our past, even our time *in utero*.

Regarding infections, long lasting effects are widely accepted facts in medicine. HPV, HepC and H. pylori may lead to respective malignomes decades after contamination.

For other diseases like osteoporosis, hypertension and diabetes, such precedent has not been known for long, however, these too can sometimes be initiated decades prior to their outbreak, mostly during pregnancy or immediately after.

The responsible molecular-biological mechanism lies in the epigenetical reorganization of the nucleosomes. Pregnancy, birth and early childhood are especially receptive to these epigenetic changes. These periods are the foundation for human health and in some ways require decision making which will concern the adult decades later.

Children with low birth weight should receive special attention. Compensatory "overfeeding" after being born premature is linked to increased probabilities of developing high blood pressure, cardiovascular disease or Type 2 diabetes some decades later [2, 3]. Therefore, pediatric care should prevent excessive upfeeding of newborns and during the early years the child's weight should be kept within the lower third. Similar treatment is required for premature children who show tendencies to increased insulin resistance and hypertension during puberty. Therefore, these parameters should be included in screening examinations of these children during early adolescence [4, 5].

Birth weight presents a surrogate parameter for countless reactions *in utero* which finally lead to growth retardation. It understandably influences clinical consequences of genetic variation. In contrast to underweight children, combined with normal birth weight the vitamin D-receptor genotype has no influence on bone density later in life. Coupled with low birth weight, carrying the BB-genotype is responsible for the outbreak of osteoporosis in later stages of life. Accordingly, low birth weight should be included in the list of risk factors for osteoporosis [6, 7].

Experimental data implies even the pre-implantatory phase of the embryo to be impressionable and to lead to consequences later in life. A low protein diet during implantation may cause hypertension [8]. The periconceptional period is especially sensitive to nutritional deficiency in Vitamin B12, folic acid and methionine. The modifying effect of these substances on the epigenetic code is well known. It makes sense to pay attention to the vitamins mentioned above in mothers to be, especially after long-term use of the pill. An imbalance in maternal vitamins (Vitamin B12, folic acid and methionine) during pregnancy may cause insulin resistance in later childhood [9].

In particular, the kidney, heart and pancreas can be subject to prenatal growth disturbances that remanifest in these organs later in life. The inuterine procedure of predefining the number of nephrones is influenced by dietary imbalance (protein malnutrition) and also by an elevated maternal level of uric acid (urat).

The Renin–Angiotensin system appears to play a decisive role, effected by a diet high in salt as well as by hyperuricemia. This has an adverse effect on nephrogenesis.

A diminished antiapototic homeobox gene product Pax 2 likewise results in the reduction of the number of nephrones. Subsequent pancreatic functional determination is controlled by the balance of beta cell proliferation and apoptosis. Glucocorticoids, which during pregnancy affect the developing pancreas, constrain pancreatic transcription factors as well as the urethral homebox 1 in beta-cell precursors, inducing increased apoptosis and postpartally a lower amount of insulin-producing cells. Combined with a protein restricted diet this entails decreased expression of insulin regulation (proteins) in skeletal muscles. Proteinkinase C betaisoform, phosphoinositol -3-kinases' p85 regulating subunit and most notably GLUT4 have lower expression. This intrauterine reduction of skeletal muscle insulin regulation later contributes to an early manifestation of (islet) beta cell depletion.

Maternal nutrition also affects fetal heart development. Low caloric diet late in gestation can cause a thickened carotid intima-media layer about a decade into the child's life. Foremost, iron deficiency and its associated oxygen undersupply appears to provoke reduction in cardiomyocytes, generating hypertrophy of the remaining cells, predetermining for cardiac hypertrophy later in life.

The determination of later stages in life through intrauterine procedures seems to be caused by epigenetic mechanisms, concerning methylation of CpG islands unspecific genes promoters, chromatin structure alteration by histone modification and micro RNAs, being involved in postranscriptional control of gene expression. These gene and cell specific epigenetic modifications are driven by DNA–RNA interaction.

Thus, renal petran50 can be epigenetically modified just as well as the adrenal angiotensine 2 receptor type 1B, impacting fetal adrenal tissue apoptosis and hyperteremic glucocorticoid receptor – accordingly affecting stress response.

Epigenetic modifications are in part reversible, PPAR alpha gene hypermethylation by omega-3 fatty acids, for instance, can be undone.

The safety of *in vitro* fertilization techniques (IVF) has been frequently discussed since its introduction by Palermo *et al.* in 1992. IVF's long term effects are becoming more and more a focus of interest. Artificial extra-corporal cultivation and invasive oocyte and fetus manipulation are held responsible for fetal growth retardation, premature birth and low birth weight [10, 11]. Fetal growth retarded girls for instance, show deficient uteroplacental function during a pregnancy of their own. Observations like recurrent abortions, nidation disorder, the appearance of pathological fertilization in the course of intracytoplasmatic fertilization (intracytoplasmatic sperm injection (ICSI)) and the incidence of immature or unfertilized oocytes are feasible only since the application of IVF. Another possibility might be epigenetic changes as a cause for these observations, maybe even for the underlying sterility itself [10].

In every developmental stage, growing organisms pass through phases of increased sensitivity, being exposed and receptive to inner (DNA) and certain external mechanisms of regulation (environmental stress). Development of human diseases can – in the long term – originate from the relation between delayed and physiological progression of biological processes [10].

Current animal-experiment-based studies show neuroendocrine behavioral patterns to be impacted epigenetically. Omitted postpartal tactile stimulation effects increased hippocampal glucocorticoid receptor gene expression in the offspring. The consequence could be subsequent overshooting stress reactions. Intrauterine exposure to glucocorticoids, which can also be caused by changes in 11-hydroxysteroid-dehydrogenase, produce similar outcomes. There are two isoforms of the aforesaid enzyme: Type 1, converting cortisone into cortisol and the inverse (cortisol into cortisone). Type 2, unidirectional converting circulating cortisol into inactive cortisone. This mechanism protects the unborn against excessive glucocorticoid concentration, which can also come from the maternal organism. Stress reactions, malnutrition and exercise repress Type 2 -17 beta hydroxysteroid dehydrogenase through epigenetic mechanisms, with ceasing cortisole inactivation. This leads to subsequent hypercorticism (later in life), changing the balance between sympathetic and parasympathetic nervous systems and could be jointly responsible for excessive stress reactions. Epidemiological explorations on humans that have been conducted so far seem to confirm assumptions of intrauterine stress influencing infants' behavior.

Hyperactivity, observed in early childhood, anxiousness and posterior depressive moods appear to have a close connection to maternal hypercorticism in the last trimester as well as to (maternal) psychosocial stress.

That the outcome of epigenetic modifications be passed on, as recent results seem to confirm, is of certain significance. This phenomenon has been investigated on biocides and fungicides, serving as endocrine disruptors (hormonally active agents) and securing methylation of genes that are crucial for spermatogenesis. Thereby sperm quality declines, as measured by quantity and motility. If pregnancy can be achieved regardless, male offspring show similar defects without this person ever being in contact with biocides. This epigenetic constraint of spermatogenesis is apparently passed on to the next generation [11, 12].

On September 12th, 1809 Darwin was born. The year 2009 marks the 200th anniversary of his birth. Based on countless observations he was of the opinion that adaption to the environment leads to development of a species. The subject of genes and epigenetics was not yet known to him. The dogma that is Charles Darwin's theory, for our destiny to be determined in the book of genes, has since fallen aside.

The clarification of connections concerning genetics and the environment (epigenetics) present two fundamental additions to our knowledge of human development. On the one hand epigenetic research sheds light on the constant discussion about the importance of the environment for our physical sickness and health. On the other hand it leads us to understand our need to not only support our own health, but also to improve our children's health.

Epigenetics frees us of the theory of a predefined destiny but in turn burdens us with much more responsibility, for our own as well as our descendants daily lives.

References

1 Egger, G. et al. (2004) *Nature*, **429**, 457–463.
2 Barker, D.J.P., Osmond, C., Forsén, T.J., Kajantie, E., and Eriksson, J.G. (2005) Trajectories of growth among children who have coronary events as adults. *N. Engl. J. Med.*, **353**, 1802–1809.
3 Barker, D.J.P. (1998) In utero programming of chronic disease. *Clin. Sci.*, **95**, 115–128.
4 Hofman, P.L., Regan, F., and Jackson, W.E. (2004) Premature birth and later insulin resistance. *N. Engl. J. Med.*, **351**, 2179–2186.
5 Hovi, P., Andersson, S., and Eriksson, J.G. (2007) Glucose regulation in young adults with very low birth weight. *N. Engl. J. Med.*, **356**, 2053–2063.
6 Javaid, M.K., and Cooper, C. (2002) Prenatal and childhood influences on osteoporosis. *Best. Pract. Res. Clin. Endocrinol. Metab.*, **16**, 349–367.
7 Dennison, E.M., Arden, N.K., and Keen, R.W. (2001) Birthweight, vitamin D receptor genotype and the programming of osteoporosis. *Paediatr. Perinat. Epidemiol.*, **15**, 211–219.
8 Kwong, W.Y., Wild, A.E., Roberts, P., Willis, A.C., and Fleming, T.P. (2000) Maternal undernutrition during the preimplantation period of rat development causes blastocyst abnormalities and programming of postnatal hypertension. *Development*, **127**, 4195–4202.
9 Yajnik, C.S., Deshpande, S.S., and Jackson, A.A. (2008) Vitamin B12 and folat concentration during pregnancy and insulin resistence in the offspring: the Pune Maternal Nutrition Study. *Diabetologia*, **51**, 29–38.
10 Niemitz, E.L. (2004) *Am. J. Hum. Genet.*, **74**, 599–609.
11 Maher, E.R., Afnan, M., and Barratt, C.L. (2003) Epigenetic risks related to assisted reproductive technologies: epigenetics, imprinting, ART and icebergs? *Human Reprod*, **18** (12), 2508–2511.
12 Rosenwaks, Z. (2007) *PNAS*, **104**, 5709–5710.

15
Epigenetics and Tumorigenesis

Heidrun Karlic and Franz Varga

Abstract

The last years have seen conceptual and technical advances uncovering epigenetic pathways which present promising targets for diagnostic and therapeutic strategies in an increasing number of solid tumors and leukemias. Epigenetic modifications which promote the malignant potential of stem cells or respective progenitor cells are gaining an increased importance in malignancies of the elderly, when effects of environmental exposures, lifestyles and metabolism have left their traces in the genome and the associated protein components of the chromatin. This explains that besides "classic" tumor suppressor genes (e.g., of the P53 pathway), expression patterns of hormone receptors such as estrogen receptor and growth factor receptors, vitamin response, DNA-repair and apoptotic pathways (death associated protein kinase) are altered by epigenetic mechanisms in tumor cells. An insight into these complex mechanisms presents the basis for improved models which outline the chances and risks of therapeutic interventions combining chromatin-targeting drugs.

15.1
Introduction

Despite major advances in understanding key molecular lesions in cellular control pathways that contribute to cancer, it remains true that microscopic examination of nuclear structure is a gold standard in cancer diagnosis. Changes in the nuclear architecture, which largely involve the state of chromatin configuration, have the potential to confirm the cancer phenotype in a single cell. The most important cues are the size of the nucleus, nuclear outline, a condensed nuclear membrane, prominent nucleoli, dense "hyperchromatic" chromatin and a high nuclear/cytoplasmic ratio. Such structural features, visible under a microscope, likely correlate with profound alterations in chromatin function and resultant changes in gene expression states and/or chromosomal stability. Linking changes observable at a microscopic level with the molecular marks discussed throughout this book remains one of the great challenges in cancer research.

In this chapter, we review epigenetic marks, typified by changes in DNA cytosine methylation at CpG dinucleotides and histone modifications, which are abnormally distributed in cancer cells. They are increasingly being linked to events such as aging that affect the stablity and function of stem-cell genomes and thus contribute significantly to the cancer phenotype (Figure 15.1). Such epigenetic modifications can have significant impact on chromatin structure and transcriptional activity and, in contrast to genetic aberrations, are reversible phenomena [1].

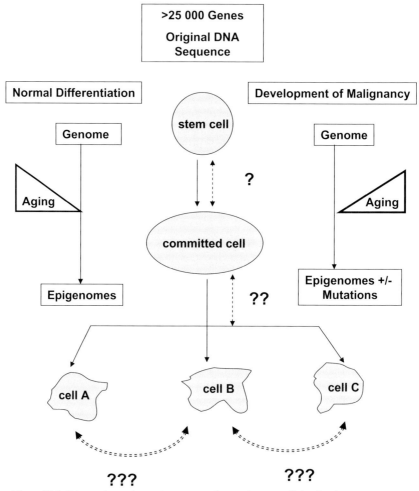

Figure 15.1 Epigenetics and genetics in normal or malignant cellular development.

15.2
Role of Metabolism Within the Epigenetic Network

The extent to which epigenetic change can also be acquired in response to external stimuli from environment, lifestyle and metabolism represents an exciting dimension in the "nature versus nurture" debate [2].

Metabolic enzymes supply acetyl groups from acetyl-CoA for histone acetylation from carbohydrate and fat metabolism [3] and methyl groups from dietary methyl donors [4]. Acetylation at lysine residues is a widespread posttranslational histone modification with examples known for histones H2A, H2B, H3, and H4. The levels of histone acetylation play a crucial role in chromatin remodeling and in the regulation of gene transcription as well as cell cycle progression and DNA repair [5–8]. The presence of acetylated lysine in histone tails is associated with a more relaxed chromatin state and gene-transcription activation, while the deacetylation of lysine residues is associated with a more condensed chromatin state and transcriptional gene silencing [9]. Various histone acetyl transferases (HATs) have been identified (for review see e.g., [10]). As typical for protein acetyltransferases, their acetylation donor is acetyl-CoA, a central molecule of both carbohydrate and fatty acid metabolism (Figure 15.2).

Acetyl-CoA–synthetases producing nucleocytosolic acetyl-CoA directly regulate global histone acetylation [3]. In contrast to histone acetylation, which is closely related to energy metabolism, methylation of histones and DNA is associated with amino acid metabolism: A pathway which is key to many of these reactions is the metabolic cycling of methionine. Briefly, methionine is converted to the methyl cofactor S-adenosylmethionine (SAM or AdoMet). Subsequent to methyl donation, the product S-adenosylhomocysteine (SAH) becomes homocysteine (Hcy), which is then either catabolized or remethylated to methionine.

The palindromic CpG dinucleotide, which is found in clusters in the regulatory region of genes, often serves as substrate for DNA methyltransferases (DNMTs) targeting the 5-carbon position of the cytosine residues. The added methyl group can interfere with transcription factor binding, thereby regulating transcription [11]. It can also designate a possible attachment site for methyl-CpG-binding proteins, which in turn effect further regulation by their association with the histone deacetylase containing chromatin remodeling complexes. DNMT1, 3a and 3b are the most thoroughly studied DNMTs, and the activity of these enzymes is often described as being either maintenance or *de novo* methylation [12]. The former process serves to maintain pre-existing epigenetic control status in dividing cells by methylating hemimethylated sites on newly synthesized strands, whereas the latter methylates sites in which both strands are unmethylated, for example, during early embryonic development [13].

It appears possible that the reduced expression of metabolism-associated genes in aged individuals is based on epigenetic mechanisms. In age-associated diseases such as myelodysplastic syndrome (MDS) epigenetic changes affect on the one hand genes which play a role in cell proliferation and differentiation and on the other hand important tumor suppressor genes, as recently reviewed by our group

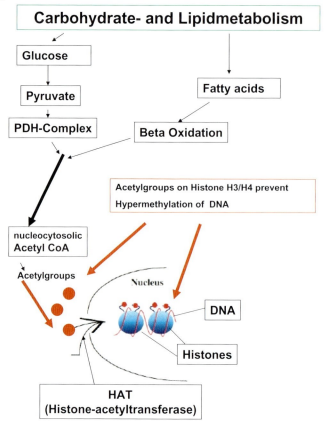

Figure 15.2 Epigenetic impact of energy metabolism.

[14]. Metabolic changes during the aging process that are associated with an increased risk of malignancy include both carbohydrate and fatty acid metabolism. The role of insulin and insulin-like growth factor (IGF-1) signaling in aging is one of the most extensively studied pathways [15]. Although IGFs might not be considered classical hematopoietic growth factors, some reports have shown that IGFs play a crucial role in hematopoiesis regulating proliferation and differentiation via the IGF-1 receptor (IGF-1R) [16]. Within MDS cases, IGF-1R expression was higher in advanced than in less advanced subgroups and correlated with blast counts. IGF-1R overexpression may predict malignant proliferation in hematopoietic cells, such as the transformation of MDS to AML [17]. The above-mentioned dysregulation of insulin-signaling also affects key enzymes of oxidative metabolism which are essential for energy production from fat. Changes of this enzyme network at the mitochondrial level are known to be associated with the aging process, apoptosis, and many diseases. A comparative study quantifying expression of these enzymes in different age groups showed expected age-dependent effects. In addition, a MDS-specific reduction of microsomal carnitine palmitoyl-

transferase is caused by promoter methylation [18, 19]. The reduction in transcription of different genes in blood cells, which is well known in different tissues, may reflect a systemic signaling process, associated with aging, apoptosis, and MDS [18]. Furthermore, data exist suggesting that changes in relative mRNA levels of these enzymes could represent the hematopoietic regenerative potential including evidence for a possible predictive value of such analyses [20].

15.3
Epigenetic Modification by DNA Methylation During Lifetime

The great fidelity with which DNA methylation patterns in mammals are inherited after each cell division is ensured by the DNMTs. However, the aging cell undergoes a DNA methylation drift: Early studies showed that global DNA methylation decreases during aging in many tissue types and it was subsequently observed that mammalian fibroblasts cultured to senescence increasingly lost DNA methylation [21]. The decrease in global DNA methylation during aging is probably mainly the result of the passive demethylation of heterochromatic DNA as a consequence of a progressive loss of DNMT1 efficacy or erroneous targeting of the enzyme by other cofactors (or both) [22]. However, this needs to be confirmed. An increased expression of the *de novo* DNA methylase DNMT3b, which acts in a rather targeted manner, could be a natural response of the cell to loss of DNA methylation in repeated DNA sequences as well. A logical outcome of DNMT3b overexpression could be that specific regions such as promoter CpG islands, which are commonly unmethylated in normal cells, become aberrantly hypermethylated, as previously reported for the genes coding for the human mutL homolog 1 (MLH1), whose mutated form defines a low penetrance risk for colorectal cancer, and for the cyclin-dependent kinase inhibitor CDKN2A (p14ARF, p16) that is known as an important tumor suppressor gene (reviewed in Ref. [23]).

Several specific regions of the genomic DNA become hypermethylated during aging [24]. For instance, there is an increase in methylcytosines in hepatic Gck promoter in livers of senescent rats which may represent an important marker for diabetogenic potential during the aging process [25]. Methylation of promoter CpG islands in nontumorigenic tissues has been reported for several genes, including estrogen receptor (ER), myogenic differentiation antigen 1 (MYOD1), insulin-like growth factor II (IGF2) and tumor suppressor candidate 33 (N33). In some cases, such as MLH1 and CDKN2A, which are frequently inactivated in colon cancer, hypermethylation was also common in normal aged tissues (reviewed in Ref. [26]). Another study found increasing promoter hypermethylation of the tumor suppressor genes lysyl oxidase (LOX), CDKN2A, runt-related transcription factor 3 (RUNX3), and TPA (tumor promoting agent)-inducible gene 1 (TIG1) in non-neoplastic gastric mucosa that was significantly correlated with aging [27]). Other examples of genes with increased promoter methylation during aging include genes associated with the structural integrity of cells and their transcriptional regulators such as collagen $\alpha1(I)$ [28], E-cadherin [29] and fos [30].

Thus, the accumulation of epigenetic alterations during aging might contribute to tumorigenic transformation. Although it is possible to associate the accumulation of methylation at the promoters of these tumor suppressor genes during aging with the predisposition to develop cancer, there is no experimental or mechanistic evidence of a direct relationship between these genes and the aging process. The regulation of the CDKN2A locus during aging and tumorgenesis deserves special attention. On the one hand, the promoter region of CDKN2A gains an increased number of methylated CpGs in normal gastric epithelia during aging [28]. The increased hypermethylation within this promoter suggests that CDKN2A (=p16INK4a) expression is reduced. On the other hand, the expression of CDKN2A is known to increase with aging in mammals in a tissue-specific manner (reviewed in Ref. [31]) and, what is more striking, it has been proposed that its upregulation is directly involved in the decrease of self-renewal potential of some mature stem cells [32].

15.4
Interaction of Genetic and Epigenetic Mechanisms in Cancer

Cancer is caused by (a mitotically heritable) deregulation of genes, which control whether cells divide, die, and move from one part of the body to another. During the process of carcinogenesis, genes can become activated in ways that enhance division or prevent cell death, or alternatively, they can become inactivated so that they no longer apply the brakes to these processes. The first type of genes is called "oncogenes" and the second "tumor suppressor" genes. It is the interaction between these two gene classes that results in the formation of cancer. Genes can be inactivated by at least three pathways, including (i) a mutation inducing a disabled function of the gene, (ii) a deletion so that the gene becomes lost and thus not available to work appropriately, (iii) an epigenetic change. This epigenetic silencing can involve histone modifications, and inappropriate methylation of cytosine residues in CpG sequence motifs that reside within control regions which govern gene expression. As outlined in this book, the basic mechanisms responsible for maintaining the silenced state are quite well understood. Consequently, we also know that epigenetics has profound implications for cancer prevention, detection and therapy. A series of approved drugs can reverse epigenetic changes and restore normal gene activity in cancer cells. In addition, because the changes in DNA methylation can be analyzed with a high degree of sensitivity, many strategies to detect cancer early rely on finding DNA methylation changes. Detection of associated changes in the expression profile of (de)methylating enzymes and the underlying metabolic pathways are gaining importance, both as diagnostic tools and as targets for therapy and prevention. These data, particularly DNA- and chromatin-methylation patterns that are fundamentally altered in cancers, have led to new opportunities for the understanding, detection and prevention of cancer [33].

15.5
DNA Methylation in Normal and Cancer Cells

DNA methylation, catalyzed by DMNTs, involves the addition of a methyl group to the carbon-5 position of the cytosine ring in the CpG dinucleotide and results in the formation of methylcytosine [34, 35]. Although CpG dinucleotides are under-represented in the mammalian genome, CpG rich regions (CpG islands) are found within the promoter regions of approximately 40–50% of human genes [36]. Cytosines within CpG islands, especially those associated with promoter regions, are less methylated in normal cells. This lack of methylation in promoter-associated CpG islands allows gene transcription to occur, providing that the appropriate transcription factors are present and the chromatin structure is open [37]. Abnormal DNA methylation patterns [38] have been recognized in cancer cells for over two decades [39]. Paradoxically, cancer cells are associated with global hypomethylation but with regional hypermethylation of CpG islands at gene promoters [39, 40]. Aberrant genome-wide hypomethylation may relate to tumorigenesis by promoting genomic instability. Methylation of promoter CpG islands is associated with a closed chromatin structure and transcriptional silencing of the associated gene [37, 38, 41]. Knudson's "two-hit" model for cancer proposed that a dominantly inherited predisposition to cancer entails a germline mutation, while tumorigenesis requires a second, somatic mutation. Non-hereditary cancer of the same type requires the same two hits but both are somatic [42, 43].

The frequency of this process, the variety of genes involved, and the large repertoire of cancers shown to harbor dense methylated promoter CpG islands all reflect the critical role of epigenetic mechanisms in cancer initiation and progression. While certain genes, such as CDKN2A, have been shown to be hypermethylated in many tumor types, in general, the pattern of genes hypermethylated in cancer cells is tissue specific [44]. Many fundamental cellular pathways are inactivated in human cancer by this type of epigenetic lesion: DNA repair (MLH1; O-6-methylguanine-DNA methyltransferase, MGMT; breast cancer 1, early onset, BRCA1), cell cycle (CDKN2A, p16INK4a, p14Arf; CDKN2B, p15INK4b), cell invasion and adherence (E-cadherin, CDH1; adenomatous polyposis coli APC; H-cadherin, CDH13, von Hippel Lindau tumor suppressor, VHL), apoptosis (death-associated protein kinase 1, DAPK1; caspase 8, CASP8; FAS; tumor necrosis factor receptor superfamily, member 10a, TNFRSF10A, TRAIL-R1) detoxification (glutathione S-transferase pi 1, GSTP1and hormonal response (retinoic acid receptor, beta, RARB; estradiol receptor α, ESR1) [44–48]. The deregulation of such pathways is likely to confer a survival advantage to the affected cell and thus contribute to the step-wise progression of a normal cell to a cancer cell. Altered DNA methylation patterns in human cancer are not only of importance to our understanding of the molecular pathogenesis of this disease but may also may serve as markers for cancer diagnosis and prognosis, and prediction of response to therapy.

15.6
Promoter Hypermethylation in Hematopoietic Malignancies

DNA hypermethylation is a common mechanism of gene inactivation in hematopoietic malignancies and gains increasing importance as a therapeutic target, especially when combined with other targeted drugs such as monoclonal antibodies or small-molecule inhibitors [49]. The spectrum of genes inactivated by hypermethylation in hematopoietic malignancies differs from solid tumors, although many cancer-related pathways are known to be deregulated in leukemia/lymphoma as a result of DNA hypermethylation [44]. Table 15.1 shows a summary of those genes commonly hypermethylated in hematopoietic malignancies. In general, the pattern of promoter methylation found in hematopoietic malignancies can be considered to be an aberrant and specific phenomenon, with disease-specific methylation patterns of key CpG islands found for particular genes. Most investigators have found that normal hematopoietic progenitors are free of such patterns of gene promoter methylation [50, 51]. However, as discussed previously, not all promoter methylation is abnormal or disease related. Dynamic changes in promoter methylation have been shown to play a role in the control of gene expres-

Table 15.1 Genes frequently methylated in hematopoietic malignancies.

	Acute myeloid leukemias	Myelodysplastic syndromes	Acute lymphoid leukemia	Lymphoma	Multiple myeloma
DNA repair	–	–	–	–	MGMT
Hormone response	Estrogen receptor	Estrogen receptor	–	–	–
Vitamin response	RARB2	–	–	RARB2	–
Cell cycle	p15	p15	p15, p16	p16	p15, p16
P53 network	–	HIC-1	p73	p73	–
Cell adherence and invasion	E-cadherin	E-cadherin, calcitonin	E-cadherin	–	–
Apoptosis	DAPK1 (sec AML)	DAPK1 (contradictory results)	DAPK1	DAPK1	DAPK1
Tyrosine kinase cascades	SOCS-1	–	–	–	SOCS-1
Other pathways		IGF1, ABL (CML)	GSTP1		

sion levels for growth factor receptors, growth factors and cytokines during normal myeloid development [52, 53].

15.7
Hypermethylated Gene Promoters in Solid Cancers

Genes commonly hypermethylated in solid tumors are summarized in Table 15.2. To understand the significance of genes for tumorigenesis and the challenges for the future in this field, the cancer-related genes which are affected by transcrip-

Table 15.2 Genes frequently methylated in solid tumors.

	Gyn. (breast, ovary, uterus)	Prostate	Gastrointest (esophagus, stomach, pancreas, liver, colon)	Lung, head-neck	Kidney, bladder	Brain
DNA repair	BRCA1		MGMT	MGMT		MGMT
Hormone response	ER Pgr (progest rec, breast)	AR	AR ER	ER		
Vitamin response						
Ras signaling		RASSF1A		RASSF1A		
Cell cycle	p16 p15		p16 p15	p16 p15	p16	p16
P53 network						
Cell adherence and invasion	hMLH1 TIMP-3 E-cadherin		hMLH1 TIMP-3 calcitonin	TIMP-3 E-cadherin	TIMP-3	TIMP-3
Wnt signaling		APC SOX7	APC IGFBP-3			
Apoptosis	DAPK		DAPK	DAPK		
Tyrosine kinase cascades	–	TGF2R	–	–	–	–
Other pathways	GSTP1 LOX	GSTP1 PTGS2 MDR1	GSTP1 IGF2 (colon) LOX (colon)	LOX		

tional inactivation may be divided into three groups: The first comprises those which were instrumental in defining promoter hypermethylation and gene silencing as an important mechanism for loss of tumor suppressor gene function in cancer. These were already recognized as classic tumor suppressor genes which, when mutated in the germline of families, cause inherited forms of cancer. They are often mutated in sporadic forms of cancers but can frequently be hypermethylated on one or both alleles in such tumors [33, 37]. In addition, for these genes, promoter hypermethylation can sometimes constitute the "second hit" in Knudson's hypothesis by being associated with loss of function of the second copy of the gene in familial tumors where the first hit is a germline mutation [54, 55]. In some instances, 5-azacytidine-induced reactivation of these genes in cultured tumor cells has been shown to restore the key tumor suppressor gene function which is lost during tumor progression [56]. The second group of epigenetically silenced genes are those previously identified as candidate tumor suppressor genes by virtue of their function, but they have not been found to have an appreciable frequency of mutational inactivation. Examples include the putative tumor suppressor gene Ras association domain family 1 (RASSF1A) and fragile histidine triad gene (FHIT, encodes a diadenosine 5′,5′′′-P1,P3-triphosphate hydrolase) on chromosome 3p in lung and other types of tumors [57, 58]. Others are those known to encode proteins which subserve functions critical to prevention of tumor progression, such as the pro-apoptotic gene death-associated protein kinase 1 (DAPK1) [59]. These genes present an important challenge for the field of cancer in that, despite their having been identified as having frequent promoter hypermethylation in tumors, it must be proven – since many of the genes are not frequently, or not at all, mutated – how the genes actually contribute to tumorigenesis.

The third group of genes is being identified through strategies applied to randomly detect aberrantly silenced genes with promoter hypermethylation [60, 61]. As compared to those genes in the second group, it is a challenge to place these genes into a functional context for cancer progression because their functions may be totally unknown.

15.8
Interaction DNA Methylation and Chromatin

Hypomethylated genes in cancer are known to have key histone modifications associated with their promoter regions [62]. DNA methyltransferases (DNMT), which catalyzes DNA methylation, also bind histone deacetylases (HDACs) and have the potential to target these enzymes to regions of gene silencing [63, 64]. Removal of acetyl groups from histone lysine tails (deacetylation) by HDACs is one of several modifications made to these proteins associated with transcriptional silencing [62]. Protein complexes of methyl-CpG-binding proteins, transcriptional corepressors, chromatin-remodelling proteins and HDACs bind to hypermethylated DNA regions, resulting in a transcriptionally repressive chromatin state in a heritable manner [64]. Histone deacetylation increases ionic interactions between the positively charged histones and negatively charged DNA, which yields a more compact chromatin structure and represses gene transcription by limiting the accessibility of the methyl transferases. For example, in DNMT1 knockout cancer cells there is an increase in the amount of acetylated forms of histone H3 and a decrease in that of the methylated forms of histone H3. These changes are associated with the loss of interaction of HDACs and the heterochromatin protein HP1 with histone H3. These data strongly indicate that histone hyperacetylation is not always the result of a loss of HDAC activity, but that it could be due to a loss of HDAC targeted to specific DNA sequences. One possible explanation is that changes in DNA methylation also cause histone modification due to direct interactions between the enzymes regulating different epigenetic modifications [65]. HDAC2 is also involved in the regulation of neuronal differentiation through a direct interaction with the N-terminal domain of DNMT3b. Treatment of the pheochromocytoma cell line PC12 with HDAC inhibitors prevents the nerve growth factor-induced differentiation of this cell line, while the overexpression of the N-terminal domain of DNMT3b facilitates differentiation [66]. Various mechanisms are involved in histone hypoacetylation. These changes can be explained by a decrease in HAT activity due to the mutations or chromosomal translocations characteristic of leukemias, or to changes that result in the increased activity of HDACs.

Focusing on the role of HDACs in cancer, the available data indicate that there is more than one mechanism by which HDACs function in cancer development. HDACs also participate in gene expression regulation mediated by nuclear receptors. Estrogen receptors (ERs) belong to a large superfamily of nuclear receptors that modulate the expression of genes regulating critical breast and ovary functions. HDAC1 interacts with ER-a and suppresses its transcriptional activity. The interaction of HDAC1 with ER-a is mediated by the activation function-2 (AF-2) domain and DNA-binding domain of ER-a, and this interaction is weakened in the presence of estrogens [67]. Furthermore, another study indicates that the ER-gene transcription is regulated by a multiprotein complex that includes HDACs, DNMTs, and retinoblastoma protein Rb [68]. There are a number of studies showing altered expression of individual HDACs in tumor samples. For example, there is an increase in HDAC1 expression in gastric [69], prostate [70], colon [71] and breast [72] carcinomas. Overexpression of HDAC2 has been found in cervical [73] and gastric [74] cancers, and in colorectal carcinoma with loss of APC expression [75]. Other studies have reported high levels of HDAC3 and HDAC6 expression in colon and breast cancer specimens, respectively [71, 76]. Several examples of the roles of chromatin-modifying activities are also known in hematologic malignancies [1]. For example, acute myeloid leukemia (AML), chronic myeloid leukemia (CML) and acute promyelocytic leukemia (PML) are caused by chromosomal translocations altering the use of HDACs and consequently DNA methylation. In PML, the PML gene is fused to the retinoic acid receptor (RAR). This receptor recruits HDAC activity and DNA methylation, and causes a state of transcriptional silencing, as shown with experimental promoter constructs. It has been suggested that this targeting of chromatin change can potentially lead to tumor suppressor gene silencing, which participates in a cellular differentiation block [77]. In AML, The DNA-binding domain of the transcription factor AML-1 is fused to a protein called ETO, which interacts with an HDAC. Repression of cellular differentiation by the mistargeted HDAC contributes to aberrant gene expression and, ultimately leukemia [78]. These are just two examples of the direct involvement of chromatin modifications in the oncogenic phenotype. It has, however, become clear that chromatin modifications can directly and indirectly alter the patterns of cytosine methylation, an epigenetic change of the DNA which can either "initiate" or "lock in" silencing of key genes leading to heritable perturbations in key cellular pathways.

Acknowledgment

Financial support for this study was given by the Ludwig Boltzmann Society (Cluster Oncology) and the Jubiläumsfonds der Österreichischen Nationalbank (Project No. 13068).

References

1 Wolffe, A.P. (2001) Chromatin remodeling: why it is important in cancer. Oncogene, 20, 2988–2990.
2 Ezzat, S. (2008) Chromatin remodeling: the interface between extrinsic cues and the genetic code? Clin. Invest. Med., 31, E272–E281.
3 Takahashi, H., McCaffery, J.M., Irizarry, R.A., and Boeke, J.D. (2006) Nucleocytosolic acetyl-coenzyme a synthetase is required for histone acetylation and global transcription. Mol. Cell, 23, 207–217.
4 Ulrey, C.L., Liu, L., Andrews, L.G., and Tollefsbol, T.O. (2005) The impact of metabolism on DNA methylation. Hum. Mol. Genet., 14 (1), R139–R147.
5 Bird, A.W., Yu, D.Y., Pray-Grant, M.G., Qiu, Q., Harmon, K.E., Megee, P.C. et al. (2002) Acetylation of histone H4 by Esa1 is required for DNA double-strand break repair. Nature, 419, 411–415.
6 Krebs, J.E., Fry, C.J., Samuels, M.L., and Peterson, C.L. (2000) Global role for chromatin remodeling enzymes in mitotic gene expression. Cell, 102, 587–598.
7 Masumoto, H., Hawke, D., Kobayashi, R., and Verreault, A. (2005) A role for cell-cycle-regulated histone H3 lysine 56 acetylation in the DNA damage response. Nature, 436, 294–298.
8 Suka, N., Suka, Y., Carmen, A.A., Wu, J., and Grunstein, M. (2001) Highly specific antibodies determine histone acetylation site usage in yeast heterochromatin and euchromatin. Mol. Cell, 8, 473–479.
9 Iizuka, M., and Smith, M.M. (2003) Functional consequences of histone modifications. Curr. Opin. Genet. Dev., 13, 154–160.
10 Roth, S.Y., Denu, J.M., and Allis, C.D. (2001) Histone acetyltransferases. Annu. Rev. Biochem., 70, 81–120.
11 James, S.J., Melnyk, S., Pogribna, M., Pogribny, I.P., and Caudill, M.A. (2002) Elevation in S-adenosylhomocysteine and DNA hypomethylation: potential epigenetic mechanism for homocysteine-related pathology. J. Nutr., 132, S2361–S2366.
12 Okano, M., Bell, D.W., Haber, D.A., and Li, E. (1999) DNA methyltransferases Dnmt3a and Dnmt3b are essential for de novo methylation and mammalian development. Cell, 99, 247–257.

13 Li, E., Bestor, T.H., and Jaenisch, R. (1992) Targeted mutation of the DNA methyltransferase gene results in embryonic lethality. *Cell*, **69**, 915–926.

14 Pfeilstocker, M., Karlic, H., Nosslinger, T., Sperr, W., Stauder, R., Krieger, O. et al. (2007) Myelodysplastic syndromes, aging, and age: correlations, common mechanisms, and clinical implications. *Leuk. Lymphoma*, **48**, 1900–1909.

15 Holzenberger, M., Kappeler, L., and De Magalhaes Filho, C. (2004) IGF-1 signaling and aging. *Exp. Gerontol.*, **39**, 1761–1764.

16 Shimon, I., and Shpilberg, O. (1995) The insulin-like growth factor system in regulation of normal and malignant hematopoiesis. *Leuk. Res.*, **19**, 233–240.

17 Qi, H., Xiao, L., Lingyun, W., Ying, T., Yi-Zhi, L., Shao-Xu, Y. et al. (2006) Expression of type 1 insulin-like growth factor receptor in marrow nucleated cells in malignant hematological disorders: correlation with apoptosis. *Ann. Hematol.*, **85**, 95–101.

18 Karlic, H., Lohninger, A., Laschan, C., Lapin, A., Bohmer, F., Huemer, M. et al. (2003) Downregulation of carnitine acyltransferases and organic cation transporter OCTN2 in mononuclear cells in healthy elderly and patients with myelodysplastic syndromes. *J. Mol. Med.*, **81**, 435–442.

19 Pfeilstocker, M. (2006) Hypermethylation and reduced carnitine availability – a possible mechanism for downregulation of microsomal Carnitine Palmitoyltransferase (mCPT) in Myelodysplastic Syndromes (MDS). *Blood*, **85**, 386–393.

20 Fillitz, M., Karlic, H., Tuchler, H., Zeibig, J., Spiegel, W., Wihlidal, P. et al. (2006) Does mRNA level of microsomal carnitine palmitoyltransferase predict yield of peripheral blood stem cell apheresis? *Ann. Hematol.*, **85**, 386–393.

21 Wilson, V.L., and Jones, P.A. (1983) DNA methylation decreases in aging but not in immortal cells. *Science*, **220**, 1055–1057.

22 Casillas, M.A. Jr., Lopatina, N., Andrews, L.G., and Tollefsbol, T.O. (2003) Transcriptional control of the DNA methyltransferases is altered in aging and neoplastically-transformed human fibroblasts. *Mol. Cell. Biochem.*, **252**, 33–43.

23 Issa, J.P. (2003) Age-related epigenetic changes and the immune system. *Clin. Immunol.*, **109**, 103–108.

24 Kim, S.K., Jang, H.R., Kim, J.H., Kim, M., Noh, S.M., Song, K.S. et al. (2008) CpG methylation in exon 1 of transcription factor 4 increases with age in normal gastric mucosa and is associated with gene silencing in intestinal-type gastric cancers. *Carcinogenesis*, **29**, 1623–1631.

25 Jiang, M.H., Fei, J., Lan, M.S., Lu, Z.P., Liu, M., Fan, W.W. et al. (2008) Hypermethylation of hepatic Gck promoter in ageing rats contributes to diabetogenic potential. *Diabetologia*, **51**, 1525–1533.

26 Shi, H., Wang, M.X., and Caldwell, C.W. (2007) CpG islands: their potential as biomarkers for cancer. *Expert. Rev. Mol. Diagn.*, **7**, 519–531.

27 So, K., Tamura, G., Honda, T., Homma, N., Waki, T., Togawa, N. et al. (2006) Multiple tumor suppressor genes are increasingly methylated with age in non-neoplastic gastric epithelia. *Cancer Sci.*, **97**, 1155–1158.

28 Takatsu, M., Uyeno, S., Komura, J., Watanabe, M., and Ono, T. (1999) Age-dependent alterations in mRNA level and promoter methylation of collagen alpha1(I) gene in human periodontal ligament. *Mech. Ageing Dev.*, **110**, 37–48.

29 Bornman, D.M., Mathew, S., Alsruhe, J., Herman, J.G., and Gabrielson, E. (2001) Methylation of the E-cadherin gene in bladder neoplasia and in normal urothelial epithelium from elderly individuals. *Am. J. Pathol.*, **159**, 831–835.

30 Choi, E.K., Uyeno, S., Nishida, N., Okumoto, T., Fujimura, S., Aoki, Y. et al. (1996) Alterations of c-fos gene methylation in the processes of aging and tumorigenesis in human liver. *Mutat. Res.*, **354**, 123–128.

31 Kim, W.Y., and Sharpless, N.E. (2006) The regulation of INK4/ARF in cancer and aging. *Cell*, **127**, 265–275.

32 Janzen, V., Forkert, R., Fleming, H.E., Saito, Y., Waring, M.T., Dombkowski, D.M. et al. (2006) Stem-cell ageing modified by the cyclin-dependent kinase

inhibitor p16INK4a. *Nature*, **443**, 421–426.
33 Esteller, M. (2007) Epigenetic gene silencing in cancer: the DNA hypermethylome. *Hum. Mol. Genet.*, **16** (1), R50–R59.
34 Holliday, R., and Grigg, G.W. (1993) DNA methylation and mutation. *Mutat. Res.*, **285**, 61–67.
35 Bird, A. (2002) DNA methylation patterns and epigenetic memory. *Genes Dev.*, **16**, 6–21.
36 Gardiner-Garden, M., and Frommer, M. (1987) CpG islands in vertebrate genomes. *J. Mol. Biol.*, **196**, 261–282.
37 Jones, P.A., and Baylin, S.B. (2002) The fundamental role of epigenetic events in cancer. *Nat. Rev. Genet.*, **3**, 415–428.
38 Baylin, S.B., and Jones, P.A. (2007) Epigenetic determinants of cancer, in *Epigenetics* (eds C.D. Allis, T. Jenuwein, D. Reinberg, and M.-L. Caparros), Cold Spring Harbor Laboratory Press, New York, pp. 457–477.
39 Feinberg, A.P., and Vogelstein, B. (1983) Hypomethylation distinguishes genes of some human cancers from their normal counterparts. *Nature*, **301**, 89–92.
40 Esteller, M. (2003) Cancer epigenetics: DNA methylation and chromatin alterations in human cancer. *Adv. Exp. Med. Biol.*, **532**, 39–49.
41 Herman, J.G., and Baylin, S.B. (2003) Gene silencing in cancer in association with promoter hypermethylation. *N. Engl. J. Med.*, **349**, 2042–2054.
42 Knudson, A.G. (1996) Hereditary cancer: two hits revisited. *J. Cancer Res. Clin. Oncol.*, **122**, 135–140.
43 Knudson, A.G. (2001) Two genetic hits (more or less) to cancer. *Nat. Rev. Cancer*, **1**, 157–162.
44 Galm, O., Herman, J.G., and Baylin, S.B. (2006) The fundamental role of epigenetics in hematopoietic malignancies. *Blood Rev.*, **20**, 1–13.
45 Esteller, M., and Herman, J.G. (2002) Cancer as an epigenetic disease: DNA methylation and chromatin alterations in human tumours. *J. Pathol.*, **196**, 1–7.
46 Hopkins-Donaldson, S., Ziegler, A., Kurtz, S., Bigosch, C., Kandioler, D., Ludwig, C. et al. (2003) Silencing of death receptor and caspase-8 expression in small cell lung carcinoma cell lines and tumors by DNA methylation. *Cell Death Differ.*, **10**, 356–364.
47 Hellwinkel, O.J., Kedia, M., Isbarn, H., Budaus, L., and Friedrich, M.G. (2008) Methylation of the TPEF- and PAX6-promoters is increased in early bladder cancer and in normal mucosa adjacent to pTa tumours. *BJU Int.*, **101**, 753–757.
48 Yang, H.J., Liu, V.W., Wang, Y., Tsang, P.C., and Ngan, H.Y. (2006) Differential DNA methylation profiles in gynecological cancers and correlation with clinico-pathological data. *BMC Cancer*, **6**, 212.
49 Bishton, M., Kenealy, M., Johnstone, R., Rasheed, W., and Prince, H.M. (2007) Epigenetic targets in hematological malignancies: combination therapies with HDACis and demethylating agents. *Expert. Rev. Anticancer Ther.*, **7**, 1439–1449.
50 Herman, J.G., Civin, C.I., Issa, J.P., Collector, M.I., Sharkis, S.J., and Baylin, S.B. (1997) Distinct patterns of inactivation of p15INK4B and p16INK4A characterize the major types of hematological malignancies. *Cancer Res.*, **57**, 837–841.
51 Melki, J.R., Vincent, P.C., Brown, R.D., and Clark, S.J. (2000) Hypermethylation of E-cadherin in leukemia. *Blood*, **95**, 3208–3213.
52 Lubbert, M., Mertelsmann, R., and Herrmann, F. (1997) Cytosine methylation changes during normal hematopoiesis and in acute myeloid leukemia. *Leukemia*, **11** (Suppl. 1), S12–S18.
53 Sakashita, K., Koike, K., Kinoshita, T., Shiohara, M., Kamijo, T., Taniguchi, S. et al. (2001) Dynamic DNA methylation change in the CpG island region of p15 during human myeloid development. *J. Clin. Invest.*, **108**, 1195–1204.
54 Grady, W.M., Willis, J., Guilford, P.J., Dunbier, A.K., Toro, T.T., Lynch, H. et al. (2000) Methylation of the CDH1 promoter as the second genetic hit in hereditary diffuse gastric cancer. *Nat. Genet.*, **26**, 16–17.
55 Esteller, M., Fraga, M.F., Guo, M., Garcia-Foncillas, J., Hedenfalk, I., Godwin, A.K. et al. (2001) DNA

methylation patterns in hereditary human cancers mimic sporadic tumorigenesis. *Hum. Mol. Genet.*, **10**, 3001–3007.

56 Herman, J.G., Umar, A., Polyak, K., Graff, J.R., Ahuja, N., Issa, J.P. et al. (1998) Incidence and functional consequences of hMLH1 promoter hypermethylation in colorectal carcinoma. *Proc. Natl. Acad. Sci. U. S. A.*, **95**, 6870–6875.

57 Dammann, R., Li, C., Yoon, J.H., Chin, P.L., Bates, S., and Pfeifer, G.P. (2000) Epigenetic inactivation of a RAS association domain family protein from the lung tumour suppressor locus 3p21.3. *Nat. Genet.*, **25**, 315–319.

58 Burbee, D.G., Forgacs, E., Zochbauer-Muller, S., Shivakumar, L., Fong, K., Gao, B. et al. (2001) Epigenetic inactivation of RASSF1A in lung and breast cancers and malignant phenotype suppression. *J. Natl. Cancer Inst.*, **93**, 691–699.

59 Katzenellenbogen, R.A., Baylin, S.B., and Herman, J.G. (1999) Hypermethylation of the DAP-kinase CpG island is a common alteration in B-cell malignancies. *Blood*, **93**, 4347–4353.

60 Suzuki, H., Gabrielson, E., Chen, W., Anbazhagan, R., van Engeland, M., Weijenberg, M.P. et al. (2002) A genomic screen for genes upregulated by demethylation and histone deacetylase inhibition in human colorectal cancer. *Nat. Genet.*, **31**, 141–149.

61 Yamashita, K., Upadhyay, S., Osada, M., Hoque, M.O., Xiao, Y., Mori, M. et al. (2002) Pharmacologic unmasking of epigenetically silenced tumor suppressor genes in esophageal squamous cell carcinoma. *Cancer Cell*, **2**, 485–495.

62 Struhl, K. (1998) Histone acetylation and transcriptional regulatory mechanisms. *Genes Dev.*, **12**, 599–606.

63 Fuks, F., Burgers, W.A., Brehm, A., Hughes-Davies, L., and Kouzarides, T. (2000) DNA methyltransferase Dnmt1 associates with histone deacetylase activity. *Nat. Genet.*, **24**, 88–91.

64 Rountree, M.R., Bachman, K.E., and Baylin, S.B. (2000) DNMT1 binds HDAC2 and a new co-repressor, DMAP1, to form a complex at replication foci. *Nat. Genet.*, **25**, 269–277.

65 Espada, J., Ballestar, E., Fraga, M.F., Villar-Garea, A., Juarranz, A., Stockert, J.C. et al. (2004) Human DNA methyltransferase 1 is required for maintenance of the histone H3 modification pattern. *J. Biol. Chem.*, **279**, 37175–37184.

66 Bai, S., Ghoshal, K., Datta, J., Majumder, S., Yoon, S.O., and Jacob, S.T. (2005) DNA methyltransferase 3b regulates nerve growth factor-induced differentiation of PC12 cells by recruiting histone deacetylase 2. *Mol. Cell. Biol.*, **25**, 751–766.

67 Kawai, H., Li, H., Avraham, S., Jiang, S., and Avraham, H.K. (2003) Overexpression of histone deacetylase HDAC1 modulates breast cancer progression by negative regulation of estrogen receptor alpha. *Int. J. Cancer*, **107**, 353–358.

68 Macaluso, M., Cinti, C., Russo, G., Russo, A., and Giordano, A. (2003) pRb2/p130-E2F4/5-HDAC1-SUV39H1-p300 and pRb2/p130-E2F4/5-HDAC1-SUV39H1-DNMT1 multimolecular complexes mediate the transcription of estrogen receptor-alpha in breast cancer. *Oncogene*, **22**, 3511–3517.

69 Choi, J.H., Kwon, H.J., Yoon, B.I., Kim, J.H., Han, S.U., Joo, H.J. et al. (2001) Expression profile of histone deacetylase 1 in gastric cancer tissues. *Jpn. J. Cancer Res.*, **92**, 1300–1304.

70 Halkidou, K., Gaughan, L., Cook, S., Leung, H.Y., Neal, D.E., and Robson, C.N. (2004) Upregulation and nuclear recruitment of HDAC1 in hormone refractory prostate cancer. *Prostate*, **59**, 177–189.

71 Wilson, A.J., Byun, D.S., Popova, N., Murray, L.B., L'Italien, K., Sowa, Y. et al. (2006) Histone deacetylase 3 (HDAC3) and other class I HDACs regulate colon cell maturation and p21 expression and are deregulated in human colon cancer. *J. Biol. Chem.*, **281**, 13548–13558.

72 Zhang, Z., Yamashita, H., Toyama, T., Sugiura, H., Ando, Y., Mita, K. et al. (2005) Quantitation of HDAC1 mRNA expression in invasive carcinoma of the breast. *Breast Cancer Res. Treat.*, **94**, 11–16.

73 Huang, B.H., Laban, M., Leung, C.H., Lee, L., Lee, C.K., Salto-Tellez, M. et al.

(2005) Inhibition of histone deacetylase 2 increases apoptosis and p21Cip1/WAF1 expression, independent of histone deacetylase 1. *Cell Death Differ.*, **12**, 395–404.

74 Song, J., Noh, J.H., Lee, J.H., Eun, J.W., Ahn, Y.M., Kim, S.Y. *et al.* (2005) Increased expression of histone deacetylase 2 is found in human gastric cancer. *APMIS*, **113**, 264–268.

75 Zhu, P., Martin, E., Mengwasser, J., Schlag, P., Janssen, K.P., and Gottlicher, M. (2004) Induction of HDAC2 expression upon loss of APC in colorectal tumorigenesis. *Cancer Cell*, **5**, 455–463.

76 Zhang, Z., Yamashita, H., Toyama, T., Sugiura, H., Omoto, Y., Ando, Y. *et al.* (2004) HDAC6 expression is correlated with better survival in breast cancer. *Clin. Cancer Res.*, **10**, 6962–6968.

77 Di Croce, L., Raker, V.A., Corsaro, M., Fazi, F., Fanelli, M., Faretta, M. *et al.* (2002) Methyltransferase recruitment and DNA hypermethylation of target promoters by an oncogenic transcription factor. *Science*, **295**, 1079–1082.

78 Amann, J.M., Nip, J., Strom, D.K., Lutterbach, B., Harada, H., Lenny, N. *et al.* (2001) ETO, a target of t(8;21) in acute leukemia, makes distinct contacts with multiple histone deacetylases and binds mSin3A through its oligomerization domain. *Mol. Cell Biol.*, **21**, 6470–6483.

16
Epigenetic Approaches in Oncology
Sabine Zöchbauer-Müller and Robert M. Mader

Abstract

Until recently, the contribution of epigenetic mechanisms to the development of cancer has been underestimated. This shortcoming was favored by the observation that the overall cancer genome is hypomethylated, whereas hypermethylation is restricted to particular regions of the cancer genome. The insight that this hypermethylation does not occur on a random base, but may selectively affect essential cellular functions, paved the way for a different approach to cancer epigenetics. In parallel to the development of appropriate laboratory methods, diagnostic tools in tumor entities such as lung cancer are currently validated focusing on biologic fluids, including easily accessible blood samples. These results may indicate future therapeutic strategies as epigenetic targeting is envisaged. For the time being, derivatives of aza(deoxy)cytidines or the histone deacetylase inhibitor vorinostat have already been approved for the treatment of myelodysplastic syndrome and cutaneous T-cell lymphoma, respectively. These achievements show that epigenetic diagnostics and therapeutics are fuelling biomedical research in parallel. In order to fully exploit the potential of epigenetic modulators in cancer therapy, compounds able to selectively target epigenetic mechanisms by interfering with specific drug–promoter interactions are a challenging but extremely promising next step.

16.1
Introduction

Epigenetics has become a hot topic in several disease entities in recent years, among them malignant diseases. The impressive amount of knowledge accumulated in this time points toward a fruitful exploitation of epigenetics in terms of both diagnostic tools and therapeutic interventions. With regard to epigenetic regulation, the inner contradiction of the malignant cell is illustrated by the coexistence of two states of DNA methylation, which differ profoundly from the normal cell. In a normal cell, the pericentromeric heterochromatin is hypermethylated, whereas controlling functions such as tumor suppressor genes are actively transcribed, that is, the CpG islands on the corresponding promoter are hypomethylated (Figure 16.1). In contrast, the pericentromeric heterochromatin becomes hypomethylated, resulting in genomic instability caused by recombinatorial events during mitosis. In parallel, epigenetic silencing at specific sites such as tumor suppressor genes contributes to tumorigenesis by relieving the physiologic mechanisms aiming at the control of hyperproliferative phenotypes [1]. This event is believed to occur very early during the development of tumors, as synthesized in the model of an epigenetic progenitor origin of cancer proposed by Andrew Feinberg and colleagues in 2006 [2]. This model is supported by a wealth of observations and may serve as a starting point for the development of innovative strategies in diagnostics and therapy.

16.2
DNA Methylation, Chromatin and Transcription

With regard to the epigenetic imbalance observed during the malignant transformation, the central question for our understanding was: How and under which conditions, such as environmental stress or a pre-malignant cellular background, does DNA methylation lead to gene silencing? The primary event, methylation of

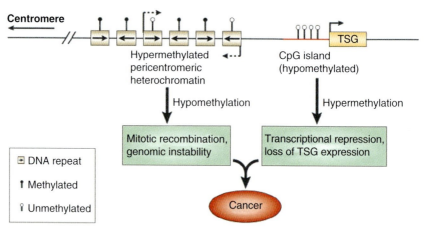

Figure 16.1 DNA methylation and cancer. In tumor cells, repeat-rich heterochromatin becomes hypomethylated and this contributes to genomic instability, a hallmark of tumor cells, through increased mitotic recombination events. De novo methylation of CpG islands also occurs in cancer cells, and can result in the transcriptional silencing of growth-regulatory genes. These changes in methylation are early events in tumorigenesis. Reprinted by permission from Macmillan Publishers Ltd: *Nat Rev Genet* (2005), **6**, 597–610, copyright 2005.

the DNA at the CpG islands, is accomplished by DNA methyltransferases (DNMT) 1, 3A, and 3B. In order to modulate gene expression, the methyl-CpG binding domain proteins (MBDs) bind to methylcytosines, which subsequently inhibit gene transcription [3, 4]. Two of these MBDs, MeCP2 and MBD2, participate in protein complexes that recruit transcriptional co-repressors and histone deacetylases (HDAC) [5–7]. Mediated by these complexes, sites of methylated DNA can target the formation of chromatin, including deacetylated sites of histones, which is typical for transcriptionally repressive domains [3–7]. These results suggest a concerted action of DNA methylation and histone deacetylation in gene silencing and may therefore explain the synergistic effect of methylation inhibitors and HDAC inhibitors on re-expression of silenced genes in tumors [8]. In addition, DNMT1 can directly bind to HDACs and to transcriptional co-repressors, resulting in a direct suppression of gene transcription in a partially HDAC-dependent manner [9–11].

As a paradigm for the central role of DNA methylation in early lesions of the colon, the mRNA expression of three DNA methyltransferases has recently been the subject of investigations in the colorectal adenoma–carcinoma sequence by our group [12]. We compared the expression of DNMT1, DNMT 3A, and DNMT 3B with that of methyl-CpG binding domain protein 2 (MBD2), recently described as the only active DNA demethylase [13]. RNA was isolated from normal colonic mucosa, aberrant crypt foci, benign adenomas, and malignant colorectal carcinomas and analyzed by reverse transcriptase-PCR with subsequent quantification by capillary gel electrophoresis. In contrast to MBD2, pairwise comparisons between tumors and matched, adjacent healthy mucosa tissue revealed that expression of all three genes encoding DNA methyltransferases were increased by two- to threefold, suggesting a relevant role of all investigated DNA methyltransferases during colorectal tumorigenesis (Figure 16.2). This increase was not counterbalanced by enhanced expression of MBD2, which emphasizes increased DNA methylating activity rather than suppressed demethylation as the primary event of epigenetic regulation in the adenoma–carcinoma sequence.

16.3
Methods for Detecting Methylation

Several laboratory methods have been described for detecting DNA methylation, for example, bisulfite genomic sequencing, methylation-specific PCR (MSP), real-time polymerase chain reaction PCR-based assays and restriction enzyme digest. In the MSP assay, originally described by Herman and coworkers [14], treatment of genomic DNA with sodium bisulfite converts unmethylated, but not methylated, cytosines to uracil. During subsequent PCR cycles, uracil bases are converted to thymidine, resulting in sequence differences between methylated and unmethylated DNA. These differences can be detected either by DNA sequencing, as the most specific way of detecting methylation, or by MSP using PCR primers which distinguish between methylated and unmethylated DNA sequences. By using this

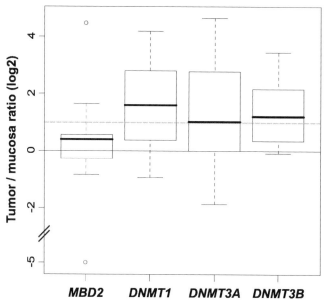

Figure 16.2 Expression of MBD2, DNMT1, DNMT3A, and DNMT3B in matched mucosa-tumor samples. Box-plot of tumor/mucosa log2-ratios, calculated from the corresponding PCR product fluorescence signals, showing ~two- to ~three-fold higher expression of DNMT encoding genes in tumor samples. The dashed gray line indicates the two-fold change threshold.

approach, 0.1% of methylated DNA present in an unmethylated DNA sample can be detected, a finding confirmed later by other authors [15]. These results suggest that MSP is a very sensitive and fast method and is suitable to detect methylation in a large number of samples. A disadvantage of this method is, however, that the obtained results are qualitative but not quantitative in nature.

Contemporary quantitative PCR approaches to detect DNA methylation often rely on real-time quantitative PCR [16–18]. By using fluorescence-based, real-time quantitative PCR the authors reported an even 10-fold higher sensitivity of this method compared with MSP. The disadvantage of using a variety of different laboratory assays to detect DNA methylation may be an assay bias, which is probably a major reason for the different frequencies of DNA methylation of certain genes reported in the literature.

16.4
The Paradigm of Lung Cancer

Causing over 1 million deaths worldwide each year, lung cancer is the leading cause of cancer deaths in the world [19], mostly provoked by tobacco smoking as the main risk factor, which is responsible for approximately 90% of lung cancer deaths [20]. The major histologic types of lung cancer are non-small cell lung cancers (NSCLC), which are further subclassified into squamous cell carcinomas, adenocarcinomas, large cell carcinomas, and mixed types as well as small cell lung cancers (SCLC).

16.4.1
Frequently Methylated Tumor Suppressor Genes and Other Cancer-Related Genes in Lung Carcinomas

In agreement with the relevant contributions of epigenetic silencing during tumorigenesis, several genes which are frequently methylated in lung carcinomas have been identified. Of note, the methylation pattern is tumor type-specific with differences between NSCLCs and SCLCs. The RAS association domain family 1A (*RASSF1A*) gene is a candidate tumor suppressor gene located at the chromosomal region 3p21.3. RASSF1A plays a role during G_1/S-phase cell cycle progression [21, 22]. The frequency of methylation of *RASSF1A* in primary NSCLCs ranges between 30 and 40% [23–26] and is of prognostic impact in NSCLC patients [24]. Patients with tumor DNA methylated at *RASSF1A* sites had a shorter overall survival compared with patients with unmethylated *RASSF1A* sites in tumor DNA. These observations have been confirmed in stage I lung adenocarcinoma patients [27]. As a phenomenon contributing to malignant transformation, starting cigarette smoking at an early age is associated with *RASSF1A* methylation, which was associated with a poor prognosis in patients suffering from NSCLC [25]. Smokers who started smoking before the age of 19 were 4.23 times more likely to have *RASSF1A* methylated lung tumors than smokers starting thereafter [25].

Another gene interacting with cell cycle progression is the *p16* gene, which is located at chromosome 9p21. Its product, $p16^{INK4a}$ binds to cyclin-dependent protein kinase 4 (CDK4) and inhibits the ability of CDK4 to interact with cyclin D1 [28]. Methylation of *p16* has been reported by several authors to occur frequently in NSCLCs [26, 29–31]. Moreover, methylation of *p16* is significantly associated with pack-years smoked, duration of smoking, and negatively correlated with the time since quitting smoking [32].

The tumor suppressor in lung cancer 1 (*TSLC1*) gene was cloned from a region where frequent allelic loss was observed in lung carcinomas, the chromosomal region 11q23.2. The regulatory function of *TSLC1* is frequently lost in lung carcinomas by DNA methylation at rates as high as 79% of primary NSCLC samples, which showed allelic loss at the *TSLC1* locus [33]. With regard to cell cell contact, the cadherin *CDH13* (H-cadherin) is a cell surface glycoprotein responsible for homophilic cell recognition and adhesion. *CDH13* is methylated in 43% of primary NSCLCs. *DAL-1* (differentially expressed in adenocarcinoma of the lung-1, *EPB41L3*), is localized in the chromosomal region 18p11.3 and is a candidate tumor suppressor gene. This chromosomal region is affected by LOH in different types of cancer, including NSCLC. We observed methylation at the *DAL-1* promoter in 55% of primary NSCLCs [34]. In parallel to disease progression, *DAL-1* methylation occurs more frequently in tumors from patients with diseases stage II–III when compared with stage I.

16.4.2
Monitoring of DNA Methylation in Blood Samples

In lung cancer patients, methylation of genes of interest can not only be detected in tumor tissues, but also in serum/plasma samples and in exfoliative material of the aerodigestive tract [35]. It has been shown that cancer patients have increased levels of free DNA in their sera, which is thought to derive from lysed cancer cells [36].

This was first shown by Esteller and coworkers who investigated the presence of DNA in serum from lung cancer patients by methylation-specific PCR [37]. This group analyzed primary NSCLCs and matched serum samples from 22 patients for the methylation pattern of 4 different tumor suppressor genes (*DAPK, GSTP1, p16, MGMT*). Methylation of at least one of these genes was detected in 68% of samples from patients with NSCLCs. When comparing primary tumors with methylation and matched serum samples, 73% of the matched serum samples were found to be methylated. In addition, none of the sera from patients with tumors not demonstrating methylation was positive. Usadel and coworkers investigated the frequency of *APC* methylation in primary NSCLCs and paired preoperative serum or plasma samples of these patients by semiquantitative methylation-specific real-time PCR with fluorophores [38]. 47% of serum and/or plasma samples from patients with *APC* methylated tumors carried detectable amounts of methylated *APC* promoter DNA. In contrast, no methylation at the *APC* promoter site was detected in serum samples from 50 healthy controls. Ramirez' group observed a close correlation between methylation of the genes *RASSF1A, DAPK* and *TMS1* in tumors and in serum samples from NSCLC patients by using methylation-specific PCR [39], which was also a feasible approach to study the methylation of *p16* in serum and plasma samples from NSCLC patients [40].

16.5
Epigenetics and Therapy

The already mentioned epigenetic progenitor model of tumorigenesis involves as series of early epigenetic events as a major driving force of the process [2]. This definition understands cancer as a polyclonal epigenetic disruption of stem/progenitor cells, involving specific tumor progenitor genes (Figure 16.3). The cooperation with genetic lesions – as, for example, proposed by the colorectal adenoma–carcinoma sequence – is driving tumor progression further. This cooperation is referred to as genetic and epigenetic plasticity emphasizing (i) the central role of early epigenetic events together with (ii) genetic instability as driving forces. If this model turns out to be correct, one might expect lines of intervention by interference with the DNA methylating machinery, either by direct inhibition of DNA methyltransferases or by inhibition of histone deacetylases. Of note is that the therapeutic interventions may target early lesions in the process of malignant transformation, paving the way for novel upfront strategies.

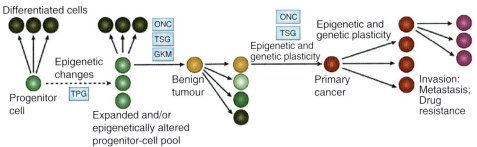

Figure 16.3 The epigenetic progenitor model of cancer. According to this model, cancer arises in three steps. First is an epigenetic alteration of stem/progenitor cells within a given tissue, which is mediated by aberrant regulation of tumor-progenitor genes (TPG) within the stem cells themselves, due to the influence of the stromal compartment, or environmental damage. Second is a gatekeeper mutation (GKM) (tumor-suppressor gene (TSG) in solid tumors, and rearrangement of oncogene (ONC) in leukemia and lymphoma). Although these GKMs are themselves monoclonal, the expanded or altered progenitor compartment increases the risk of cancer when such a mutation occurs and the frequency of subsequent primary tumors (shown as separately arising tumors). Third is genetic and epigenetic instability, which leads to increased tumor evolution. Note that many of the properties of advanced tumors (invasion, metastasis and drug resistance) are inherent properties of the progenitor cells that give rise to the primary tumor and do not require other mutations (highlighting the importance of epigenetic factors in tumor progression). Reprinted by permission from Macmillan Publishers Ltd: *Nat Rev Genet* (2006), **7**, 21–33, copyright 2006.

16.6
Epigenetic Alterations Under Cytotoxic Stress

There is little evidence that cytotoxic treatment interferes directly with the methylation status at DNA promoter sites. Theoretically, one might expect epigenetic silencing as one of the prompt mechanisms to react to cytotoxic stress in order to acquire drug resistance. For 6-thioguanine, this has been demonstrated by bisulfite sequencing of CpG-rich promoter regions with regard to two genes associated with response to the compound, that is, HPRT and CDX1 [41]. The authors observed a significant increase of the DNA methylation in the HPRT promoter region (control: 1.7%; after exposure to 6-thioguanine: 4.7%). With regard to CDX1, the increase was not significant when considering the complete promoter region of 570 CpG-sites. Remarkably, a four CpG-subregion was particularly involved with a highly significant increase from 7% methylation to 27% after exposure to the drug. This CDX1 promoter subregion is strongly correlated with gene expression [42], which suggests a directed DNA methylation mechanism. This investigation leaves us with several open issues which need to be addressed in the near future: Is this a general mechanism, which refers to other metabolic genes too? Is this *in vitro* observation also relevant *in vivo*? If so, what is the predictive power of this phenomenon for anticancer treatment?

16.7
Therapeutic Applications of Inhibitors of DNA Methylation

The well known compounds 5-azacytidine (5-aza-CR) and 5-aza-2′-deoxycytidine (5-aza-CdR) are potent inhibitors of DNA methylation as well as inducers of cellular differentiation [43]. While 5-azacytidine is incorporated into both DNA and RNA, 5-aza-2′-deoxycytidine is incorporated in DNA only. Incorporation of these compounds into cellular DNA leads to rapid loss of DNMT1 activity, because the enzyme becomes irreversibly bound to the 5-aza-nucleotide [44]. As a consequence of the inhibition of the maintenance DNA methyltransferase DNMT1, genes silenced by methylation can be re-expressed *in vitro* after treatment with these agents [24, 45–47]. This concept has been developed to treatment strategies in humans, resulting in the successful application and approval of 5-azacytidine (Vidaza®) by the FDA in 2005 for the treatment of myelodysplastic syndrome. This was primarily based on the encouraging results of the landmark studies performed by the Cancer and Leukemia Group B (better known as CALGB). In these and other clinical trials, 5-azacytidine was well tolerated, even by elderly patients, mild nausea, vomiting, and reactions at the injection site being the most common adverse effects. The sister compound 5-aza-2′-deoxycytidine has received its approval for the treatment of the same disease by the FDA one year later and is marketed under the brand name Dacogen® [48].

Besides the myelodysplastic syndrome, 5-azacytidine has been administered to patients with chronic and acute myeloid leukemia and selected solid tumors. However, the number of clinical trials of 5-azacytidine and 5-aza-2′-deoxycytidine in patients with solid tumors is small and responses are low when directly compared with the efficacy of conventional therapy. Momparler and coworkers obtained very interesting results in a pilot clinical trial on 5-aza-2′-deoxycytidine in patients with stage IV NSCLC [49]. Patients without prior chemotherapy were administered an 8h i.v. infusion of 5-aza-2′-deoxycytidine for one or more cycles at intervals of 5 to 7 weeks, depending on recovery of the granulocyte count. Although only seven patients were included, for all but one the survival time increased in parallel to the number of cycles of 5-aza-2′-deoxycytidine administered. Interestingly, one patient survived more than 6 years. In this study, the authors observed a delayed action of 5-aza-2′-deoxycytidine on tumor growth, which became manifested after one or more cell divisions. If this observation turns out to be a general principle, it suggests that patients require several treatment cycles before antitumor activity becomes evident.

Because side effects of 5-azacytidine and 5-aza-2′-deoxycytidine such as nausea, vomiting, diarrhea, myelosuppression include also the ability to form mutagenic lesions, the search for new methylation inhibitors with less toxic side effects is underway. Alternative inhibitors of DNMT1 have been described, but not reached the state of maturity necessary for approval in the use in humans [50, 51]. Among them, antisense oligonucleotides directed against DNMT1 mRNA and hairpin-structured oligonucleotide substrate mimics may be suitable approaches.

The pipeline for epigenetic interference seems to be definitely richer with regard to the inhibitors of histone deacetylases. Among them, suberoylanilide hydroxamic acid or vorinostat has proven to be efficient in the treatment of cutaneous T-cell lymphoma, followed by FDA approval in 2007 (Zolinza®). Vorinostat is an oral inhibitor of both classes of histone deacetylases, namely class I and class II [48]. Other examples of clinical studies include depsipeptide or sodium phenylbutyrate as single agents or these agents in combination with other antineoplastic therapies. For example, valproic acid is under investigation in hematologic malignancies as well as in advanced solid tumors.

Of interest is that a synergistic effect in reactivating gene expression between 5-aza-2'-deoxycytidine and the histone deacetylase inhibitor trichostatin A has been demonstrated [8]. The combination of these agents has shown promising results on both cell kill and cancer-related gene reactivation in cancer cell lines [8, 52]. The exact underlying mechanism for this synergy has not yet been unraveled. Nevertheless, these data offer the opportunity for strategies with increased therapeutic efficiency with concomitant reduction of the dose of 5-aza-2'-deoxycytidine used in clinical studies. When proven in humans, this approach could establish the combination of histone deacetylase inhibitors with 5-aza-2'-deoxycytidine, resulting in potentially less side effects. However, clinical trials are urgently needed to prove this concept.

16.8
How May Methylation Become Relevant to Clinical Applications?

So far, several authors have reported a prognostic impact of methylation of selected genes which are detected in primary tumors from lung cancer patients, such as *RASSF1A, APC, CDH1,* and *MGMT* [16, 24, 27, 53, 54]. With regard to lung cancer patients, the definition of genes with prognostic impact methylated in tumors would allow for the individualization of therapy for those patients. These patients might potentially benefit from the epigenetic diagnostic strategy by several means: either by early detection of the malignant process at stages where curative strategies are likely to succeed or from a more aggressive treatment strategy in advanced disease. In terms of personalized medicine, the use of methylation inhibitors, either alone or in combination with cytotoxic therapy, may be envisaged. In the future, the combination of methylation inhibitors with histone deacetylase inhibitors will be tested in cancer patients, thus further enlarging our therapeutic options. Completely out of reach for the time being is the development of targeted epigenetic strategies, which lead to the selective re-expression of silenced genes by specific interactions with silenced promoter subregions, either by interaction with methyl-CpG binding domain proteins or active demethylation of CpG-islands.

Because methylation is tissue-specific, the methylation pattern of several genes may serve to distinguish different tumor histologies [29]. Moreover, a different methylation pattern is not only found in various tumor entities but also between histologic subtypes of lung carcinomas [55, 56].

Besides that, methylation may also be used in lung cancer risk assessment and early lung cancer detection. Methylation can be detected in the bronchial epithelium from cancer-free smokers and has already been observed up to three years prior to the diagnosis of overt lung cancer [57]. As the frequencies of gene methylation in the bronchial epithelium differ, the identification of the best prognostic battery of genes associated with increased lung cancer risk will take some more time. By this approach, it would be possible to assess the "individual" disease risk instead of that of cohorts. The data currently available suggest the requirement for rigorous testing in order to identify individuals at high risk of developing lung cancer.

Once individualized, this subgroup of people could undergo diagnostic procedures for early detection of lung cancer, such as spiral computer tomography scan, which would help to identify the ideal candidates for chemoprevention trials. Within therapeutic settings, methylation patterns of relevant genes could serve as surrogate biomarkers of the efficacy of chemoprevention. In overt disease, methylation patterns may be indicative for response to therapy in advanced disease, because the material tested for epigenetic aberrations in smokers has to be available in an easy, non-invasive and cheap way. In accordance with these requirements, we reported that the frequencies of methylation of certain genes were comparable between bronchial brushes and sputum samples obtained from heavy smokers, suggesting that sputum samples are an ideal specimen for this kind of study [35].

16.9
Conclusions

Epigenetics are here to stay in the field of experimental and clinical oncology. There is already vast evidence for an epigenetic involvement in the early stages of tumorigenesis. The understanding of this involvement needs to be detailed in the coming years. In contrast to the usual scenario, it is the therapeutic application that has preceded diagnostic tools in the clinic. Nevertheless, we expect relevant contributions in this field in the very near future. As a possible route of development and forward-looking approach in cancer therapy, compounds able to target selected epigenetic mechanisms by specific drug–promoter interactions are definitely a worthwhile goal to achieve.

References

1 Robertson, K.D. (2005) DNA methylation and human disease. *Nat. Rev. Genet.*, **6**, 597–610.

2 Feinberg, A.P., Ohlsson, R., and Henikoff, S. (2006) The epigenetic progenitor origin of human cancer. *Nat. Rev. Genet.*, **7**, 21–33.

3 Bird, A.P., and Wolffe, A.P. (1999) Methylation-induced repression – belts, braces, and chromatin. *Cell*, **99**, 451–454.

4 Ng, H.H., and Bird, A. (1999) DNA methylation and chromatin modification. *Curr. Opin. Genet. Dev.*, **9**, 158–163.

5 Jones, P.L., Veenstra, G.J., Wade, P.A., Vermaak, D., Kass, S.U., Landsberger, N., Strouboulis, J., and Wolffe, A.P. (1998) Methylated DNA and MeCP2 recruit histone deacetylase to repress transcription. *Nat. Genet.*, **19**, 187–191.

6 Nan, X., Ng, H.H., Johnson, C.A., Laherty, C.D., Turner, B.M., Eisenman, R.N., and Bird, A. (1998) Transcriptional repression by the methyl-CpG-binding protein MeCP2 involves a histone deacetylase complex. *Nature*, **393**, 386–389.

7 Wade, P.A., Gegonne, A., Jones, P.L., Ballestar, E., Aubry, F., and Wolffe, A.P. (1999) Mi-2 complex couples DNA methylation to chromatin remodelling and histone deacetylation. *Nat. Genet.*, **23**, 62–66.

8 Cameron, E.E., Bachman, K.E., Myohanen, S., Herman, J.G., and Baylin, S.B. (1999) Synergy of demethylation and histone deacetylase inhibition in the re-expression of genes silenced in cancer. *Nat. Genet.*, **21**, 103–107.

9 Fuks, F., Burgers, W.A., Brehm, A., Hughes-Davies, L., and Kouzarides, T. (2000) DNA methyltransferase Dnmt1 associates with histone deacetylase activity. *Nat. Genet.*, **24**, 88–91.

10 Robertson, K.D., Ait-Si-Ali, S., Yokochi, T., Wade, P.A., Jones, P.L., and Wolffe, A.P. (2000) DNMT1 forms a complex with Rb, E2F1 and HDAC1 and represses transcription from E2F-responsive promoters. *Nat. Genet.*, **25**, 338–342.

11 Rountree, M.R., Bachman, K.E., and Baylin, S.B. (2000) DNMT1 binds HDAC2 and a new co-repressor, DMAP1, to form a complex at replication foci. *Nat. Genet.*, **25**, 269–277.

12 Schmidt, W.M., Sedivy, R., Forstner, B., Steger, G.G., Zochbauer-Muller, S., and Mader, R.M. (2007) Progressive up-regulation of genes encoding DNA methyltransferases in the colorectal adenoma-carcinoma sequence. *Mol. Carcinog.*, **46**, 766–772.

13 Detich, N., Theberge, J., and Szyf, M. (2002) Promoter-specific activation and demethylation by MBD2/demethylase. *J. Biol. Chem.*, **277**, 35791–35794.

14 Herman, J.G., Graff, J.R., Myohanen, S., Nelkin, B.D., and Baylin, S.B. (1996) Methylation-specific PCR: a novel PCR assay for methylation status of CpG islands. *Proc. Natl. Acad. Sci. U. S. A.*, **93**, 9821–9826.

15 Virmani, A.K., Muller, C., Rathi, A., Zoechbauer-Mueller, S., Mathis, M., and Gazdar, A.F. (2001) Aberrant methylation during cervical carcinogenesis. *Clin. Cancer Res.*, **7**, 584–589.

16 Brabender, J., Usadel, H., Danenberg, K.D., Metzger, R., Schneider, P.M., Lord, R.V., Wickramasinghe, K., Lum, C.E., Park, J., Salonga, D., Singer, J., Sidransky, D., Holscher, A.H., Meltzer, S.J., and Danenberg, P.V. (2001) Adenomatous polyposis coli gene promoter hypermethylation in non-small cell lung cancer is associated with survival. *Oncogene*, **20**, 3528–3532.

17 Eads, C.A., Danenberg, K.D., Kawakami, K., Saltz, L.B., Blake, C., Shibata, D., Danenberg, P.V., and Laird, P.W. (2000) MethyLight: a high throughput assay to measure DNA methylation. *Nucleic. Acids. Res.*, **28**, E32.

18 Eads, C.A., Danenberg, K.D., Kawakami, K., Saltz, L.B., Danenberg, P.V., and Laird, P.W. (1999) CpG island hypermethylation in human colorectal tumors is not associated with DNA methyltransferase overexpression. *Cancer Res.*, **59**, 2302–2306.

19 Parkin, D.M., Bray, F.I., and Devesa, S.S. (2001) Cancer burden in the year 2000.

The global picture. *Eur. J. Cancer*, **37** (Suppl. 8), S4–S66.

20 Wingo, P.A., Ries, L.A., Giovino, G.A., Miller, D.S., Rosenberg, H.M., Shopland, D.R., Thun, M.J., and Edwards, B.K. (1999) Annual report to the nation on the status of cancer, 1973–1996, with a special section on lung cancer and tobacco smoking. *J. Natl. Cancer Inst.*, **91**, 675–690.

21 Lerman, M.I., and Minna, J.D. (2000) The 630-kb lung cancer homozygous deletion region on human chromosome 3p21.3: identification and evaluation of the resident candidate tumor suppressor genes. The International Lung Cancer Chromosome 3p21.3 Tumor Suppressor Gene Consortium. *Cancer Res.*, **60**, 6116–6133.

22 Shivakumar, L., Minna, J., Sakamaki, T., Pestell, R., and White, M.A. (2002) The RASSF1A tumor suppressor blocks cell cycle progression and inhibits cyclin D1 accumulation. *Mol. Cell. Biol.*, **22**, 4309–4318.

23 Agathanggelou, A., Honorio, S., Macartney, D.P., Martinez, A., Dallol, A., Rader, J., Fullwood, P., Chauhan, A., Walker, R., Shaw, J.A., Hosoe, S., Lerman, M.I., Minna, J.D., Maher, E.R., and Latif, F. (2001) Methylation associated inactivation of RASSF1A from region 3p21.3 in lung, breast and ovarian tumours. *Oncogene*, **20**, 1509–1518.

24 Burbee, D.G., Forgacs, E., Zochbauer-Muller, S., Shivakumar, L., Fong, K., Gao, B., Randle, D., Kondo, M., Virmani, A., Bader, S., Sekido, Y., Latif, F., Milchgrub, S., Toyooka, S., Gazdar, A.F., Lerman, M.I., Zabarovsky, E., White, M., and Minna, J.D. (2001) Epigenetic inactivation of RASSF1A in lung and breast cancers and malignant phenotype suppression. *J. Natl. Cancer Inst.*, **93**, 691–699.

25 Kim, D.H., Kim, J.S., Ji, Y.I., Shim, Y.M., Kim, H., Han, J., and Park, J. (2003) Hypermethylation of RASSF1A promoter is associated with the age at starting smoking and a poor prognosis in primary non-small cell lung cancer. *Cancer Res.*, **63**, 3743–3746.

26 Toyooka, S., Maruyama, R., Toyooka, K.O., McLerran, D., Feng, Z., Fukuyama, Y., Virmani, A.K., Zochbauer-Muller, S., Tsukuda, K., Sugio, K., Shimizu, N., Shimizu, K., Lee, H., Chen, C.Y., Fong, K.M., Gilcrease, M., Roth, J.A., Minna, J.D., and Gazdar, A.F. (2003) Smoke exposure, histologic type and geography-related differences in the methylation profiles of non-small cell lung cancer. *Int. J. Cancer*, **103**, 153–160.

27 Tomizawa, Y., Kohno, T., Kondo, H., Otsuka, A., Nishioka, M., Niki, T., Yamada, T., Maeshima, A., Yoshimura, K., Saito, R., Minna, J.D., and Yokota, J. (2002) Clinicopathological significance of epigenetic inactivation of RASSF1A at 3p21.3 in stage I lung adenocarcinoma. *Clin. Cancer Res.*, **8**, 2362–2368.

28 Sherr, C.J. (1996) Cancer cell cycles. *Science*, **274**, 1672–1677.

29 Esteller, M., Corn, P.G., Baylin, S.B., and Herman, J.G. (2001) A gene hypermethylation profile of human cancer. *Cancer Res.*, **61**, 3225–3229.

30 Kashiwabara, K., Oyama, T., Sano, T., Fukuda, T., and Nakajima, T. (1998) Correlation between methylation status of the p16/CDKN2 gene and the expression of p16 and Rb proteins in primary non-small cell lung cancers. *Int. J. Cancer*, **79**, 215–220.

31 Zochbauer-Muller, S., Fong, K.M., Virmani, A.K., Geradts, J., Gazdar, A.F., and Minna, J.D. (2001) Aberrant promoter methylation of multiple genes in non-small cell lung cancers. *Cancer Res.*, **61**, 249–255.

32 Kim, D.H., Nelson, H.H., Wiencke, J.K., Zheng, S., Christiani, D.C., Wain, J.C., Mark, E.J., and Kelsey, K.T. (2001) p16(INK4a) and histology-specific methylation of CpG islands by exposure to tobacco smoke in non-small cell lung cancer. *Cancer Res.*, **61**, 3419–3424.

33 Kuramochi, M., Fukuhara, H., Nobukuni, T., Kanbe, T., Maruyama, T., Ghosh, H.P., Pletcher, M., Isomura, M., Onizuka, M., Kitamura, T., Sekiya, T., Reeves, R.H., and Murakami, Y. (2001) TSLC1 is a tumor-suppressor gene in human non-small-cell lung cancer. *Nat. Genet.*, **27**, 427–430.

34 Heller, G., Fong, K.M., Girard, L., Seidl, S., End-Pfutzenreuter, A., Lang, G.,

Gazdar, A.F., Minna, J.D., Zielinski, C.C., and Zochbauer-Muller, S. (2006) Expression and methylation pattern of TSLC1 cascade genes in lung carcinomas. *Oncogene*, **25**, 959–968.

35 Zochbauer-Muller, S., Lam, S., Toyooka, S., Virmani, A.K., Toyooka, K.O., Seidl, S., Minna, J.D., and Gazdar, A.F. (2003) Aberrant methylation of multiple genes in the upper aerodigestive tract epithelium of heavy smokers. *Int. J. Cancer*, **107**, 612–616.

36 Fujiwara, K., Fujimoto, N., Tabata, M., Nishii, K., Matsuo, K., Hotta, K., Kozuki, T., Aoe, M., Kiura, K., Ueoka, H., and Tanimoto, M. (2005) Identification of epigenetic aberrant promoter methylation in serum DNA is useful for early detection of lung cancer. *Clin. Cancer Res.*, **11**, 1219–1225.

37 Esteller, M., Sanchez-Cespedes, M., Rosell, R., Sidransky, D., Baylin, S.B., and Herman, J.G. (1999) Detection of aberrant promoter hypermethylation of tumor suppressor genes in serum DNA from non-small cell lung cancer patients. *Cancer Res.*, **59**, 67–70.

38 Usadel, H., Brabender, J., Danenberg, K.D., Jeronimo, C., Harden, S., Engles, J., Danenberg, P.V., Yang, S., and Sidransky, D. (2002) Quantitative adenomatous polyposis coli promoter methylation analysis in tumor tissue, serum, and plasma DNA of patients with lung cancer. *Cancer Res.*, **62**, 371–375.

39 Ramirez, J.L., Sarries, C., de Castro, P.L., Roig, B., Queralt, C., Escuin, D., de Aguirre, I., Sanchez, J.M., Manzano, J.L., Margeli, M., Sanchez, J.J., Astudillo, J., Taron, M., and Rosell, R. (2003) Methylation patterns and K-ras mutations in tumor and paired serum of resected non-small-cell lung cancer patients. *Cancer Lett.*, **193**, 207–216.

40 Bearzatto, A., Conte, D., Frattini, M., Zaffaroni, N., Andriani, F., Balestra, D., Tavecchio, L., Daidone, M.G., and Sozzi, G. (2002) p16(INK4A) Hypermethylation detected by fluorescent methylation-specific PCR in plasmas from non-small cell lung cancer. *Clin. Cancer Res.*, **8**, 3782–3787.

41 Bredberg, A., and Bodmer, W. (2007) Cytostatic drug treatment causes seeding of gene promoter methylation. *Eur. J. Cancer*, **43**, 947–954.

42 Wong, N.A., Britton, M.P., Choi, G.S., Stanton, T.K., Bicknell, D.C., Wilding, J.L., and Bodmer, W.F. (2004) Loss of CDX1 expression in colorectal carcinoma: promoter methylation, mutation, and loss of heterozygosity analyses of 37 cell lines. *Proc. Natl. Acad. Sci. U. S. A.*, **101**, 574–579.

43 Momparler, R.L., Bouffard, D.Y., Momparler, L.F., Dionne, J., Belanger, K., and Ayoub, J. (1997) Pilot phase I–II study on 5-aza-2′-deoxycytidine (Decitabine) in patients with metastatic lung cancer. *Anticancer Drugs*, **8**, 358–368.

44 Ferguson, A.T., Vertino, P.M., Spitzner, J.R., Baylin, S.B., Muller, M.T., and Davidson, N.E. (1997) Role of estrogen receptor gene demethylation and DNA methyltransferase.DNA adduct formation in 5-aza-2′deoxycytidine-induced cytotoxicity in human breast cancer cells. *J. Biol. Chem.*, **272**, 32260–32266.

45 Toyooka, K.O., Toyooka, S., Virmani, A.K., Sathyanarayana, U.G., Euhus, D.M., Gilcrease, M., Minna, J.D., and Gazdar, A.F. (2001) Loss of expression and aberrant methylation of the CDH13 (H-cadherin) gene in breast and lung carcinomas. *Cancer Res.*, **61**, 4556–4560.

46 Virmani, A.K., Rathi, A., Zochbauer-Muller, S., Sacchi, N., Fukuyama, Y., Bryant, D., Maitra, A., Heda, S., Fong, K.M., Thunnissen, F., Minna, J.D., and Gazdar, A.F. (2000) Promoter methylation and silencing of the retinoic acid receptor-beta gene in lung carcinomas. *J. Natl. Cancer Inst.*, **92**, 1303–1307.

47 Zochbauer-Muller, S., Fong, K.M., Maitra, A., Lam, S., Geradts, J., Ashfaq, R., Virmani, A.K., Milchgrub, S., Gazdar, A.F., and Minna, J.D. (2001) 5′ CpG island methylation of the FHIT gene is correlated with loss of gene expression in lung and breast cancer. *Cancer Res.*, **61**, 3581–3585.

48 Donepudi, S., Mattison, R.J., and Kihslinger, J.E., and Godley, L.A. (2007) Modulators of DNA methylation and histone deacetylation. *Update Cancer Ther.*, **2**, 157–169.

49 Momparler, R.L., Eliopoulos, N., and Ayoub, J. (2000) Evaluation of an inhibitor of DNA methylation, 5-aza-2′-deoxycytidine, for the treatment of lung cancer and the future role of gene therapy. *Adv. Exp. Med. Biol.*, **465**, 433–446.

50 Bigey, P., Knox, J.D., Croteau, S., Bhattacharya, S.K., Theberge, J., and Szyf, M. (1999) Modified oligonucleotides as bona fide antagonists of proteins interacting with DNA. Hairpin antagonists of the human DNA methyltransferase. *J. Biol. Chem.*, **274**, 4594–4606.

51 Fournel, M., Sapieha, P., Beaulieu, N., Besterman, J.M., and MacLeod, A.R. (1999) Down-regulation of human DNA-(cytosine-5) methyltransferase induces cell cycle regulators p16(ink4A) and p21(WAF/Cip1) by distinct mechanisms. *J. Biol. Chem.*, **274**, 24250–24256.

52 Bovenzi, V., and Momparler, R.L. (2001) Antineoplastic action of 5-aza-2′-deoxycytidine and histone deacetylase inhibitor and their effect on the expression of retinoic acid receptor beta and estrogen receptor alpha genes in breast carcinoma cells. *Cancer Chemother. Pharmacol.*, **48**, 71–76.

53 Brabender, J., Usadel, H., Metzger, R., Schneider, P.M., Park, J., Salonga, D., Tsao-Wei, D.D., Groshen, S., Lord, R.V., Takebe, N., Schneider, S., Holscher, A.H., Danenberg, K.D., and Danenberg, P.V. (2003) Quantitative O (6)-methylguanine DNA methyltransferase methylation analysis in curatively resected non-small cell lung cancer: associations with clinical outcome. *Clin. Cancer Res.*, **9**, 223–227.

54 Tang, X., Khuri, F.R., Lee, J.J., Kemp, B.L., Liu, D., Hong, W.K., and Mao, L. (2000) Hypermethylation of the death-associated protein (DAP) kinase promoter and aggressiveness in stage I non-small-cell lung cancer. *J. Natl. Cancer Inst.*, **92**, 1511–1516.

55 Toyooka, S., Toyooka, K.O., Maruyama, R., Virmani, A.K., Girard, L., Miyajima, K., Harada, K., Ariyoshi, Y., Takahashi, T., Sugio, K., Brambilla, E., Gilcrease, M., Minna, J.D., and Gazdar, A.F. (2001) DNA methylation profiles of lung tumors. *Mol. Cancer Ther.*, **1**, 61–67.

56 Virmani, A.K., Tsou, J.A., Siegmund, K.D., Shen, L.Y., Long, T.I., Laird, P.W., Gazdar, A.F., and Laird-Offringa, I.A. (2002) Hierarchical clustering of lung cancer cell lines using DNA methylation markers. *Cancer Epidemiol. Biomarkers Prev.*, **11**, 291–297.

57 Palmisano, W.A., Divine, K.K., Saccomanno, G., Gilliland, F.D., Baylin, S.B., Herman, J.G., and Belinsky, S.A. (2000) Predicting lung cancer by detecting aberrant promoter methylation in sputum. *Cancer Res.*, **60**, 5954–5958.

17
Epigenetic Dysregulation in Aging and Cancer

Despina Komninou and John P. Richie

Abstract

The strongest risk factor for most types of cancer is aging. Numerous aging-dependent genetic and epigenetic events as well as metabolic alterations are likely to be major contributors to increased cancer susceptibility in old age promoting a pro-tumorigenic tissue microenvironment. Epigenetic-mediated changes in response to dietary and other environmental triggers throughout the course of life dictate epigenetic "signatures" that alter metabolic pathways linking the process of aging with cancer risk. A wide spectrum of multiple age-related epigenetic "hits", from promoter hypermethylation, histone deacetylation and gene silencing to impaired NF-κB signaling with up-regulation of cytokines and chronic inflammation, may account for predisposition to age-associated malignancies such as colon, breast (postmenopausal) and prostate cancer. The cancer-prone phenotype of old age is associated with insulin resistance and its related metabolic syndrome which elicits many of the signs of early aging. In fact, longevity studies have often implicated genes that regulate lifespan through insulin or insulin-like signaling pathways. Interestingly, a family of epigenetic enzymes with histone deacetylase (HDAC) activity, known as sirtuins, are involved in the extension of lifespan mediated by caloric restriction (CR) regulating gene expression, DNA repair, insulin sensitivity, NF-κB signaling and apoptosis. CR exerts protective effects on carcinogenesis by maintaining an insulin-sensitive phenotype characterized by lifelong maintenance of optimal levels of key fat-derived cytokines (lipokines). Mechanistic studies with models of CR or with "CR-mimetics" may provide insight on how the aging-related insulin resistance and subsequent chronic inflammation might modulate epigenetic processes that affect age-associated diseases such as the development and progression of cancer. A greater understanding of the interrelations among diet, aging, insulin resistance, inflammation and cancer at the epigenetic level of regulation will help us design new dietary/pharmacological preventive strategies specifically targeting key epigenetic signaling networks.

17.1
Introduction

The American Cancer Society has reported that cancer is by far the leading cause of death among men and women aged 60 to 79 [1]. This striking relationship between cancer and age is often overlooked and understudied. It has been estimated that the rates of spontaneous mutations (genetic changes) over a lifetime could not account for the cancer-prone phenotype of old age unless other specific age-related (epigenetic) changes are brought into the equation as they become critical in the control of cell growth, constructing a pro-tumorigenic tissue microenvironment [2, 3].

Epigenetic events result in changes that do not affect the gene's sequence of DNA (genotype), but alter the gene expression (phenotype), affecting susceptibility to complex diseases including cancer and other aging-related conditions such as cardiovascular disease, type II diabetes and obesity [4–6]. Epigenetic alterations are propagated through cell division, accumulate over lifetime and are influenced by environmental/dietary triggers dictating epigenetic "signatures" with important implications for human biology and disease process [7]. Diet-induced changes in epigenetic regulators during the course of life may precipitate the metabolic/inflammatory dysregulation governing the strong association between aging and cancer development [8].

One epigenetic modification that has been postulated to play a role in both processes, carcinogenesis and aging, is hypermethylation, the addition of methyl groups in certain regions of DNA, which reduces the expression of genes with subsequent disruption of normal gene function [9]. Specifically, hypermethylation of promoter-associated CpG islands is a gene silencing epigenetic modification which is common during aging and often involves genes that are hypermethylated in cancer [10]. For example, the estrogen receptor (ER) gene is not methylated in young individuals, but it is partially methylated in older individuals, and hypermethylated in colonic, breast and prostate tumors [11].

In this chapter, we focus on aging-related epigenetic alterations and their biological significance in the cancer-prone phenotype of old age which is characterized by metabolic decline and dysregulation of key fat-derived cytokines resulting in impaired NF-κB and insulin signaling that promote an inflammatory microenvironment predisposed to neoplasia. Anti-aging models, such as caloric restriction (CR) and methionine restriction (MR), along with CR "mimetics" have been useful experimental tools to explore the interconnection between diet, insulin resistance, chronic inflammation and disease progression through modulation of epigenetic signaling common in aging and cancer development.

17.2
The Cancer-Prone Metabolic Phenotype of Aging

Aging is associated with a metabolic decline characterized by the development of changes in fat distribution, obesity, mainly abdominal obesity, and insulin resistance [12–14]. It is becoming clear that the increased adipose tissue is not simply a reservoir for energy, but rather an active and dynamic endocrine organ capable of expressing several cytokines and other fat-derived peptides (lipokines). Increased accumulation of this metabolically active fat mass (visceral adiposity) is a common and typical change in body composition during aging and is associated with increased plasma insulin levels, impaired glucose tolerance and chronic low-grade inflammation [15–17]. Visceral/abdominal obesity occurs despite the decreased subcutaneous fat and progressive sarcopenia typical of aging, and has been associated with greater risks for age-related diseases. The lipokines produced by this type of adipose tissue include TNF-α, IL-6, leptin and adiponectin, which are implicated in the pathogenesis of diabetes and cancer [18–20].

These adipocytokines are being studied as inflammatory regulators as well as possible modifiable markers of the increased risk and even predictors of cancers associated with the metabolic phenotype of aging. In addition, factors such as diets high in fat, simple carbohydrates and total calories, physical inactivity and abdominal obesity, are associated with increased risk for aging-associated conditions such as insulin resistance and cancer development [21–23].

Insulin resistance elicits many of the signs of early aging and is a common condition in older animals and humans [14, 24], suggesting that insulin-signaling pathways mediate aging processes. Longevity studies often implicate genes that regulate lifespan through insulin or insulin-like signaling pathways [25, 26]. Several key players, such as insulin, IGF-1, leptin and adiponectin, have been investigated for their involvement in cancer development [15, 27–29].

An important mediator of insulin sensitivity with anti-inflammatory and anti-cancer properties is adiponectin, the most abundant gene product secreted by fat cells, which is significantly reduced in obese/diabetic animals and humans [30–32]. Leptin, which exerts effects opposite to those of adiponectin, is an important factor for the growth of normal and tumor cells, but it enhances cell invasion and migration only in malignant cells [33]. Expression of leptin is induced by insulin and both hormones are proposed as biomarkers of breast cancer progression [34]. While leptin and TNF-alpha activate the inflammation/oxidative stress responsive transcriptional factor nuclear factor kappaB (NF-kappaB), adiponectin inhibits NF-κB signaling [35]. Adiponectin production is stimulated by agonists of peroxisome proliferator-activated receptor gamma (PPAR), such as thiazolidinediones (a class of antidiabetic drugs), which improve insulin action by enhancing the expression and secretion of adiponectin and antagonizing the suppressive effect of TNF-α on adiponectin production [36]. PPARs are nuclear receptors that dimerize with retinoid receptors to act as transcription factors and have been implicated in lipid, glucose and energy homeostasis providing a molecular link between nutrition and gene expression [37]. They also seem to exert antiproliferative, proapoptotic and antiinflammatory effects that may be mediated, in part, by antagonizing the activities of NF-kappaB [38]. Furthermore, compounds such as hypolipidemic drugs, certain NSAIDs and antioxidants, which are found to activate PPARs, can also correct the abnormal NF-κB signaling associated with age-related inflammatory dysregulations [39].

Although the cancer-prone metabolic phenotype of old age involves a diversity of genes which become differentially expressed during the process of aging, the underlying mechanisms are not well understood. As this phenotype contributes to the age-related increase in morbidity and mortality [40], the importance of epigenetic regulations represents a research area of growing interest in the postgenomic era. Several investigations suggest that environmental factors, including diet, play a role as key regulators of epigenetic processes linking the aging-related insulin resistance and metabolic syndrome with cancer susceptibility. Accumulating evidence provides support for the effects of diet on epigenetic modulations such as DNA methylation and histone acetylation [8].

17.3
Age-Related Epigenetic Silencing Via DNA Methylation

The best-known epigenetic modification that gradually and selectively leads to gene silencing during aging is the enhanced methylation of CpG-rich areas (CpG islands) at the promoters of the involved genes [41]. It seems that some cells in older individuals progressively lose the expression of certain genes via increased frequency of methylation in their promoter area. It is estimated that several hundreds of genes are affected by this age-related hypermethylation. The type and number of silencing genes are likely influenced by different environmental, lifestyle, dietary and metabolic effectors throughout the life course of an individual. When the affected genes include tumor-suppressor genes, DNA-repair and apoptosis genes, which are important regulators of the neoplastic phenotype, a potentially critical mechanism for cancer may have been triggered. Indeed, promoter hypermethylation is a common event in human neoplasia that affects several hundred genes of a tumor in a "multiple hit" and stepwise fashion contributing to the multistep carcinogenesis model proposed by Fearon and Vogelstein [42].

There is strong evidence that promoter hypermethylation of several genes, including ER, IGF-2, MYOD1, MLH1, MGMT (O^6-methylguanine-DNA methyltransferase) and CDKN2A (the gene encoding p16), occurs early in neoplasia, preceding malignant transformation. This cancer-prone epigenetic alteration is detected in normal tissues as an age-related trend, in normal-appearing tissues of cancer patients, and also in pre-malignant and malignant tissues to a higher degree. Specifically, the MGMT (DNA-repair enzyme) gene silencing by promoter hypermethylation is found in up to 40% of colorectal cancers [43] and also in normal-appearing colorectal mucosa adjacent to the tumor, suggesting a marker of field effects useful for early detection and risk assessment in colon cancer [44]. In addition, the ER gene is not methylated in young individuals, but it appears to be partially methylated in older individuals, and hypermethylated in 100% of colonic tumors [10].

In breast cancer studies, the lack of evidence for mutational inactivation of ER led to investigations examining the role of epigenetic alterations in ER loss of expression [45, 46]. The ER is unmethylated in normal breast tissue and most ER positive breast tumor cell lines, while it is methylated in approximately 50% of unselected primary breast cancers and most ER negative breast cancer cell lines [47]. Loss of ER expression during breast cancer progression has been associated with hypermethylation of the ER gene that was evident in 34% of ductal carcinoma *in situ* lesions and increased significantly, to nearly 60%, in metastatic lesions [48]. Furthermore, demethylation of the ER gene in ER negative human breast cancer cells treated with DNA methylation inhibitors resulted in ER re-expression and activation [49, 50]. Hypermethylation of other genes involved in breast cancer progression is also reported, including E-cadherin and retinoic acid receptor beta [48, 50].

A number of genes are reported to be hypermethylated in prostate cancer, including the ER gene [51] as well as genes essential for apoptosis, suggesting an epigenetic "signature of apoptotic silencing" for this cancer [52]. DNA methylation and its role in aberrant apoptosis as part of the survival mechanisms in cancer represent a significant finding with potential use in diagnostic, prognostic and therapeutic settings [52].

Another interesting observation is that aberrant hypermethylation has been detected in conditions of chronic inflammation such as ulcerative colitis [41], Barrett's esophagus [53] and chronic hepatitis [54, 55], which are associated with cancer development in the affected tissues over time. Epigenetic alterations are also involved in other pathologies such as atherosclerosis [56] and rheumatoid arthritis [57], characterized by uncontrolled cell proliferation in an inflammatory micro-environment. It has been proposed that chronic inflammation dictates "accelerated" silencing via DNA methylation changes in an age-related fashion affecting onset of disease and disease progression [41].

Diet may affect DNA methylation patterns through availability of methyl donors (folate, choline, methionine) and related cofactors (vitamins B12, B6, B2), as well as through changes in the activity of DNA methyltransferases [58, 59]. The contribution of dietary factors in combination with methylene tetrahydrofolate reductase (*MTHFR*) gene polymorphisms to cancer susceptibility [60] represents an interesting area of research to explore interactions between nutrients, genetics and epigenetics that modulate cancer development and progression in high-risk populations, particularly when they get older. Since epigenetic changes are potentially reversible, dietary interventions in high-risk groups may reduce the age-related DNA methylation and its associated cancer risk.

As mentioned above, DNA methylation is an early event in carcinogenesis, contributing to continuous positive selection for cancerous and metastatic phenotypes in an age-related fashion. It can also be easily detected in human samples and has been suggested as a useful marker for early detection and prognosis, as well as for monitoring patient response to treatment [61]. The practical idea of reversing age-related DNA hypermethylation has a great impact on cancer p revention and other age-associated diseases. Besides dietary interventions, demethylating agents such as 5-azacytidine, decitabine and antisense DNA methyltransferase, may eventually be effective in cancer prevention and disease progression [62, 63].

17.4
Inflammatory Control of Age-Related Epigenetic Regulators

The "oxidative stress/inflammation" hypothesis of aging, stating that overproduced or uncontrolled reactive oxygen species contribute to the proinflammatory states of the aging process [64], represents a significant advance in aging research as it suggests that interventions targeting the inhibition or delay of these processes can ameliorate age-related pathologies, including cancer. For example, accumulating evidence indicates that the upregulation of a redox responsive transcriptional factor, nuclear factor kappa B (NF-κB), which activates the expression of proinflammatory genes, is involved in both cancer and aging processes [39, 65]. NF-κB is also found to be activated by increased levels of insulin, IGF-1 and leptin and inhibited by increased levels of GSH and adiponectin [35, 66–68].

Consistent with the "oxidative stress/inflammation" hypothesis of aging, ROS generation is found to be gradually increased with age in the rat heart, liver and brain [69, 70]. In addition, the DNA-binding activity of the NF-κB factor is increased during aging in all tissues of the rats and mice tested [71–73], and COX-2, an NF-κB responsive inflammatory gene product, is found to be upregulated in rat kidney during aging [74].

The cancer-prone metabolic phenotype of aging promotes insulin resistance and chronic, low-grade inflammation via fat-derived cytokine production, such as TNF-α and IL-6, which is stimulated by progressive macrophage infiltration of adipose tissue [17]. Diets high in energy, fat and high glycemic index carbohydrates enhance insulin resistance which deteriorates metabolic homeostasis and promotes an inflammatory microenvironment in tissues predisposed to neoplasia [27]. One mechanism through which these factors may elicit their effects is activation of NF-κB. Therefore, cell growth and antiapoptotic effects of insulin are apparently mediated through both IGF-1 and NF-κB signaling. The notion that NF-κB is a key mediator in insulin resistance was strongly supported by studies demonstrating that high doses of salicylates, which inhibit its activity, reversed hyperglycemia, hyperinsulinemia and dyslipidemia in obese rodents and humans with type-2 diabetes, by sensitizing insulin signaling [66, 75]. Notably, this effect of salicylates was independent of COX inhibition by salicylates. Interestingly, aspirin and other non-steroidal anti-inflammatory agents (NSAIDs) which reduce the risk of colon cancer by about half are also found to mediate their chemopreventive action through similar signaling pathways [76].

An interesting, new bioinformatic approach systematically examined fourteen predicted motifs of major regulators of age-dependent gene expression in nine human and mouse tissue types [77]. The regulator most strongly associated with aging was NF-κB. Inhibition of NF-κB in the skin of old mice changed the tissue characteristics and gene expression patterns to those of young mice. This study shows that NF-κB regulates age-related gene expression signatures (aging epigenetics) and its constitutive activation is required for many of the aging phenotypes. Of special interest, Vanden Berghe *et al.* discuss the crucial role of NF-κB in inflammation-triggered epigenetics as epigenetic regulators themselves become susceptible to inflammatory control, and vice versa [78]. Indeed, NF-κB-induced IL-6 was found to elicit epigenetic changes via regulation of DNA methyltransferases [79, 80], while impaired DNA methylation was found to increase IL-6 levels [81].

It seems that NF-κB activation and DNA hypermethylation are both important mediators of aging, inflammation and carcinogenesis. The role of these two mediators in the inhibition of apoptosis makes them key players in positive selection for the survival of cells that maintain the metabolic/inflammation phenotype of aging, which gradually accumulates additional epigenetic "hits" (alterations) accelerating cancer development and progression in a tissue-specific manner. This phenomenon of antiapoptosis is enforced by two distinct mechanisms. One is the suppression of the TNF-α-induced apoptosis by NF-κB [82], and the other is the silencing of key apoptotic genes by DNA hypermethylation, leading to alterations in the "apoptotic methylation signatures" [52]. Sabotaging the process of apoptosis is one of the principal mechanisms in the evolution of cancer that characterizes the aggressive nature of the malignant phenotype. Therefore, it will be interesting to explore whether dietary and/or pharmacological manipulations can modulate the epigenetics of impaired DNA methylation and NF-κB signaling during aging and prevent cancer development.

17.5
Lessons from Anti-Aging Modalities

In experimental animal systems, strategies that prolong life span prevent both the metabolic and the tumorigenic phenotypes of aging. For example, caloric restriction (CR), a well-accepted aging-delay modality, retards or inhibits aging-associated changes, including increased insulin sensitivity [83] body fat accumulation [84], cholesterol and triglyceride levels [85]. When, in a non-insulin-dependent diabetes model, rats were subjected to 30% food restriction, their body weight, intraabdominal fat, plasma triglycerides, insulin and glucose, and tissue triglyceride accumulation were all decreased [86]. CR also inhibits carcinogenesis and the development of both, spontaneous and chemically induced tumors (including colon and mammary) in experimental models [87, 88]. A possible mechanism for the CR anti-aging and anti-tumor effects is a decrease in ROS production and an enhancement of antioxidant defense systems [89]. It has been reported that CR suppressed ROS generation during aging and NF-κB upregulation in the kidney of old rats [64]. CR also limited oxidative stress, as assessed by rapid recovery in GSH levels and inhibited NF-κB DNA binding activity in rat myocardial ischemia-reperfusion injury model [90].

Another dietary intervention that enhances maximum longevity in rats and mice is methionine restriction (MR), which shares many of the metabolic phenotypes of CR without a restriction in energy intake [91–93]. MR prevents the development of age-related pathologies such as colon carcinogenesis, spontaneous testicular tumors and chronic progressive nephropathy [94]. MR also leads to remarkably increased blood glutathione levels and prevents its depletion during aging [92]. Interestingly, decreased blood levels of glutathione, a major regulator of oxidative stress, are often found in the elderly [95] and have been associated with the pathogenesis of diabetes [96] and cancer [97]. In addition, MR profoundly decreases mitochondrial ROS production and oxidative damage to mitochondria DNA and proteins, without decreasing oxygen consumption [98]. These changes are similar to those observed in CR and emphasize the role of mitochondrial function and efficiency in anti-aging mechanisms.

The formation of efficient mitochondria is linked to sirtuin 1 (SIRT1), the mammalian ortholog of Sir2 (silent information regulator 2) and a member of a family of epigenetic enzymes with NAD^+-dependent histone deacetylase (HDAC) activity, known as sirtuins [99]. Sirtuins play a key role in the extension of lifespan mediated by CR, regulating gene expression and cell signaling involved in insulin sensitivity, metabolism, mitochondria function, stress responses, inflammation and aging [100]. SIRT1 deacetylates and activates PPARγ coactivator 1α (PGC-1α), a nutrient sensing factor that increases mitochondria biogenesis and efficiency shifting fat oxidation from incomplete to complete with decreased leak and ROS generation. Interestingly, 25% caloric restriction in overweight men and women (25–50 year) for six months resulted in increased expression of SIRT1 and PGC-1α, the two genes contributing to mitochondria efficiency and nutrient sensing [101]. This finding raises the exciting possibility for diet–epigenetic interactions that can dictate the beneficial metabolic signatures of CR in humans, during adulthood and within a short period of time.

Another important consideration is the role of adiponectin in mediating diet–epigenetic regulation. It seems that adiponectin signaling is involved in the regulation of the SIRT1–PGC-1α pathway by inducing SIRT1 expression [101]. Notably, MR increased plasma adiponectin levels twofold and CR twofold [102]. This remarkable diet-induced increase in adiponectin, a major regulator of insulin sensitivity, was observed in MR rats as early as 8 weeks on the dietary intervention. It is an intriguing finding, considering that the main phenotypic change in CR and MR is a substantial decrease in fat mass. Thus, mechanistic studies exploring the specific diet-induced epigenetic changes in adiponectin expression may lead to a better understanding of the metabolic benefits of these anti-aging models.

As was mentioned above, NF-κB activity is continually required to maintain the cancer-prone metabolic phenotype of aging. In this regard, anti-aging strategies interrupt cross-talk pathways where NF-κB signaling is significantly involved. For example, upregulation of SIRT1 may, in part, explain the cancer preventive effects of CR through inhibition of NF-B activity and subsequent decrease in cell survival signals restoring TNF-α-induced apoptosis [103]. SIRT1-induced deacetylation and inhibition of NF-κB may also be responsible for the CR-like effects of resveratrol (the red wine polyphenol) which was found to enhance metabolism and survival of mice on a high-caloric diet [104]. This study showed that resveratrol shifted expression patterns of mice on a high-caloric diet toward those on a standard diet, suggesting a nutrient–epigenetic interaction. In addition, an interesting, recently published, mechanistic study revealed that the function of another sirtuin, SIRT6, prevents hyperinduction of NF-κB-dependent gene expression (including those involved in aging), as it interacts with chromatin-bound NF-κB, deacetylating it and destabilizing NF-κB binding to chromatin [105].

Altogether, these findings suggest that attenuation of NF-κB signaling via SIRT1, or SIRT6, or even adiponectin-enhancing manipulations, by dietary or/and pharmacological interventions, has the potential to delay the aging process and prevent cancer and other age-related pathologies. To this end, a new area of research has evolved that focuses on the development of "CR-mimetics", identifying compounds that mimic CR effects, including PPARs and sirtuins activators, as well as plant-derived polyphenols (resveratrol) and insulin action enhancers.

17.6 Conclusions

During the process of aging, a metabolically declined phenotype accelerates epigenomic dysregulations and selectively maintains those that promote cancer risk. There is strong evidence for the involvement of NF-κB signaling as a major regulator of the cancer-prone metabolic phenotype of aging. DNA hypermethylation, sirtuins and lipokines are also important modulators of this phenotype (Figure 17.1). A clear view of the busy crossroads between metabolism, chronic inflammation, aging and cancer and a better understanding of the underlying mechanisms for epigenetic regulations/dysregulations are essential for the design of targeted therapies to prevent, or even reverse, the pathologic phenotypes that attack the rapidly growing, overweight and obese aging population.

Strikingly, the proportion of all deaths from cancer that is attributed to overweight and obesity in the United States among people at age 50 or older may be as high as 14% in men and 20% in women [106]. It is estimated that 90 000 cancer deaths could be prevented each year if people could maintain normal weight (BMI < 25) throughout life. Along these lines, there is sufficient evidence of a

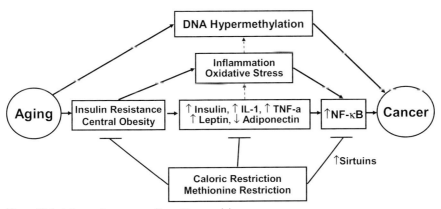

Figure 17.1 Aging and cancer predisposition model.

cancer-preventive effect of avoidance of weight gain for aging-related cancers, including cancer of the breast (in postmenopausal women), colon, endometrium, prostate, kidney and esophagus. The magnitude of this major public health problem is likely to grow as the percentage of elderly people is expected to double by the year 2025 [107].

Therefore, it will be a great challenge for future anti-aging and anticancer therapies to selectively target key signaling networks and intervene with safe epigenetic modulators, including dietary factors and/or pharmaceutical agents, protecting against multiple "hits" common in aging and cancer development. Since these are long-term processes, specific dietary manipulations would present the most suitable, mainstream approaches for lifelong and non-toxic practices recommended to the general public and enhanced by selective epigenetic-altering agents for high-risk populations.

References

1 Jemal, A., Siegel, R., Ward, E., Hao, Y., Xu, J., Murray, T., and Thun, M.J. (2008) Cancer statistics. *CA Cancer J. Clin.*, **58**, 71–96.

2 DePinho, R.A. (2000) The age of cancer. *Nature*, **408**, 248–254.

3 Anisimov, V.N. (1989) Age-related mechanisms of susceptibility to carcinogenesis. *Semin. Oncol.*, **16**, 10–19.

4 Bjornsson, H.T., Fallin, M.D., and Feinberg, A.P. (2004) An integrated epigenetic and genetic approach to common human disease. *Trends Genet.*, **20**, 350–358.

5 Bird, A. (2007) Perceptions of epigenetics. *Nature*, **447**, 396–398.

6 Jones, R.S. (2007) Epigenetics: reversing the "irreversible". *Nature*, **450**, 357–359.

7 Fraga, M.F., and Esteller, M. (2007) Epigenetics and aging: the targets and the marks. *Trends Genet.*, **23**, 413–418.

8 Ross, S.A., and Milner, J.A. (2007) Epigenetic modulation and cancer: effect of metabolic syndrome? *Am. J. Clin. Nutr.*, **86**, s872–s877.

9 Esteller, M., Fraga, M.F., Paz, M.F., Campo, E., Colomer, D., Novo, F.J., Calasanz, M.J., Galm, O., Guo, M., Benitez, J., and Herman, J.G. (2002) Cancer epigenetics and methylation. *Science*, **297**, 1807–1808, discussion 1807–1808.

10 Issa, J.P., Ottaviano, Y.L., Celano, P., Hamilton, S.R., Davidson, N.E., and Baylin, S.B. (1994) Methylation of the oestrogen receptor CpG island links ageing and neoplasia in human colon. *Nat. Genet.*, **7**, 536–540.

11 Issa, J.P. (2008) Cancer prevention: epigenetics steps up to the plate. *Cancer Prev. Res. (Phila. Pa.)*, **1**, 219–222.

12 Barzilai, N., and Gupta, G. (1999) Interaction between aging and syndrome X: new insights on the pathophysiology of fat distribution. *Ann. N. Y. Acad. Sci.*, **892**, 58–72.

13 Kotz, C.M., Billington, C.J., and Levine, A.S. (1999) Obesity and aging. *Clin. Geriatr. Med.*, **15**, 391–412.

14 Reaven, G.M. (1988) Banting lecture 1988. Role of insulin resistance in human disease. *Diabetes*, **37**, 1595–1607.

15 Shimokata, H., Tobin, J.D., Muller, D.C., Elahi, D., Coon, P.J., and Andres, R. (1989) Studies in the distribution of body fat: I. Effects of age, sex, and obesity. *J. Gerontol.*, **44**, M66–73.

16 Bjorntorp, P. (1991) Adipose tissue distribution and function. *Int. J. Obes.*, **15** (Suppl. 2), 67–81.

17 Wellen, K.E., and Hotamisligil, G.S. (2003) Obesity-induced inflammatory changes in adipose tissue. *J. Clin. Invest.*, **112**, 1785–1788.

18 Hotamisligil, G.S., Shargill, N.S., and Spiegelman, B.M. (1993) Adipose expression of tumor necrosis factor-alpha: direct role in obesity-linked insulin resistance. *Science*, **259**, 87–91.

19 Ahren, B., Mansson, S., Gingerich, R.L., and Havel, P.J. (1997) Regulation of plasma leptin in mice: influence of age, high-fat diet, and fasting. *Am. J. Physiol.*, **273**, R113–R120.

20 Scherer, P.E., Williams, S., Fogliano, M., Baldini, G., and Lodish, H.F. (1995) A novel serum protein similar to C1q, produced exclusively in adipocytes. *J. Biol. Chem.*, **270**, 26746–26749.

21 Giovannucci, E. (1995) Insulin and colon cancer. *Cancer Causes Control*, **6**, 164–179.

22 McKeown-Eyssen, G.E., Bright-See, E., Bruce, W.R., Jazmaji, V., Cohen, L.B., Pappas, S.C., and Saibil, F.G. (1994) A randomized trial of a low fat high fibre diet in the recurrence of colorectal polyps. Toronto Polyp Prevention Group. *J. Clin. Epidemiol.*, **47**, 525–536.

23 Komninou, D., Ayonote, A., Richie, J.P. Jr., and Rigas, B. (2003) Insulin resistance and its contribution to colon carcinogenesis. *Exp. Biol. Med. (Maywood)*, **228**, 396–405.

24 Barzilai, N., and Rossetti, L. (1996) Age-related changes in body composition are associated with hepatic insulin resistance in conscious rats. *Am. J. Physiol.*, **270**, E930–E936.

25 Bluher, M., Kahn, B.B., and Kahn, C.R. (2003) Extended longevity in mice lacking the insulin receptor in adipose tissue. *Science*, **299**, 572–574.

26 Holzenberger, M., Dupont, J., Ducos, B., Leneuve, P., Geloen, A., Even, P.C., Cervera, P., and Le Bouc, Y. (2003) IGF-1 receptor regulates lifespan and resistance to oxidative stress in mice. *Nature*, **421**, 182–187.

27 Bruce, W.R., Wolever, T.M., and Giacca, A. (2000) Mechanisms linking diet and colorectal cancer: the possible role of insulin resistance. *Nutr. Cancer*, **37**, 19–26.

28 Gabriely, I., Ma, X.H., Yang, X.M., Rossetti, L., and Barzilai, N. (2002) Leptin resistance during aging is independent of fat mass. *Diabetes*, **51**, 1016–1021.

29 Rose, D.P., Komninou, D., and Stephenson, G.D. (2004) Obesity, adipocytokines, and insulin resistance in breast cancer. *Obes. Rev.*, **5**, 153–165.

30 Yamauchi, T., Kamon, J., Waki, H., Terauchi, Y., Kubota, N., Hara, K., Mori, Y., Ide, T., Murakami, K., Tsuboyama-Kasaoka, N., Ezaki, O., Akanuma, Y., Gavrilova, O., Vinson, C., Reitman, M.L., Kagechika, H., Shudo, K., Yoda, M., Nakano, Y., Tobe, K., Nagai, R., Kimura, S., Tomita, M., Froguel, P., and Kadowaki, T. (2001) The fat-derived hormone adiponectin reverses insulin resistance associated with both lipoatrophy and obesity. *Nat. Med.*, **7**, 941–946.

31 Arita, Y., Kihara, S., Ouchi, N., Takahashi, M., Maeda, K., Miyagawa, J., Hotta, K., Shimomura, I., Nakamura, T., Miyaoka, K., Kuriyama, H., Nishida, M., Yamashita, S., Okubo, K., Matsubara, K., Muraguchi, M., Ohmoto, Y., Funahashi, T., and Matsuzawa, Y. (1999) Paradoxical decrease of an adipose-specific protein, adiponectin, in obesity. *Biochem. Biophys. Res. Commun.*, **257**, 79–83.

32 Barb, D., Williams, C.J., Neuwirth, A.K., and Mantzoros, C.S. (2007) Adiponectin in relation to malignancies: a review of existing basic research and clinical evidence. *Am. J. Clin. Nutr.*, **86**, s858–866.

33 Garofalo, C., and Surmacz, E. (2006) Leptin and cancer. *J. Cell Physiol.*, **207**, 12–22.

34 Stephenson, G.D., and Rose, D.P. (2003) Breast cancer and obesity: an update. *Nutr. Cancer*, **45**, 1–16.

35 Ouchi, N., Kihara, S., Arita, Y., Okamoto, Y., Maeda, K., Kuriyama, H., Hotta, K., Nishida, M., Takahashi, M., Muraguchi, M., Ohmoto, Y., Nakamura, T., Yamashita, S., Funahashi, T., and Matsuzawa, Y. (2000) Adiponectin, an adipocyte-derived plasma protein, inhibits endothelial NF-{kappa}B signaling through a cAMP-dependent pathway. *Circulation*, **102**, 1296–1301.

36 Maeda, N., Takahashi, M., Funahashi, T., Kihara, S., Nishizawa, H., Kishida,

K., Nagaretani, H., Matsuda, M., Komuro, R., Ouchi, N., Kuriyama, H., Hotta, K., Nakamura, T., Shimomura, I., and Matsuzawa, Y. (2001) PPARgamma ligands increase expression and plasma concentrations of adiponectin, an adipose-derived protein. *Diabetes*, **50**, 2094–2099.

37 Gelman, L., Fruchart, J.C., and Auwerx, J. (1999) An update on the mechanisms of action of the peroxisome proliferator-activated receptors (PPARs) and their roles in inflammation and cancer. *Cell Mol. Life Sci.*, **55**, 932–943.

38 Pineda Torra, I., Gervois, P., and Staels, B. (1999) Peroxisome proliferator-activated receptor alpha in metabolic disease, inflammation, atherosclerosis and aging. *Curr. Opin. Lipidol.*, **10**, 151–159.

39 Spencer, N., Poynter, M., Im, S., and Daynes, R. (1997) Constitutive activation of NF-kappa B in an animal model of aging. *Int. Immunol.*, **9**, 1581–1588.

40 Trevisan, M., Liu, J., Bahsas, F.B., and Menotti, A. (1998) Syndrome X and mortality: a population-based study. Risk Factor and Life Expectancy Research Group. *Am. J. Epidemiol.*, **148**, 958–966.

41 Issa, J.P., Ahuja, N., Toyota, M., Bronner, M.P., and Brentnall, T.A. (2001) Accelerated age-related CpG island methylation in ulcerative colitis. *Cancer Res.*, **61**, 3573–3577.

42 Fearon, E.R., and Vogelstein, B. (1990) A genetic model for colorectal tumorigenesis. *Cell*, **61**, 759–767.

43 Nagasaka, T., Sharp, G.B., Notohara, K., Kambara, T., Sasamoto, H., Isozaki, H., MacPhee, D.G., Jass, J.R., Tanaka, N., and Matsubara, N. (2003) Hypermethylation of O6-methylguanine-DNA methyltransferase promoter may predict nonrecurrence after chemotherapy in colorectal cancer cases. *Clin. Cancer Res.*, **9**, 5306–5312.

44 Shen, L., Kondo, Y., Rosner, G.L., Xiao, L., Hernandez, N.S., Vilaythong, J., Houlihan, P.S., Krouse, R.S., Prasad, A.R., Einspahr, J.G., Buckmeier, J., Alberts, D.S., Hamilton, S.R., and Issa, J.P. (2005) MGMT promoter methylation and field defect in sporadic colorectal cancer. *J. Natl. Cancer Inst.*, **97**, 1330–1338.

45 Ottaviano, Y.L., Issa, J.P., Parl, F.F., Smith, H.S., Baylin, S.B., and Davidson, N.E. (1994) Methylation of the estrogen receptor gene CpG island marks loss of estrogen receptor expression in human breast cancer cells. *Cancer Res.*, **54**, 2552–2555.

46 Yang, X., Phillips, D.L., Ferguson, A.T., Nelson, W.G., Herman, J.G., and Davidson, N.E. (2001) Synergistic activation of functional estrogen receptor (ER)-alpha by DNA methyltransferase and histone deacetylase inhibition in human ER-alpha-negative breast cancer cells. *Cancer Res.*, **61**, 7025–7029.

47 Lapidus, R.G., Nass, S.J., and Davidson, N.E. (1998) The loss of estrogen and progesterone receptor gene expression in human breast cancer. *J. Mammary Gland Biol. Neoplasia*, **3**, 85–94.

48 Nass, S.J., Herman, J.G., Gabrielson, E., Iversen, P.W., Parl, F.F., Davidson, N.E., and Graff, J.R. (2000) Aberrant methylation of the estrogen receptor and E-cadherin 5′ CpG islands increases with malignant progression in human breast cancer. *Cancer Res.*, **60**, 4346–4348.

49 Ferguson, A.T., Lapidus, R.G., Baylin, S.B., and Davidson, N.E. (1995) Demethylation of the estrogen receptor gene in estrogen receptor-negative breast cancer cells can reactivate estrogen receptor gene expression. *Cancer Res.*, **55**, 2279–2283.

50 Bovenzi, V., and Momparler, R.L. (2001) Antineoplastic action of 5-aza-2′-deoxycytidine and histone deacetylase inhibitor and their effect on the expression of retinoic acid receptor beta and estrogen receptor alpha genes in breast carcinoma cells. *Cancer Chemother. Pharmacol.*, **48**, 71–76.

51 Li, L.C., Chui, R., Nakajima, K., Oh, B.R., Au, H.C., and Dahiya, R. (2000) Frequent methylation of estrogen receptor in prostate cancer: correlation with tumor progression. *Cancer Res.*, **60**, 702–706.

52 Murphy, T.M., Perry, A.S., and Lawler, M. (2008) The emergence of DNA methylation as a key modulator of

aberrant cell death in prostate cancer. *Endocr. Relat. Cancer*, **15**, 11–25.
53 Jones, P.A., and Laird, P.W. (1999) Cancer epigenetics comes of age. *Nat. Genet.*, **21**, 163–167.
54 Gao, W., Kondo, Y., Shen, L., Shimizu, Y., Sano, T., Yamao, K., Natsume, A., Goto, Y., Ito, M., Murakami, H., Osada, H., Zhang, J., Issa, J.-P.J., and Sekido, Y. (2008) Variable DNA methylation patterns associated with progression of disease in hepatocellular carcinomas. *Carcinogenesis*, **29**, 1901–1910.
55 Yutaka Kondo, Y.K., Sakamoto, M., Mizokami, M., Ueda, R., and Hirohashi, S. (2000) Genetic instability and aberrant DNA methylation in chronic hepatitis and cirrhosis – a comprehensive study of loss of heterozygosity and microsatellite instability at 39 loci and DNA hypermethylation on 8 CpG islands in microdissected specimens from patients with hepatocellular carcinoma. *Hepatology*, **32**, 970–979.
56 Dong, C., Yoon, W., Goldschmidt-Clermont, P.J. (2002) DNA methylation and atherosclerosis. *J. Nutr.*, **132**, 2406S–2409.
57 Sánchez-Pernaute, O., Ospelt, C., Neidhart, M., and Gay, S. (2008) Epigenetic clues to rheumatoid arthritis. *J. Autoimmunol*, **30**, 12–20.
58 Ross, S.A. (2003) Diet and DNA methylation interactions in cancer prevention. *Ann. N. Y. Acad. Sci.*, **983**, 197–207.
59 Davis, C.D., and Uthus, E.O. (2004) DNA methylation, cancer susceptibility, and nutrient interactions. *Exp. Biol. Med. (Maywood)*, **229**, 988–995.
60 Curtin, K., Bigler, J., Slattery, M.L., Caan, B., Potter, J.D., and Ulrich, C.M. (2004) MTHFR C677T and A1298C polymorphisms: diet, estrogen, and risk of colon cancer. *Cancer Epidemiol. Biomarkers Prev.*, **13**, 285–292.
61 Lofton-Day, C., and Lesche, R. (2003) DNA methylation markers in patients with gastrointestinal cancers. Current understanding, potential applications for disease management and development of diagnostic tools. *Dig. Dis.*, **21**, 299–308.
62 Das, P.M., Singal, R. (2004) DNA methylation and cancer. *J. Clin. Oncol.*, **22**, 4632–4642.
63 Yoo, C.B., Chuang, J.C., Byun, H.-M., Egger, G., Yang, A.S., Dubeau, L., Long, T., Laird, P.W., Marquez, V.E., and Jones, P.A. (2008) Long-term epigenetic therapy with oral zebularine has minimal side effects and prevents intestinal tumors in mice. *Cancer Prev. Res.*, **1**, 233–240.
64 Chung, H.Y., Kim, H.J., Kim, J.W., and Yu, B.P. (2001) The inflammation hypothesis of aging: molecular modulation by calorie restriction. *Ann. N. Y. Acad. Sci.*, **928**, 327–335.
65 Mayo, M.W., Wang, C.-Y., Cogswell, P.C., Rogers-Graham, K.S., Lowe, S.W., Der, C.J., and Baldwin, A.S. Jr. (1997) Requirement of NF-{kappa}B activation to suppress p53-independent apoptosis induced by oncogenic ras. *Science*, **278**, 1812–1815.
66 Yuan, M., Konstantopoulos, N., Lee, J., Hansen, L., Li, Z.W., Karin, M., and Shoelson, S.E. (2001) Reversal of obesity- and diet-induced insulin resistance with salicylates or targeted disruption of Ikkbeta. *Science*, **293**, 1673–1677.
67 Remacle-Bonnet, M.M., Garrouste, F.L., Heller, S., Andre, F., Marvaldi, J.L., and Pommier, G.J. (2000) Insulin-like growth factor-I protects colon cancer cells from death factor-induced apoptosis by potentiating tumor necrosis factor alpha-induced mitogen-activated protein kinase and nuclear factor kappaB signaling pathways. *Cancer Res.*, **60**, 2007–2017.
68 Staal, F.J., Anderson, M.T., Staal, G.E., Herzenberg, L.A., and Gitler, C. (1994) Redox regulation of signal transduction: tyrosine phosphorylation and calcium influx. *Proc. Natl. Acad. Sci. U. S. A.*, **91**, 3619–3622.
69 Bejma, J., Ramires, P., and Ji, L.L. (2000) Free radical generation and oxidative stress with ageing and exercise: differential effects in the myocardium and liver. *Acta. Physiol. Scand.*, **169**, 343–351.

70 Baek, B.S., Kwon, H.J., Lee, K.H., Yoo, M.A., Kim, K.W., Ikeno, Y., Yu, B.P., and Chung, H.Y. (1999) Regional difference of ROS generation, lipid peroxidation, and antioxidant enzyme activity in rat brain and their dietary modulation. *Arch. Pharm. Res.*, **22**, 361–366.

71 Helenius, M., Hanninen, M., Lehtinen, S.K., and Salminen, A. (1996) Aging-induced up-regulation of nuclear binding activities of oxidative stress responsive NF-kB transcription factor in mouse cardiac muscle. *J. Mol. Cell. Cardiol.*, **28**, 487–498.

72 Korhonen, P., Helenius, M., and Salminen, A. (1997) Age-related changes in the regulation of transcription factor NF-kappa B in rat brain. *Neurosci. Lett.*, **225**, 61–64.

73 Komninou, D. (2000) Oxidative Stress and Aging in Colon Carcinogenesis. Pathology, Ph.D. Thesis. New York Medical College, Valhalla, New York.

74 Kim, H.J., Kim, K.W., Yu, B.P., and Chung, H.Y. (2000) The effect of age on cyclooxygenase-2 gene expression: NF-kappaB activation and IkappaBalpha degradation. *Free Radic. Biol. Med.*, **28**, 683–692.

75 Hundal, R.S., Petersen, K.F., Mayerson, A.B., Randhawa, P.S., Inzucchi, S., Shoelson, S.E., and Shulman, G.I. (2002) Mechanism by which high-dose aspirin improves glucose metabolism in type 2 diabetes. *J. Clin. Invest.*, **109**, 1321–1326.

76 Tegeder, I., Pfeilschifter, J., and Geisslinger, G. (2001) Cyclooxygenase-independent actions of cyclooxygenase inhibitors. *FASEB J.*, **15**, 2057–2072.

77 Adler, A.S., Sinha, S., Kawahara, T.L., Zhang, J.Y., Segal, E., and Chang, H.Y. (2007) Motif module map reveals enforcement of aging by continual NF-kappaB activity. *Genes Dev.*, **21**, 3244–3257.

78 Vanden Berghe, W., Ndlovu, M.N., Hoya-Arias, R., Dijsselbloem, N., Gerlo, S., and Haegeman, G. (2006) Keeping up NF-kappaB appearances: epigenetic control of immunity or inflammation-triggered epigenetics. *Biochem. Pharmacol.*, **72**, 1114–1131.

79 Peng, B., Hodge, D.R., Thomas, S.B., Cherry, J.M., Munroe, D.J., Pompeia, C., Xiao, W., and Farrar, W.L. (2005) Epigenetic silencing of the human nucleotide excision repair gene, hHR23B, in interleukin-6-responsive multiple myeloma KAS-6/1 cells. *J. Biol. Chem.*, **280**, 4182–4187.

80 Hodge, D.R., Peng, B., Cherry, J.C., Hurt, E.M., Fox, S.D., Kelley, J.A., Munroe, D.J., and Farrar, W.L. (2005) Interleukin 6 supports the maintenance of p53 tumor suppressor gene promoter methylation. *Cancer Res.*, **65**, 4673–4682.

81 Milutinovic, S., Zhuang, Q., Niveleau, A., and Szyf, M. (2003) Epigenomic stress response. Knockdown of DNA methyltransferase 1 triggers an intra-S-phase arrest of DNA replication and induction of stress response genes. *J. Biol. Chem.*, **278**, 14985–14995.

82 Van Antwerp, D.J., Martin, S.J., Kafri, T., Green, D.R., and Verma, I.M. (1996) Suppression of TNF-alpha-induced apoptosis by NF-kappaB. *Science*, **274**, 787–789.

83 Barzilai, N., and Rossetti, L. (1995) Relationship between changes in body composition and insulin responsiveness in models of the aging rat. *Am. J. Physiol.*, **269**, E591–E597.

84 Greenberg, J.A., and Boozer, C.N. (1999) The leptin-fat ratio is constant, and leptin may be part of two feedback mechanisms for maintaining the body fat set point in non-obese male Fischer 344 rats. *Horm. Metab. Res.*, **31**, 525–532.

85 Masoro, E.J. (2005) Overview of caloric restriction and ageing. *Mech. Ageing. Dev.*, **126**, 913–922.

86 Man, Z.W., Hirashima, T., Mori, S., and Kawano, K. (2000) Decrease in triglyceride accumulation in tissues by restricted diet and improvement of diabetes in Otsuka Long-Evans Tokushima fatty rats, a non-insulin-dependent diabetes model. *Metabolism*, **49**, 108–114.

87 Steinbach, G., Kumar, S.P., Reddy, B.S., Lipkin, M., and Holt, P.R. (1993) Effects of caloric restriction and dietary fat on epithelial cell proliferation in rat colon. *Cancer Res.*, **53**, 2745–2749.

88 Klurfeld, D.M., Welch, C.B., Lloyd, L.M., and Kritchevsky, D. (1989) Inhibition of DMBA-induced mammary tumorigenesis by caloric restriction in rats fed high-fat diets. *Int. J. Cancer*, **43**, 922–925.

89 Hursting, S.D., Lavigne, J.A., Berrigan, D., Perkins, S.N., and Barrett, J.C. (2003) Calorie restriction, aging, and cancer prevention: mechanisms of action and applicability to humans. *Annu. Rev. Med.*, **54**, 131–152.

90 Chandrasekar, B., Nelson, J.F., Colston, J.T., and Freeman, G.L. (2001) Calorie restriction attenuates inflammatory responses to myocardial ischemia-reperfusion injury. *Am. J. Physiol. Heart Circ. Physiol.*, **280**, H2094–H2102.

91 Orentreich, N., Matias, J.R., DeFelice, A., and Zimmerman, J.A. (1993) Low methionine ingestion by rats extends life span. *J. Nutr.*, **123**, 269–274.

92 Richie, J.P. Jr., Leutzinger, Y., Parthasarathy, S., Malloy, V., Orentreich, N., and Zimmerman, J.A. (1994) Methionine restriction increases blood glutathione and longevity in F344 rats. *FASEB J.*, **8**, 1302–1307.

93 Miller, R.A., Buehner, G., Chang, Y., Harper, J.M., Sigler, R., and Smith-Wheelock, M. (2005) Methionine-deficient diet extends mouse lifespan, slows immune and lens aging, alters glucose, T4, IGF-I and insulin levels, and increases hepatocyte MIF levels and stress resistance. *Aging Cell*, **4**, 119–125.

94 Komninou, D., Leutzinger, Y., Reddy, B.S., and Richie, J.P. Jr. (2006) Methionine restriction inhibits colon carcinogenesis. *Nutr. Cancer*, **54**, 202–208.

95 Lang, C.A., Naryshkin, S., Schneider, D.L., Mills, B.J., and Lindeman, R.D. (1992) Low blood glutathione levels in healthy aging adults. *J. Lab. Clin. Med.*, **120**, 720–725.

96 Murakami, K., Kondo, T., Ohtsuka, Y., Fujiwara, Y., Shimada, M., and Kawakami, Y. (1989) Impairment of glutathione metabolism in erythrocytes from patients with diabetes mellitus. *Metabolism*, **38**, 753–758.

97 Ames, B.N., and Gold, L.S. (1991) Endogenous mutagens and the causes of aging and cancer. *Mutat. Res.*, **250**, 3–16.

98 Sanz, A., Caro, P., Ayala, V., Portero-Otin, M., Pamplona, R., and Barja, G. (2006) Methionine restriction decreases mitochondrial oxygen radical generation and leak as well as oxidative damage to mitochondrial DNA and proteins. *FASEB J.*, **20**, 1064–1073.

99 Nemoto, S., Fergusson, M.M., and Finkel, T. (2005) SIRT1 functionally interacts with the metabolic regulator and transcriptional coactivator PGC-1{alpha}. *J. Biol. Chem.*, **280**, 16456–16460.

100 Guarente, L. (2006) Sirtuins as potential targets for metabolic syndrome. *Nature*, **444**, 868–874.

101 Civitarese, A.E., Carling, S., Heilbronn, L.K., Hulver, M.H., Ukropcova, B., Deutsch, W.A., Smith, S.R., and Ravussin, E. (2007) Calorie restriction increases muscle mitochondrial biogenesis in healthy humans. *PLoS Medicine*, **4**, e76.

102 Malloy, V.L., Krajcik, R.A., Bailey, S.J., Hristopoulos, G., Plummer, J.D., and Orentreich, N. (2006) Methionine restriction decreases visceral fat mass and preserves insulin action in aging male Fischer 344 rats independent of energy restriction. *Aging Cell*, **5**, 305–314.

103 Yeung, F., Hoberg, J.E., Ramsey, C.S., Keller, M.D., Jones, D.R., Frye, R.A., and Mayo, M.W. (2004) Modulation of NF-kappaB-dependent transcription and cell survival by the SIRT1 deacetylase. *EMBO J.*, **23**, 2369–2380.

104 Baur, J.A., Pearson, K.J., Price, N.L., Jamieson, H.A., Lerin, C., Kalra, A., Prabhu, V.V., Allard, J.S., Lopez-Lluch, G., Lewis, K., Pistell, P.J., Poosala, S., Becker, K.G., Boss, O., Gwinn, D., Wang, M., Ramaswamy, S., Fishbein, K.W., Spencer, R.G., Lakatta, E.G., Le Couteur, D., Shaw, R.J., Navas, P., Puigserver, P., Ingram, D.K., de Cabo, R., and Sinclair, D.A. (2006) Resveratrol improves health and survival of mice on a high-calorie diet. *Nature*, **444**, 337–342.

105 Kawahara, T.L., Michishita, E., Adler, A.S., Damian, M., Berber, E., Lin, M., McCord, R.A., Ongaigui, K.C., Boxer, L.D., Chang, H.Y., and Chua, K.F. (2009) SIRT6 links histone H3 lysine 9 deacetylation to NF-kappaB-dependent gene expression and organismal life span. *Cell*, **136**, 62–74.

106 Calle, E.E., Rodriguez, C., Walker-Thurmond, K., and Thun, M.J. (2003) Overweight, obesity, and mortality from cancer in a prospectively studied cohort of U.S. adults. *N. Engl. J. Med.*, **348**, 1625–1638.

107 Campbell, P.R. (1996) Population Projections for States by Age, Sex, Race, and Hispanic Origin: 1995 to 2025. U.S. Bureau of the Census. Population Division: PPL-47.

18
The Impact of Genetic and Environmental Factors in Neurodegeneration: Emerging Role of Epigenetics

Lucia Migliore and Fabio Coppedè

Abstract

In this chapter we provide an overview of recent advances in our understanding of genetic and environmental factors in complex neurodegenerative diseases such as Alzheimer's disease, Parkinson's disease and Amyotrophic Lateral Sclerosis. The discovery of several genes responsible for the familial forms has lent new insights into the molecular pathways involved in the selective neuronal degeneration which is specific for each of these disorders. Nevertheless, the vast majority of the cases occur as sporadic forms, likely resulting from complex gene–gene and gene–environment interplay. Several environmental factors, including metals, pesticides, head injuries, lifestyles and dietary habits have been associated with increased disease risk or with protection. Hundreds of genetic polymorphisms have been investigated as possible risk factors for the sporadic forms, but results are often conflicting, not confirmed or inconclusive. It is likely that the level of expression of genes that have a fundamental role in age-related diseases, including neurodegenerative ones, can be altered due to the methylation status of their promoters. Until now only a limited number of environmental agents affecting the epigenome have been identified. Dietary modification can indeed have a profound effect on DNA methylation and genomic imprinting. Recent evidence supports the importance of modifications of our epigenome by environmental agents acting as ROS producers. Many of the processes with a key role in neurodegeneration, can be now analyzed in the light of the new epigenetic knowledge to facilitate the implementation of future disease prevention strategies.

18.1
Neurodegenerative Diseases

Neurodegenerative diseases are a heterogeneous group of pathologies of the nervous system which includes complex multifactorial diseases, such as Alzheimer's disease (AD), Parkinson's disease (PD) and amyotrophic lateral

Epigenetics and Human Health
Edited by Alexander G. Haslberger, Co-edited by Sabine Gressler
Copyright © 2010 WILEY-VCH Verlag GmbH & Co. KGaA, Weinheim
ISBN: 978-3-527-32427-9

sclerosis (ALS), monogenic disorders such as Huntington's disease (HD), and others for whom inherited, sporadic and transmissible forms are known.

AD, PD and ALS are the three major neurodegenerative diseases, affecting several million people worldwide. They are defined as complex multifactorial disorders since both familial and sporadic forms are known. Familial forms represent a minority of the cases (ranging from 5 to 10% of the total), whereas the vast majority of AD, PD and ALS occurs as sporadic forms, likely resulting from the contribution of complex interactions between genetic and environmental factors superimposed on slow, sustained neuronal dysfunction due to aging. Several causative genes for the familial forms have been discovered in recent years, they are inherited as Mendelian traits and their discovery has led to a better comprehension of the molecular pathways responsible for the selective neuronal degeneration which is specific for each of these disorders (Table 18.1).

AD represents the most common form of dementia in the elderly, characterized by progressive loss of memory and cognitive capacity severe enough to interfere with daily functioning and the quality of life. The cardinal histopathologic lesions of AD are senile plaques, composed of extracellular deposits of amyloid beta (Aβ) peptides and neurofibrillary tangles, composed of intraneuronal tau protein aggregates [1]. PD is the second most common neurodegenerative disorder after AD. Pathologically, PD is characterized by progressive and profound loss of neuromelanin containing dopaminergic neurons in the substantia nigra with the presence of cytoplasmic inclusions termed Lewy bodies (LB) and containing aggregates of α-synuclein as well as other substances [2]. ALS, also known as motor neuron disease (MND), is a progressive disorder characterized by the degeneration of motor neurons of the motor cortex, brainstem and spinal cord. The course of ALS is inexorably progressive, with 50% of the patients dying within 3 years of onset [3]. HD is a monogenic disorder transmitted as an autosomal dominant trait, meaning that all the cases result from mutations of a single gene. However, a contribution from other genes and environmental factors to age at onset and progression of the disease is indicated by several studies. The disease is characterized by selective degeneration of medium spiny GABAergic neurons in the striatum, resulting in a progressive atrophy of the caudate nucleus, putamen and globus pallidus [4].

18.2
The Role of Causative and Susceptibility Genes in Neurodegenerative Diseases

AD is a genetically complex and heterogeneous disorder. Rare, fully penetrant mutations in three genes (*APP*, *PSEN1* and *PSEN2*) are responsible for familial early onset (<65 years) autosomal dominant forms (EOAD) (Table 18.1). The amyloid precursor protein gene (*APP*) encodes for the amyloid precursor protein (APP). APP is an integral membrane protein and its cleavage mediated by β- and γ-secretases results in the production of Aβ peptides denoted as Aβ40 and Aβ42.

Table 18.1 Causative genes for familial forms of neurodegenerative diseases.

Designation	Locus	Gene	Inheritance	Function or probable function
AD1	21q21.2	Amyloid precursor protein	AD	Precursor protein of Aβ peptides
AD3	14q24.3	Presenilin 1	AD	Component of the γ-secretase complex
AD4	1q31–q42	Presenilin 2	AD	Component of the γ-secretase complex
PARK1 and PARK4	4q21	α-Synuclein	AD	Presynaptic protein, component of Lewy Bodies
PARK2	6q25.2–q27	Parkin	AR	Ubiquitin E3 ligase
PARK3	2p13	Unknown	AD	Unknowm
PARK5	4p14	UCH-L1	AD	Ubiquitin C-terminal hydrolase
PARK6	1p35–36	PINK1	AR	Mitochondrial kinase
PARK7	1p36	DJ-1	AR	Mitochondrial protein, antioxidant defence
PARK8	12p11.2	LRRK2	AD	Protein kinase
PARK9	1p36	ATP13A2	AR	Lysosomal 5 P-type ATPase
PARK10	1p32	Unknown	AD	Unknown
PARK11	2q36–37	Unknown	AD	Unknown
PARK12	Xq21–q25	Unknown	Unknown	Unknown
PARK13	2p12	OMI/HTRA2	Unknown	Mitochondrial serine protease
ALS1	21q21	SOD1	AD	Superoxide dismutase, Antioxidant defense
ALS2	2q33	Alsin	AR	Guanine nucleotide exchange factor for RAB5A
ALS3	18q21	Unknown	AD	Unknown
ALS4	9q34	Senataxin	AD	DNA/RNA helicase, DNA repair
ALS5	15q15.1–q21.1	Unknown	AR	Unknown

Table 18.1 Continued.

Designation	Locus	Gene	Inheritance	Function or probable function
ALS6	16q12.1–q12.2	Unknown	AD	Unknown
ALS7	20pter	Unknown	AD	Unknown
ALS8	20q13.3	VAPB	AD	Vesicle associated membrane protein
ALS and FTDP	17q21	MAPT	AD	Assembly of microtubules
MND, dynactin type	2p13	Dynactin 1	AD	Promotion of synapse stability
HD	4p16.3	IT15	AD	Huntingtin, proposed role in vesicular trafficking

Mutations in the *APP* gene either increase total Aβ levels, or just Aβ42 alone, which is the major component of senile plaques. The presenilin genes (*PSEN1* and *PSEN2*) encode for presenilin proteins 1 and 2, respectively. Presenilins are components of the γ-secretase complex and mediate the cleavage of APP that leads to the production of Aβ peptides. *PSEN1* and *PSEN2* mutations all result in an increased production of the Aβ42 peptide [5]. In a minority of cases PD is inherited as a Mendelian trait (Table 18.1). Studies in PD families have led to the identification of eight causative genes (*α-synuclein, parkin, UCH-L1, PINK1, DJ-1, LRRK2, ATP13A2,* and *OMI/HTRA2*) and four additional loci of linkage across the genome (PARK3, PARK10, PARK11 and PARK12) pending characterization and/or replication. The understanding of the function of proteins encoded by PD causative genes suggests that the selective loss of dopaminergic neurons and the accumulation of α-synuclein are influenced by defects in the ubiquitin-proteasomal system, mitochondrial dysfunction, and the impairment of mechanisms protecting from oxidative stress and apoptosis [2].

Studies in ALS families have led to the identification of different genes responsible for familial or atypical forms (*SOD1, alsin, SETX, VAPB, DCTN1, MAPT*) and potential ALS loci (ALS3, ALS5, ALS6 and ALS7) still pending characterization (Table 18.1). Even if the exact mechanisms leading to selective motor neuron degeneration are still not completely clear, the understanding of the function of ALS causative genes has led to the comprehension that the compromising of antioxidant defense and DNA repair mechanisms, as well as aberrant vesicle trafficking and recycling and synapse stability, are critical to ALS development [3].

HD is caused by a CAG repeat expansion within exon 1 of the gene encoding for huntingtin (*IT15*). In the normal population the number of CAG repeats is maintained below 35, while in individuals affected by HD it ranges from 35 to more than 100, resulting in an expanded polyglutamine segment in the protein.

The age of onset of HD is inversely correlated with the CAG repeat length, and it has been hypothesized that the expanded polyglutamine segment confers a dominant "gain of function" to the protein, ultimately leading to neurodegeneration [6]. Significant variance remains, however, in residual age of onset, even after CAG repeat length is factored out. Many polymorphic genes have previously shown evidence of association with age of onset of HD in several different populations, among them the GluR6 kainate glutamate receptor (*GRIK2*), *APOE*, the transcriptional coactivator CA150 (*TCERG1*), *UCHL1*, *TP53*, caspase-activated DNase (*DFFB*), and the NR2A and NR2B glutamate receptor subunits (*GRIN2A, GRIN2B*) [7].

Despite the discovery of several causative genes for the familial forms, the majority of AD, PD and ALS occur as sporadic forms resulting from the contribution of several interactions between exogenous environmental factors and the individual genetic background. Over one thousand polymorphisms in almost three hundreds different genes have been analyzed in recent years as candidate AD susceptibility factors, but only the ε4 allele of the *Apolipoprotein E* (*APOE*) gene has clearly emerged as an AD risk factor. The *APOE*-ε4 variant is associated with higher plasma cholesterol levels, and is supposed to enhance Aβ deposition and the formation of neuritic plaques [8]. For the remaining hundreds of putative AD susceptibility genes results are often conflicting, obtained in small sample-sized groups or limited to one or two papers reporting association. A recent pooled analysis of those polymorphisms which had been studied in at least three independent association studies [9], revealed few of them as possible AD risk or protective factors (Table 18.2). As for AD, several hundreds of association studies have been published in recent years claiming or denying association between variants in candidate genes and the risk of PD. Results published so far are often conflicting and inconclusive, reflecting the genetic heterogeneity of the studied populations, inadequate sample size and the possible contribution of environmental factors. The major genes which have been analyzed in PD association studies are those related to dopamine transport and metabolism (e.g., *DAT, DRD2, COMT, MAO-B*), detoxification of xenobiotics (e.g., *CYP2D6, GSTs, NAT2*) and oxidative stress (e.g., *NOS, SOD2*). Moreover, common variants of PD causative genes (e.g., *SNCA, LRRK2, UCHL1*) have been largely studied for their role as possible PD susceptibility factors. Details are shown in Table 18.2.

Almost 95% of ALS occurs as sporadic forms; however, although several genes have been studied in recent years as possible ALS susceptibility factors, no single gene has been definitively shown to be consistently associated with disease risk. Recent data support a role for the DNA repair genes *APE1* and *hOGG1* in sporadic ALS based on their protective roles against oxidative stress [10, 11]. Conflicting or inconclusive results have been obtained for angiogenesis genes *ANG* and *VEGF* [10]. Other candidate genes are those coding for neurofilaments (*NEFL, NEFM* and *NEFH*), paraoxonases (*PON1, PON2* and *PON3*), survival motor neuron (*SMN1* and *SMN2*) and the hemocromatosis (*HFE*) gene. Recent pooled analyses suggest a role for the *HFE* H63D variant and for increased copy numbers of the *SMN1* gene [10]. Details are shown in Table 18.2.

Table 18.2 Some of the proposed susceptibility genes for neurodegenerative diseases.

Genetic variant(s)	Associated with
Alzheimer's disease	
APOE-ε4	Increased risk
SORL1 variants	Increased risk
ACE intron 16 (ins/del)	Increased risk
ACE rs1800764, rs4291, rs4343	Decreased risk
CHRNB2 rs4845378	Decreased risk
CST3 5'UTR-157, 5'UTR-72	Increased risk
CST3 A25T	Increased risk
ESR1 PvuII, XbaI	Increased risk
GAPDHS rs12984928, rs4806173	Decreased risk
IDE rs2251101	Decreased risk
MTHFR A1298C	Decreased risk
NCSTR 119 intron 16	Increased risk
PRPN M129V	Decreased risk
PSEN1 rs165932	Decreased risk
TF P570S	Increased risk
TFAM rs2306604	Decreased risk
TNF rs4647198 (-1031)	Increased risk
GOLPH2 rs10868366[a], rs7019241[a]	Decreased risk
Rs 9886784 (Chromosome 9)	Increased risk
Rs 10519262[a]	Increased risk
Parkinson's disease	
SNCA Rep1	Increased risk
LRRK2 G2385R	Increased risk[b]
MAPT H1 haplotype	Increased risk
UCHL1 S18Y	Decreased risk[c]
GSTM1 null genotype	Increased risk[d1]
GSTP1 variants	Increased risk[d2]
CYP2D6 variants	Increased risk[d3]
FAM 79B Rs 1000291[a]	Increased risk
UNC5C Rs 2241743[a]	Increased risk
Rs 3018626 (Chromosome 11)[a]	Increased risk
Amyotrophic lateral sclerosis	
APE1 D148E	Increased risk[c]
ANG G110G	Increased risk[c]
hOGG1 Ser326Cys	Increased risk in males
VEGF variants	Inconclusive results
HFE H63D	Increased risk
SMN1 variable copy number	Increased risk
DPP6 variant[a]	Increased risk

a From WGA studies.
b Only in Asiatic populations.
c Conflicting results.
d In combination with environmental factors ([d1] = solvents, [d2] = pesticides and herbicides, [d3] = pesticides, tobacco smoking).

Despite hundreds of association studies based on the "candidate gene" approach, the hottest new tool in genetics is whole-genome association (WGA) studies; geneticists scan patient's DNA for half a million or more single nucleotide polymorphisms (SNPs), and then compare the results with those from a healthy control group. Unfortunately, almost none of them has highlighted genes already under suspicion by the "candidate-gene" approach, moreover, results from WGA studies are often conflicting and not replicating [12, 13]. Some variants associated with AD, PD or ALS by WGA studies are listed in Table 18.2.

18.3
The Contribution of Environmental Factors to Neurodegenerative Diseases

Several environmental factors have been largely studied in recent years as possible risk factors for neurodegeneration; among them metals, solvents, pesticides, electromagnetic fields, brain injuries and physical activity, as well as drugs and dietary factors (Table 18.3).

Metals have been extensively studied as potential AD risk factors and even if a direct causal role for aluminum or other transition metals such as zinc, copper, iron and mercury in AD has not yet been definitively demonstrated, epidemiological evidence suggests that elevated levels of these metals in the brain may be linked to the development or the progression of the disease. Ingestion of aluminum in drinking water was associated with an increased risk of AD; however, other studies failed to find such association [14, 15]. Another risk factor for AD is inorganic mercury, often present in dental amalgam applications, and a role for *APOE* as a mediator of the toxic effect of mercury has been largely suggested [16]. Human exposure to metals has been the focus of several epidemiological studies aimed at evaluating their possible contribution as PD risk factors. A recent large sample-sized study failed to find association between iron, copper and manganese exposure and PD risk [17]. However another consistent report based on 110 000 individuals in two Canadian cities suggests that environmental manganese air pollution might contribute to neuronal loss in PD [18]. Occupational lead exposure also seems to be a risk factor for PD [19]. Increased ALS risk was observed among individuals occupationally exposed to lead [20].

A recent analysis of 24 published studies assessing the role of occupational AD risk factors revealed a statistically consistent association only for pesticides [21]. The available evidence indicates that rural environment and pesticide exposures are associated with PD, however, no one agent has been consistently identified, likely because associations with specific agents may be confounded by exposure to other pesticides, making it difficult to identify the causative agent [22]. There is also evidence suggesting that human exposures to agricultural chemicals, such as pesticides, are at increased ALS risk [23]. To support this hypothesis there is a recent report of a motor neuron disorder simulating ALS induced by chronic inhalation of pyrethroid insecticides [24]. Moreover, increased post-war risk of ALS has been observed in military personnel who were deployed to the Gulf Region

Table 18.3 Some of the proposed environmental factors for neurodegenerative diseases.

Environmental factor(s)	Associated with
Alzheimer's disease	
Metals (iron, copper, zinc, mercury, aluminum)	Increased risk, inconclusive results
Pesticides	Increased risk
Solvents	Increased risk, inconclusive results
Electromagnetic fields	Increased risk, inconclusive results
Caloric restriction	Protection
Antioxidants	Protection
Mediterranean diet, fruit and vegetables	Protection
Fish and omega-3 fatty acids	Protection
Traumatic brain injuries	Increased risk
Infections and inflammation	Increased risk
Parkinson's disease	
Metals (iron, copper, manganese, lead)	Increased risk, conflicting results
Rural environment (pesticides, herbicides)	Increased risk
Tobacco smoking	Protection
Caffeine (coffee and tea drinking)	Protection
Fruit and vegetables, legumes, nuts	Protection
Fish	Protection
Head injuries with loss of consciousness	Increased risk
Amyotrophic lateral sclerosis	
Metals (lead)	Increased risk
Pesticides and insecticides	Increased risk
Electromagnetic fields	Increased risk
Some sports (soccer, football)	Increased risk
Head injuries	Increased risk
Tobacco smoking	Increased risk, in women

during the first Gulf War period, suggesting exposure to neurotoxins as an environmental risk factor [25]. There is also evidence for an increased ALS risk among welders and other workers exposed to electromagnetic fields [26, 27].

Among other factors, accumulating evidence implicates traumatic brain injury and inflammation as a possible predisposing factor in AD development [28]. Repeated traumatic loss of consciousness is also associated with increased PD risk [17, 29]. An increased ALS risk for Italian professional soccer players [30] and also for National Football League players in the United States, was observed [31]. Several hypotheses have been formulated trying to explain the causative agent of ALS among soccer players, including as possible candidates excessive physical activity, drugs and doping, dietary supplements, pesticides used on the playgrounds, and traumas to the head and to other body parts [32]. Recent evidence suggests that repeated head injuries might increase ALS risk [33].

Dietary factors have been largely studied as possible contributors of neurodegeneration; among them antioxidant compounds seem to exert a neuroprotective role (Table 18.3).

In transgenic HD mice models the environmental enrichment with several new different objects seems to delay the onset of motor symptoms [4].

18.4
Epigenetics, Environment and Susceptibility to Human Diseases

Epigenetics deals with the heritable modifications of DNA that can influence the phenotype through changing gene expression without altering primary DNA sequence. The epigenetic modifications include DNA methylation, histone modifications, and RNA-mediated pathways from non-coding RNAs, notably silencing RNA (siRNA) and microRNA (miRNA). Epigenetic modifications are key regulators of important developmental events, including X-inactivation, genomic imprinting and neuronal development. Accumulating evidence indicates that variations in gene expression due to variable modifications in DNA methylation and chromatin structure in response to the environment also play a role in differential susceptibility to diseases. Consistent with these fundamental aspects, an increasing number of human pathologies have been found to be associated with aberrant epigenetic regulation, such as mental retardation, syndromes involving chromosomal instabilities, obesity, infertility, respiratory diseases, allergies, and a great number of age-related diseases including cancer, hearing loss and neurodegenerative diseases [34–38].

Epigenetic modifications have been compared, in terms of phenotypic consequences, to genetic polymorphisms resulting in variations in gene function [39]. Recent data suggest that the epigenome is dynamic and is, therefore, responsive to environmental signals not only during the critical periods in development but also later in life. It is postulated also that not only chemicals but also exposure to social behavior, such as maternal care, could affect the epigenome [40]. Exposures to different environmental agents could lead to interindividual phenotypic diversity as well as differential susceptibility to disease and behavioral pathologies [39].

The common disease genetic and epigenetic hypothesis [35] argues that, in addition to genetic variation, epigenetics provides an added layer of variation that might mediate the relationship between genotype and internal and external environmental factors. This epigenetic component could help in understanding the marked increase in common diseases with age, as well as the phenotypic discordance between monozygotic twins [41]. It is likely that the activity of proteins that have been proved to be involved in epigenetic modifications, e.g. DNA methyltransferases, could be potentially modulated by environmental factors such as diet, alcohol, cigarette smoke or environmental toxins such as heavy metals, known to disrupt DNA methylation and chromatin [42]. The fungicide vinclozolin, an endocrine disruptor that decreases male fertility, alters DNA methylation, and changes are inherited by subsequent generations [43].

Despite a growing consensus on the importance of epigenetics in the etiology of chronic human diseases, the genes most prone to epigenetic dysregulation are incompletely defined. Moreover, until now only a few environmental agents affecting the epigenome have been identified (for a review, see Ref. [44]) and much remains to do to adequately characterize environmentally induced epigenetic alterations [45, 46].

18.5
Epigenetics and Neurodegenerative Diseases

DNA methylation is dynamically regulated in the brain throughout the lifespan, a genome-wide decline in DNA methylation occurs during normal aging, which coincides with a functional decline in learning and memory with age [47, 48]. The emerging field of studies on DNA methylation of specific brain regions may help account for region-specific functional specialization [49]. Although AD manifests in late adult life, it is not clear when the disease actually starts and how long the neuropathological processes take to develop AD. To explain the etiology of AD from an epigenetic point of view, one should consider the neuropathological features, such as neuronal cell death, tau tangles, and amyloid plaque formation, as a function of epigenome variations induced by environmental factors that have until now been associated with AD, such as diet components, or toxicological exposure (see Table 18.3).

In Alzheimer's disease, as previously discussed, Aβ peptides or fragments are the major components of amyloid plaques and are produced by the amyloidogenic cutting of the amyloid precursor protein APP. APP can be alternatively processed by γ-secretases [presenilin1 (PSEN1) and 2 (PSEN2)] and α- secretases (ADAM10 and TACE) producing non-amyloidogenic peptides, or by γ- and β-secretases (BACE) producing Aβ peptides [50]. Therefore, the balance between different secretase activities is very important in the maintenance of the physiologic levels of non-amyloidogenic and amyloidogenic fragments.

We know that accumulation of oxidative stress-induced damage in brain tissue plays an important role in the pathogenesis of normal aging and neurodegenerative diseases, including AD. Because of its high metabolic rate the brain is believed to be particularly susceptible to reactive oxygen species (ROS), and the effects of oxidative stress on neurons might be cumulative. At the time oxidative damage was observed in AD, it was supposed that amyloid aggregates were the main source of oxidative stress; however, recent evidence suggests that oxidative stress is one of the earliest events in AD [51, 52] and that Aβ peptides might be produced to function as scavengers of reactive oxidative species. Only with the persistence of oxidative stress, does the production of Aβ peptides overcome their cellular turnover, so that they start to aggregate and their anti-oxidant function evolve into pro-oxidant, ultimately leading to neuronal death [53]. The connection between epigenetic mechanisms of transcriptional silencing of genes important to ROS such as MnSOD has been firmly established [54]. Increases in ROS can also effect glutathione levels which in turn can change S-adenosylmethionine (SAM) synthesis and hence DNA methylation patterns. There are additional ROS-related mechanisms involving hydrogen peroxide that can lead to further changes of the chromatin structure. Interestingly Hitchler and Domann [55] proposed an epigenetic perspective on the free radical theory of development. This theory proposes that oxygen has a key role in development by influencing the production of metabolic oxidants that would influence in turn the antioxidant capacity of cells throughout the production of glutathione (GSH). Increased GSH production influences epigenetic processes including DNA and histone methylation by limiting the availability of S-adenosylmethionine, the cofactor utilized during epigenetic control of gene expression by DNA and histone methyltransferases [55]. Glutathione is thus an important endogenous antioxidant, found in millimolar concentrations in the brain. GSH levels have been shown to decrease with aging and, in particular, are decreased in affected brain regions and peripheral cells from AD and also PD patients. Tabaton and Tamagno [56] reviewed the role of oxidative stress as a molecular link between the β- and the γ-secretase activities, and provided a mechanistic explanation of the pathogenesis of sporadic late-onset AD: the overproduction of Aβ, dependent on the upregulation of BACE1 induced by oxidative stress, would contribute to the pathogenesis of the common, sporadic, late-onset form of AD, a major risk factor for which is aging. These authors suggest that an increase in the γ-secretase cleavage of APP mediated by oxidative stress (sporadic AD), or by *PSEN1* mutations (FAD), fosters BACE1 expression and activity.

In general, genes involved in several pathways including antioxidant defense, detoxification, inflammation, etc., are induced in response to oxidative stress and in AD. However, genes that are associated with energy metabolism, which is necessary for normal brain function, are mostly down-regulated. The PGC-1alpha role in regulation of ROS metabolism makes it a potential candidate player between ROS, mitochondria, and neurodegenerative diseases: down-regulated expression of *PGC-1alpha* has been implicated in Huntington disease and in several Huntington disease animal models [57]. Lahiri et al. [58] proposed a "Latent Early-Life Associated Regulation" model, which postulates a latent expression of specific genes triggered at the developmental stage. According to this model, environmental agents (e.g., heavy metals), intrinsic factors (e.g., cytokines), and dietary factors (e.g., cholesterol) perturb gene regulation in a long-term fashion, beginning in the early developmental stages, but with pathological outcomes significantly later in life. For example, such actions would perturb *APP* gene regulation at a very early stage via its transcriptional machinery, leading to delayed overexpression of *APP* and subsequently of Aβ deposition. According to this model, promoter activity of specific genes, such as methyl-CpG-binding protein 2 (*MeCP2*) and the transcription factors Sp1, can be altered by changes in the primary DNA sequence and by epigenetic changes through mechanisms such as DNA methylation at CpG dinucleotides or oxidation of guanosine residues [58].

By genome scan studies increased levels of gene expression are now being discovered within specific classes of genes. This can be linked to a possible modulation of the methylation of promoters, such as in a study on differential expression of the ornithine transcarbamylase (*OTC*) gene, a key enzyme of the urea cycle, found expressed in AD but not in controls [59]. Currently, genome-wide technologies are available and have been utilized to examine the methylation state of cytosine bases throughout a genome (methylome). Studies involving several physiological and disease states, mainly cancer, have been performed. Although early in the process, DNA methylation is being explored as a biomarker to be used in clinical practice for early detection of disease, tumor classification and for predicting disease outcome or recurrence [60]. It has become increasingly evident in recent years that development is under epigenetic control. Prenatally or early life dietary and environmental exposures can have a profound effect on our epigenome, resulting in birth defects and diseases developed later in life [61]. Studies in rodents have shown that exposure to lead (Pb) during brain development predetermined the expression and regulation of the amyloid precursor protein and its amyloidogenic Aβ product in old age. The expression of AD-related genes (*APP*, *BACE1*) as well as their transcriptional regulator (*Sp1*) was elevated in aged monkeys exposed to Pb as infants. Developmental exposure to Pb altered the levels, characteristics, and intracellular distribution of Aβ staining and amyloid plaques in the frontal association cortex, furthermore, it induced a decrease in DNA methyltransferase activity and higher levels of oxidative damage to DNA, indicating that epigenetic imprinting in early life influenced the expression of AD-related genes and promoted DNA damage and pathogenesis [62].

Wu *et al.* [63] propose that environmental influences occurring during brain development alter the methylation pattern of the *APP* promoter which results in a latent increase in APP and Aβ levels. Increased Aβ levels promote the production of ROS which damage DNA. Epigenetic changes in DNA methylation impact both gene transcription and the ability to repair damaged DNA and thus imprint susceptibility to DNA damage. This susceptibility plus the programmed increase in Aβ levels, via a transcriptional pathway programmed by environmental exposures in early life, exacerbates the normal process of amyloidogenesis in the aging brain, thus accelerating the onset of AD.

Few attempts have been so far made to demonstrate the occurrence of epigenetic silencing of genes that have a fundamental role in other neurodegenerative diseases, such as ALS and HD. In an epigenetic context it is likely that the level of expression of several genes is altered in age-related diseases, including neurodegenerative ones, due to the methylation status of their promoters. In particular, methylation levels dysregulation have been involved to explain the variable phenotypic espressivity (age of onset, the severity and or penetrance of the pathological phenotype) [48]. Sporadic amyotrophic lateral sclerosis (SALS) results from the death of motor neurons in the brain and spinal cord. It has been proposed that epigenetic silencing of genes vital for motor neuron function could underlie SALS. Oates and Pamphlett [64] therefore examined the methylation status of two genes, *SOD1* and *VEGF*, which are implicated in ALS. Methylation in the promoters of these genes was determined in white cell DNA and brain DNA of ALS patients. However the promoter regions were found to be largely unmethylated in all patients [64]. The metallothionein (MT) family of proteins are the primary detoxification mechanism for heavy metals and MT-Ia and MT-IIa are the most common human isoforms. It was hypothesized that inappropriate methylation at the promoters of these genes could lead to silencing of transcription and reduce the availability of MTs. The level of methylation in the promoters of genes encoding MT-Ia and MT-IIa in leukocyte and brain DNA samples from SALS patients was measured and compared with controls, but again no promoter methylation of these genes was evident in any SALS or control samples [65].

18.6
The Epigenetic Role of the Diet in Neurodegenerative Diseases

Various environmental and dietary agents and lifestyles are suspected to be implicated in the development of a wide range of human cancers through epigenetic changes (for a review see Ref. [44]). Very few data are available in this regard in the field of neurodegeneration. Dietary modification can indeed have a profound effect on DNA methylation and genomic imprinting. DNA methylation is regulated through cellular levels of S-adenosyl-methionine. The conversion of homocysteine (HCY) to methionine requires folate metabolites and is an essential step in the production of SAM. Recent studies of fundamental importance have shown that a variation of the diet can lead to an alteration of the phenotype in mice or in their offspring. Deficiency in folate and methionine, necessary for normal biosynthesis of SAM, the methyl donor for methylcytosine, leads to aberrant imprinting of insulin-like growth factor 2 in mice [66], maternal methyl donor supplementation during gestation can alter the offspring phenotype by methylating a transposable element in mice with silencing of the nearby agouti coat-color gene [67]. Moreover, the same maternal dietary supplementation, with either methyl donors like folic acid or the phytoestrogen genistein, showed a protective role in counteracting the DNA hypomethylating effect of bisphenol A, a chemical with carcinogenic properties, used in the manufacture of polycarbonate plastic [68]. Genetic polymorphisms can alter the response to dietary components (nutrigenetic effect) by influencing the absorption, metabolism, or site of action. Analogously, variation in DNA methylation patterns and other epigenetic events that influence overall gene expression can influence the biological response to food components and vice versa [69].

AD is characterized by high HCY and low folate blood levels, meaning that the conversion of HCY to methionine is altered in AD, as is the production of SAM. DNA methylation is regulated through cellular levels of SAM. The conversion of homocysteine to methionine requires folate metabolites and is an essential step in the production of SAM. Fuso and collaborators [50] studied the levels of methylation of CpG islands in the promoters of the *APP* and the *PSEN1* gene (PSEN1 is one of the components of the γ-secretases which cleave APP, producing amyloid fragments), on human neuroblastoma cell lines, observing that, in conditions of folate and vitamin B12 deprivation from the media, the status of methylation of the promoter of the *PSEN1* gene was changed, with a subsequent deregulation of the production of PS1, BACE (the β-secretase) and APP proteins [50]. This experiment has provided evidence that some of the genes responsible for the production of Aβ fragments in AD can be regulated through epigenetic mechanisms which are regulated by the cellular availability of folates and B12 vitamins, and involve the production of SAM and the status of methylation of CpG islands in the DNA. More recent results by the same authors indicate that homocysteine accumulation induced through vitamin B deprivation could impair the "methylation potential" with consequent presenilin 1, BACE and amyloid-beta upregulation. Moreover, they found that homocysteine alterations had an effect on neuroblastoma but not on glioblastoma cells; this suggested a possible differential role of the two cell types in Alzheimer's disease [70]. Studies *in vivo*, on a murine model of Alzheimer's disease, confirmed that a combined folate, B12 and B6 dietary deficiency induced hyperhomocysteinemia and imbalance of S-adenosylmethionine and S-adenosylhomocysteine. This effect was associated with PSEN1 and BACE up-regulation and amyloid-β deposition [71].

Folate deficiency seems to contribute to a variety of age-related neurological and psychological disorders, including amyotrophic lateral sclerosis. Key nutritional deficiencies could potentiate the impact of enrivonmental neurotoxins. The environmental neurotoxin arsenic has recently been linked with decreased neurofilament (NF) content in peripheral nerves. Supplementation with S-adenosyl methionine (SAM) attenuated the impact of folate deprivation on arsenic neurotoxicity, consistent with the decrease in SAM following folate deprivation and the requirement for SAM-mediated methylation for arsenic bioelimination [72].

18.7
Concluding Remarks

Many of the processes with a key role in neurodegeneration, such as the formation of senile plaques, the accumulation of ROS, the cleavage of APP by neuroscretases, can now be analyzed in the light of the new epigenetic knowledge, to facilitate the implementation of future disease prevention strategies. Since epigenetic alterations are reversible, modifying epigenetic marks contributing to disease development may provide an approach to designing new therapies such as the use of inhibitors of enzymes controlling epigenetic modifications [73].

However, we have to take into account that epigenetic therapy has its limitations, such as the non-specific activation of genes and transposable elements in normal cells, and also has the potential for mutagenicity and carcinogenicity. It is also possible that corrected epigenetic modifications may revert to their previous state because of the reversible nature of DNA methylation and histone modification patterns, although this may be prevented with continued treatment, or corrected again with retreatment [34].

The emerging field of environmental epigenomics deals with the study of metastable epialleles as epigenetically labile genomic targets [74]. Among those alleles

Table 18.4 Environmental factors with epigenetic mechanisms for neurodegenerative diseases.

Environmental factor(s)	Associated with
Folate and vitamin B12 deprivation	Increase of methylation of key genes for AD (*PS1, BACE, APP*) in neuronal cells *in vitro* and in mice
Metals (lead)	Increased risk amiloid plaques (rats, monkeys)

See the text for references

it is predictable that the role of key genes responsible for neurodegenerative diseases will soon be identified and this will allow new insights into the etiology of these diseases. Comparing Tables 18.3 and 18.4 it is evident that there is still a lot to do in the advancement of the knowledge of environmental factors that, through an epigenetic mechanism, can influence the individual susceptibility to develop a neurodegenerative disease.

References

1 Lambert, J.C., and Amouyel, P. (2007) Genetic heterogeneity of Alzheimer's disease: complexity and advances. *Psychoneuroendocrinology*, **32**, S62–S70.

2 Rosner, S., Giladi, N., and Orr-Urtreger, A. (2008) Advances in the genetics of Parkinson's disease. *Acta. Pharmacol. Sin.*, **29**, 21–34.

3 Mitchell, J.D., and Borasio, G.D. (2007) Amyotrophic lateral sclerosis. *Lancet*, **369**, 2031–2041.

4 van Dellen, A., Grote, H.E., and Hannan, A.J. (2005) Gene-environment interactions, neuronal dysfunction and pathological plasticity in Huntington's disease. *Clin. Exp. Pharmacol. Physiol.*, **32**, 1007–1019.

5 Findeis, M.A. (2007) The role of amyloid beta peptide 42 in Alzheimer's disease. *Pharmacol. Ther.*, **116**, 266–286.

6 Gusella, J.F., and Macdonald, M. (2007) Genetic criteria for Huntington's disease pathogenesis. *Brain Res. Bull.*, **72**, 78–82.

7 Andresen, J.M., Gayán, J., Djoussé, L., Roberts, S., Brocklebank, D., Cherny, S.S., US-Venezuela Collaborative Research Group, HD MAPS Collaborative Research Group, Cardon, L.R., Gusella, J.F., MacDonald, M.E., Myers, R.H., Housman, D.E., and Wexler, N.S. (2007) The relationship between CAG repeat length and age of onset differs for Huntington's disease patients with juvenile onset or adult onset. *Ann. Hum. Genet.*, **71**, 295–301.

8 Farrer, L.A., Cupples, L.A., Haines, J.L., Hyman, B., Kukull, W.A., Mayeux, R., Myers, R.H., Pericak-Vance, M.A., Risch, N., and van Duijn, C.M. (1997) Effects of age, sex, and ethnicity on the association between apolipoprotein E genotype and Alzheimer disease. A meta-analysis APOE and Alzheimer Disease Meta Analysis Consortium. *JAMA*, **278**, 1349–1356.

9 Bertram, L., McQueen, M.B., Mullin, K., Blacker, D., and Tanzi, R.E. (2007) Systematic meta-analyses of Alzheimer disease genetic association studies: the AlzGene database. *Nat. Genet.*, **39**, 17–23.

10 Schymick, J.C., Talbot, K., and Traynor, B.J. (2007) Genetics of sporadic amyotrophic lateral sclerosis. *Hum. Mol. Genet.*, **16**, R233–R242.

11 Coppedè, F., Mancuso, M., Lo Gerfo, A., Carlesi, C., Piazza, S., Rocchi, A., Petrozzi, L., Nesti, C., Micheli, D., Bacci, A., Migliore, L., Murri, L., and Siciliano, G. (2007) Association of the hOGG1 Ser326Cys polymorphism with sporadic amyotrophic lateral sclerosis. *Neurosci. Lett.*, **420**, 163–168.

12 Evangelou, E., Maraganore, D.M., and Ioannidis, J.P. (2007) Meta-analysis in genome-wide association datasets: strategies and application in Parkinson disease. *PLoS ONE*, **2**, e196.

13 Garber, K. (2008) The elusive ALS genes. *Science*, **319**, 20.

14 McLachlan, D.R., Bergeron, C., Smith, J.E., Boomer, D., and Rifat, S.L. (1996) Risk for neuropathologically confirmed Alzheimer's disease and residual aluminum in municipal drinking water employing weighted residential histories. *Neurology*, **46**, 401–405.

15 Martyn, C.N., Coggon, D.N., Inskip, H., Lacey, R.F., and Young, W.F. (1997) Aluminum concentrations in drinking water and risk of Alzheimer's disease. *Epidemiology*, **8**, 281–286.

16 Mutter, J., Naumann, J., Sadaghiani, C., Schneider, R., and Walach, H. (2004) Alzheimer disease: mercury as pathogenetic factor and apolipoprotein E as a moderator. *Neuro. Endocrinol. Lett.*, **25**, 331–339.

17 Dick, F.D., De Palma, G., Ahmadi, A., Scott, N.W., Prescott, G.J., Bennett, J., Semple, S., Dick, S., Counsell, C., Mozzoni, P., Haites, N., Wettinger, S.B., Mutti, A., Otelea, M., Seaton, A., Söderkvist, P., Felice, A., and Geoparkinson Study Group (2007) Environmental risk factors for Parkinson's disease and parkinsonism: the Geoparkinson study. *Occup. Environ. Med.*, **64**, 666–672.

18 Finkelstein, M.M., and Jerrett, M. (2007) A study of the relationships between Parkinson's disease and markers of traffic-derived and environmental manganese air pollution in two Canadian cities. *Environ. Res.*, **104**, 420–432.

19 Coon, S., Stark, A., Peterson, E., Gloi, A., Kortsha, G., Pounds, J., Chettle, D., and Gorell, J. (2006) Whole-body lifetime occupational lead exposure and risk of Parkinson's disease. *Environ. Health Perspect.*, **14**, 1872–1876.

20 Kamel, F., Umbach, D.M., Hu, H., Munsat, T.L., Shefner, J.M., Taylor, J.A., and Sandler, D.P. (2005) Lead exposure as a risk factor for amyotrophic lateral sclerosis. *Neurodegener. Dis.*, **2**, 195–201.

21 Santibáñez, M., Bolumar, F., and García, A.M. (2007) Occupational risk factors in Alzheimer's disease: a review assessing the quality of published epidemiological studies, *Occup. Environ. Med.*, **64**, 723–732.

22 Elbaz, A. (2007) Parkinson's disease and rural environment. *Rev. Prat.*, **57**, 37–39.

23 Govoni, V., Granieri, E., Fallica, E., and Casetta, I. (2005) Amyotrophic lateral sclerosis, rural environment and agricultural work in the Local Health District of Ferrara, Italy, in the years 1964–1998. *J. Neurol.*, **252**, 1322–1327.

24 Doi, H., Kikuchi, H., Murai, H., Kawano, Y., Shigeto, H., Ohyagi, Y., and Kira, J. (2006) Motor neuron disorder simulating ALS induced by chronic inhalation of pyrethroid insecticides. *Neurology*, **67**, 1894–1895.

25 Horner, R.D., Kamins, K.G., Feussner, J.R., Grambow, S.C., Hoff-Lindquist, J., Harati, Y., Mitsumoto, H., Pascuzzi, R., Spencer, P.S., Tim, R., Howard, D., Smith, T.C., Ryan, M.A.K., Coffman, C.J., and Kararskis, E.J. (2003) Occurrence of amyotrophic lateral sclerosis among Gulf War veterans. *Neurology*, **61**, 742–749.

26 Li, C.Y., and Sung, F.C. (2003) Association between occupational exposure to power frequency electromagnetic fields and amyotrophic lateral sclerosis: a review. *Am. J. Ind. Med.*, **43**, 212–220.

27 Håkansson, N., Gustavsson, P., Johansen, C., and Floderus, B. (2003) Neurodegenerative diseases in welders and other workers exposed to high levels of magnetic fields. *Epidemiology*, **14**, 420–426.

28 Van Den Heuvel, C., Thornton, E., and Vink, R. (2007) Traumatic brain injury and Alzheimer's disease: a review. *Prog. Brain Res.*, **161**, 303–316.

29 Goldman, S.M., Tanner, C.M., Oakes, D., Bhudhikanok, G.S., Gupta, A., and Langston, J.W. (2006) Head injury and Parkinson's disease risk in twins. *Ann. Neurol.*, **60**, 65–72.

30 Chiò, A., Benzi, G., Dossena, M., Mutani, R., and Mora, G. (2005) Severely increased risk of amyotrophic lateral sclerosis among Italian professional football players. *Brain*, **128**, 472–476.

31 Abel, E.L. (2007) Football increases the risk for Lou Gehrig's disease, amyotrophic lateral sclerosis. *Percept. Mot. Skills*, **104**, 1251–1254.

32 Belli, S., and Vanacore, N. (2005) Proportionate mortality in italian soccer players: is amyotrophic lateral sclerosis an occupational disease? *Eur. J. Epidemiol.*, **20**, 237–242.

33 Chen, H., Richard, M., Sadler, D.P., Umbach, D.M., and Kamel, F. (2007) Head injury and amyotrophic lateral sclerosis. *Am. J. Epidemiol.*, **166**, 810–816.

34 Lu, Q., Qiu, X., Hu, N., Wen, H., Su, Y., and Richardson, B.C. (2006) Epigenetics, disease, and therapeutic interventions. *Ageing Res. Rev.*, **5**, 449–467.

35 Feinberg, A.P. (2007) Phenotypic plasticity and the epigenetics of human disease. *Nature*, **447**, 433–440.

36 Jirtle, R.L., and Skinner, M.K. (2007) Environmental epigenomics and disease susceptibility. *Nat. Rev. Genet.*, **8**, 253–262.

37 Provenzano, M.J., and Domann, F.E. (2007) A role for epigenetics in hearing: establishment and maintenance of auditory specific gene expression patterns. *Hear Res.*, **233**, 1–13.

38 Santos-Rebouças, C.B., and Pimentel, M.M. (2007) Implication of abnormal epigenetic patterns for human diseases. *Eur. J. Hum. Genet.*, **15**, 10–17.

39 Szyf, M. (2007) The dynamic epigenome and its implications in toxicology. *Toxicol. Sci.*, **100**, 7–23.

40 Szyf, M., Weaver, I., and Meaney, M. (2007) Maternal care, the epigenome and phenotypic differences in behaviour. *Reprod. Toxicol.*, **24**, 9–19.

41 Poulsen, P., Esteller, M., Vaag, A., and Fraga, M.F. (2007) The epigenetic basis of twin discordance in age-related diseases. *Pediatr. Res.*, **61**, 38R–42R

42 Sutherland, J.E., and Costa, M. (2003) Epigenetics and the environment. *Ann. N. Y. Acad. Sci.*, **983**, 151–160.

43 Anway, M.D., Rekow, S.S., and Skinner, M.K. (2008) Transgenerational epigenetic programming of the embryonic testis transcriptome. *Genomics*, **91**, 30–40.

44 Herceg, Z. (2007) Epigenetics and cancer: towards an evaluation of the impact of environmental and dietary factors. *Mutagenesis*, **22**, 91–103.

45 Edwards, T.M., and Myers, J.P. (2007) Environmental exposures and gene regulation in disease etiology. *Environ. Health Perspect.*, **115**, 1264–1270.

46 Weidman, J.R., Dolinoy, D.C., Murphy, S.K., and Jirtle, R.L. (2007) Cancer susceptibility: epigenetic manifestation of environmental exposures. *Cancer J.*, **13**, 9–16.

47 Liu, L., van Groen, T., Kadish, I., and Tollefsbol, T.O. (2008) DNA methylation impacts on learning and memory in aging. *Neurobiol. Aging*, **30**, 549–560.

48 Siegmund, K.D., Connor, C.M., Campan, M., Long, T.I., Weisenberger, D.J., Biniszkiewicz, D., Jaenisch, R., Laird, P.W., and Akbarian, S. (2007) DNA methylation in the human cerebral cortex is dynamically regulated throughout the life span and involves differentiated neurons. *PLoS ONE*, **2**, e895.

49 Ladd-Acosta, C., Pevsner, J., Sabunciyan, S., Yolken, R.H., Webster, M.J., Dinkins, T., Callinan, P.A., Fan, J.B., Potash, J.B., and Feinberg, A.P. (2007) DNA methylation signatures within the human brain. *Am. J. Hum. Genet.*, **81**, 1304–1315

50 Fuso, A., Seminara, L., Cavallaro, R.A., D'Anselmi, F., and Scarpa, S. (2005) S-adenosylmethionine/homocysteine cycle alterations modify DNA methylation status with consequent deregulation of PS1 and BACE and beta-amyloid production. *Mol. Cell Neurosci.*, **28**, 195–204.

51 Migliore, L., Fontana, I., Trippi, F., Colognato, R., Coppedè, F., Tognoni, G., Nucciarone, B., and Siciliano, G. (2005) Oxidative DNA damage in peripheral leukocytes of mild cognitive impairment

and AD patients. *Neurobiol. Aging*, **26**, 567–573.

52 Zhu, X., Su, B., Wang, X., Smith, M.A., and Perry, G. (2007) Causes of oxidative stress in Alzheimer disease. *Cell Mol. Life Sci.*, **64**, 2202–2210.

53 Moreira, P.I., Nunomura, A., Honda, K., Aliev, G., Casadenus, G., Zhu, X., Smith, M.A., and Perry, G. (2007) The key role of oxidative stress in Alzheimer's disease, in *Oxidative Stress and Neurodegenerative Disorders* (eds G. Ali Qureshi, and S. Hassan Parvez), Elsevier, Amsterdam, pp. 451–466.

54 Hitchler, M.J., Wikainapakul, K., Yu, L., Powers, K., Attatippaholkun, W., and Domann, F.E. (2006) Epigenetic regulation of manganese superoxide dismutase expression in human breast cancer cells. *Epigenetics*, **1**, 163–171.

55 Hitchler, M.J., and Domann, F.E. (2007) An epigenetic perspective on the free radical theory of development. *Free Radic. Biol. Med.*, **43**, 1023–1036.

56 Tabaton, M., and Tamagno, E. (2007) The molecular link between beta- and gamma-secretase activity on the amyloid beta precursor protein. *Cell Mol. Life Sci.*, **64**, 2211–2218.

57 McGill, J.K., and Beal, M.F. (2006) PGC-1alpha, a new therapeutic target in Huntington's disease? *Cell*, **127**, 465–468.

58 Lahiri, D.K., Maloney, B., Basha, M.R., Ge, Y.W., and Zawia, N.H. (2007) How and when environmental agents and dietary factors affect the course of Alzheimer's disease: the "LEARn" model (latent early-life associated regulation) may explain the triggering of AD. *Curr. Alzheimer Res.*, **4**, 219–228.

59 Bensemain, F., Hot, D., Ferreira, S., Dumont, J., Bombois, S., Maurage, C.A., Huot, L., Hermant, X., Levillain, E., Hubans, C., Hansmannel, F., Chapuis, J., Hauw, J.J., Schraen, S., Lemoine, Y., Buée, L., Berr, C., Mann, D., Pasquier, F., Amouyel, P., and Lambert, J.C. (2008) Evidence for induction of the ornithine transcarbamylase expression in Alzheimer's disease. *Molecular Psychiatry*, **14**, 106–116.

60 Zilberman, D. (2007) The human promoter methylome. *Nat. Genet.*, **39**, 442–443.

61 Reamon-Buettner, S.M., and Borlak, J. (2007) A new paradigm in toxicology and teratology: altering gene activity in the absence of DNA sequence variation. *Reprod. Toxicol.*, **24**, 20–30.

62 Wu, J., Basha, M.R., Brock, B., Cox, D.P., Cardozo-Pelaez, F., McPherson, C.A., Harry, J., Rice, D.C., Maloney, B., Chen, D., Lahiri, D.K., and Zawia, N.H. (2008) Alzheimer's disease (AD)-like pathology in aged monkeys after infantile exposure to environmental metal lead (Pb): evidence for a developmental origin and environmental link for AD. *J. Neurosci.*, **28**, 3–9.

63 Wu, J., Basha, M.R., and Zawia, N.H. (2008) The environment, epigenetics and amyloidogenesis. *J. Mol. Neurosci.*, **34**, 1–7.

64 Oates, N., and Pamphlett, R. (2007) An epigenetic analysis of SOD1 and VEGF in ALS. *Amyotroph. Lateral Scler.*, **8**, 83–86.

65 Morahan, J.M., Yu, B., Trent, R.J., and Pamphlett, R. (2007) Are metallothionein genes silenced in ALS? *Toxicol. Lett.*, **168**, 83–87.

66 Waterland, R.A., Lin, J.R., Smith, C.A., and Jirtle, R.L. (2006) Post-weaning diet affects genomic imprinting at the insulin-like growth factor 2 (Igf2) locus. *Hum. Mol. Genet.*, **15**, 705–716.

67 Waterland, R.A., and Jirtle, R.L. (2003) Transposable elements: targets for early nutritional effects on epigenetic gene regulation. *Mol. Cell Biol.*, **23**, 5293–5300.

68 Dolinoy, D.C., Huang, D., and Jirtle, R.L. (2007) Maternal nutrient supplementation counteracts bisphenol A-induced DNA hypomethylation in early development. *Proc. Natl. Acad. Sci. U. S. A.*, **104**, 13056–13061.

69 Milner, J.A. (2006) Diet and cancer: facts and controversies. *Nutr. Cancer*, **56**, 216–224.

70 Fuso, A., Cavallaro, R.A., Zampelli, A., D'Anselmi, F., Piscopo, P., Confaloni, A., and Scarpa, S. (2007) Gamma-Secretase is differentially modulated by alterations of homocysteine cycle in neuroblastoma and glioblastoma cells. *J. Alzheimers Dis.*, **1** (3), 275–290.

71 Fuso, A., Nicolia, V., Cavallaro, R.A., Ricceri, L., D'Anselmi, F., Coluccia, P.,

Calamandrei, G., and Scarpa, S. (2008) B-vitamin deprivation induces hyperhomocysteinemia and brain S-adenosylhomocysteine, depletes brain S-adenosylmethionine, and enhances PS1 and BACE expression and amyloid-beta deposition in mice. *Mol. Cell Neurosci.*, **37**, 731–746.

72 Dubey, M., and Shea, T.B. (2007) Potentiation of arsenic neurotoxicity by folate deprivation: protective role of S-adenosyl methionine. *Nutr. Neurosci.*, **10**, 199–204.

73 Allen, A. (2007) Epigenetic alterations and cancer: new targets for therapy. *IDrugs*, **10**, 709–712.

74 Dolinoy, D.C., and Jirtle, R.T. (2008) Environmental epigenomics in human health and disease. *Environ. Mol. Mutagen.*, **49**, 4–8.

19
Epigenetic Biomarkers in Neurodegenerative Disorders
Borut Peterlin

Abstract

Neurodegenerative disorders like Parkinson's, Alzheimer's, and Huntington's diseases generally begin late in life and are characterized by a presymptomatic phase. Therefore, biomarkers are needed for early, accurate and specific diagnosis and to provide surrogate endpoints for new drug efficacy testing. Inherited neurological and neurodegenerative disorders with mutation in genes that modify epigenetic marks, the current evidence of epigenetic dysregulation in neurodegenerative disorders and the beneficial effect of histone deacetylases (HDACs) inhibitors in models of neurodegenerative disorders imply that epigenetics abnormalities contribute to neurodegeneration. Bioinformatic analysis of differentially transcribed genes predicted to have CpG islands in the promoter regions support this concept. It could be therefore argued that epigenetic biomarkers might be useful for diagnostic purposes and the development/testing of new drugs as well as for improvement of our understanding of the complex interaction between genetic predisposition and environment in neurodegenerative disorders.

19.1
Introduction

Neurodegenerative disorders are caused by the death and dysfunction of brain cells and range from common to rare diseases. Indeed, diseases like Parkinson's disease (PD), Alzheimer's disease (AD) and Huntington's disease (HD) pose serious public health challenges and are a major burden to patients, families and society. These diseases generally begin late in life and are characterized by a presymptomatic phase which may last for several years before clinical symptoms appear. Additionally, the first symptoms of neurodegenerative disorders are rarely specific and may therefore imply several diseases with different prognosis. Finally, development and testing of new drugs is a high risk endeavor and failed clinical trials litter the path to success. Therefore, major goals of clinical research are to improve early detection of diseases, to enhance the accuracy and specificity of diagnosis and to provide surrogate endpoints for new drug efficacy testing. Biomarkers are tools to achieve these goals and could be defined as characteristics that can be objectively measured and evaluated as indicators of normal biological processes, pathological processes, or pharmacological responses to a therapeutic intervention [1]. Several categories of biomarkers are emerging in the neurodegeneration field including genomic, neuroimaging, clinical and biochemical biomarkers. Genomic biomarkers, such as mutation testing in genes responsible for monogenic forms of PD, AD or HD, already enable specific symptomatic or pre-symptomatic diagnosis of these disorders. However, mutation identification in an individual has a very limited implication for prediction of prognosis or response to new drugs. Various global "omic" approaches to the development of new biomarkers such as transcriptomics or proteomics have already proved useful in the cancer area and raised expectations in the neurodegenerative field as well. Epigenomics is the genome-wide study of epigenetic features which is beginning to provide new understanding of the global role of epigenetic changes during development, differentiation and disease processes. Epigenetic marks span a spectrum from covalent modification of DNA (DNA methylation) to modification of DNA-packaging histones (acetylation, phosphorylation) and finally to the reflection of these modifications in altered chromatin structure. Another level of epigenetic mechanism is associated with gene regulation by small non-coding RNA molecules. These include long-chain non-coding RNA playing a role in cis-regulation of gene clusters as well as whole chromosomes and small interfering RNA (siRNA) and miRNA, which mainly exert their influence in the post-transcriptional level of gene expression regulation [2]. Epigenetic marks may, in contrast to genotype, vary in three principal dimensions: within a cell type; as a function of time (e.g., during development and aging) and between cell types. In addition, epigenetic marks are susceptible to modification by environmental factors including pharmacological agents. As the field of epigenomic biomarkers in neurodegenerative disorders is still in an early phase of development, it is our purpose to present some current evidence which supports the need for further research.

19.2
Epigenetic Marks in Inherited Neurological and Neurodegenerative Disorders

The etiology of several neurological inherited disorders has been directly linked to epigenetic abnormalities [3]. Fragile X syndrome, the most common hereditary form of mental retardation, is associated with tri-nucleotide expansions. The mutation, expansion of CGG (>200) repeats in the 5'-UTR of the FMR1 gene, results in gene silencing [4] through DNA methylation of the expanded CGG tract and binding of methyl CpG binding proteins as well as recruitment of histone deacetylases and other transcriptional repressors [5]. Moreover, pre-mutation (expansion of CGG in the range of 60–200 repeats) in the FMR1 gene is associated with a neurodegenerative disease, fragile X tremor and ataxia syndrome (FXTAS), which is characterized by tremor and ataxia [6]. FRAXE is caused by expansion of CCG (>200) repeats in the 5'-UTR or within the FMR2 gene, while another form of mental retardation is associated with a deletion of the FMR3 gene, a putative non-coding RNA (ncRNA) transcribed from the opposite DNA strand to the FMR1 and FMR2 genes [7]. Several other non-syndromic and syndromic forms (Rett syndrome, Krabbe disease, Werner syndrome, Rubinstein–Taybi syndrome, ICF syndrome) of mental retardation have been linked to mutations affecting genes involved in the epigenetic control of gene expression [3] including DNA helicases (ATRX, CHD7), Kruppel-type zinc finger proteins, transcription regulators (BCOR, PQBP1) genes involved in chromatin remodeling and methylation (MeCP2, RSK2, DNMT3B, H3K9, FTSJ1, SMCX). Angelman and PraderWilli syndromes are the best known imprinting syndromes, being caused by either deletion or epigenetic abnormalities of the same chromosomal region (15q11–q13) involving IPW, ZNF127 and UBE3A genes.

19.3
Epigenetic Dysregulation in Neurodegenerative Disorders

A common theme in neurodegenerative disorders such as HD, PD and AD is the concept that intraneuronal aggregates, such as plaques, interfere with transcription and cause deficits in plasticity and cognition [8].

Transcriptional dysregulation and aberrant chromatin remodeling are central features in the pathology of Huntington disease. There is evidence of altered transcriptional homeostasis associated with increased histone methylation and decreased acetylation status in both human and animal models [9, 10]. The mutant huntingtin gene localizes primarily to the nucleus where it forms aggregates of mutant polyQ protein, which bind and functionally impair transcription factors and coactivators such as CREB-binding protein [11]. Moreover bioinformatic analysis of microarray datasets of brain tissue and blood has shown that transcription is deregulated in large genomic regions and that altered chromosomal clusters in the two tissues are remarkably similar [12].

In Parkinson disease it was shown that nuclear targeting of alpha-synuclein promotes its toxicity and that sequestration of alpha-synuclein to the cytoplasm is protective in the Drosophila model of PD. It was further demonstrated that alpha synuclein binds directly to histones, reduces levels of acetylated histone H3, and inhibits histone acetytransferases (HAT)-mediated acetyltransferase activity [13].

Similarly it has been shown that amyloid-beta, which is the core component of senile plaques, the pathological markers for Alzheimer's disease and cerebral amyloid angiopathy, may induce global hypomethylation while increasing neprilysin (NEP) methylation and further suppressing the NEP expression in the cerebral endothelial cells model [14]. As NEP is one of the enzymes responsible for amyloid-beta degradation these results support the concept of the epigenetically associated vicious cycle.

Further evidence on the epigenetic contribution to neurodegenerative diseases comes from the experiments with histone deacetylases (HDACs) inhibitors. HDACs remove acetyl groups from lysine/arginine residues in the amino-terminal tails of core histones and other proteins, thus shifting the balance toward chromatin condensation and consequently silencing of gene expression.

It was demonstrated in HD that HDAC inhibitors rescue lethality and photoreceptor neurodegeneration in the Drosophila model of polyglutamine disease as well as attenuate neuronal loss, increase motor function and extend survival in the R6/2 mouse model of HD [15–17].

Administration of HDAC inhibitors rescued alpha-synuclein toxicity and protected against dopaminergic cell death both *in vitro* and in the Drosophila model of PD [13, 18].

In the mouse model of AD and tauopathies, SIRT1 and andresveratrol, a SIRT1-activating molecule reduced neurodegeneration in the hippocampus, prevented learning impairment, and decreased the acetylation of the known SIRT1 substrates PGC-1alpha and p53. SIRT1, which belongs to the Silent information regulator 2 (Sir2) family of sirtuin class III histone deacetylases (HDACs) was shown to be upregulated in mouse models for AD [19].

It was also demonstrated that HDAC inhibitors restore acetylation status as well as learning and memory in a mouse model of neurodegeneration [20].

Finally, increased levels of plasma homocysteine (hyperhomocysteinemia) have been reported in several neurodegenerative disorders, including PD, AD and HD [21]. Homocysteine is metabolized to methionine after activation to S-adenosyl-methionine, which is known to act as a methyl donor. Furthermore, homocysteine itself influences global and gene promoter-specific deoxyribonucleic acid (DNA) methylation [22]. Therefore hyperhomocysteinemia and, consequently, dysregulation of epigenetic-DNA methylation may be one important pathophysiological mechanism of neurodegenerative disorders.

19.4
Gene Candidates for Epigenetic Biomarkers

Available evidence suggests that epigenetic dysregulation contributes to differences in gene expression and consequently to the etiology of neurodegenerative disorders. Differential gene expression could result directly from differences of methylation in promoter-associated CpG islands of the respective genes. CpG islands located in the promoter regions of genes can play important roles in the regulation of gene expression.

Global gene expression profiling with DNA microarrays measures the transcriptional activity of thousands of genes. There are several reports demonstrating the difference in brain gene expression in neurodegenerative disorders [23–25]. However, findings from brain genomic profiling cannot be directly considered as a potential biomarker due to the difficulty in obtaining brain tissue for diagnostic purposes. It is therefore of great interest that transcriptomic changes have been

Table 19.1 Number of genes differentially expressed and predicted to have CpG islands.

Type of experiment/comparison	Differentially transcribed	Differentially transcribed/CpG islands
HD brain	130	48
HD blood	983	326
PD brain	830	244
PD blood	8	3
HD brain/blood	13	5
PD brain/blood	2	1
HD/PD brain	17	9
HD/PD blood	1	1

reported when blood was analyzed in patients with neurodegenerative disorders. These changes could reflect dysregulation of gene expression in the brain (common pathological pathways), involvement of blood cells in the pathophysiology of neurodegenration (e.g., immune system) or non-specific reaction to damage in the brain.

We therefore hypothesize that we can identify potential epigenetic biomarkers on the global scale by intersection of genes identified in transcriptomic studies of neurodegeneration with genes predicted to have CpG islands in promotor regions and which are therefore susceptible to methylation dysregulation.

To identify genes with CpG islands we mined the data management system BioMart (CG content ≥65%, observed CpG, expected CpG ≥65%, percentage CpG ≥20%) and found 4.314 genes.

Selected transcriptomic studies in the brain and blood of HD, PD (at $p < 0.001$) [23, 24, 26, 27] are shown in Table 19.1. Our analysis show that about one third of differentially expressed genes in HD and PD in both analyzed tissues (brain and blood) are predicted to have CpG islands. There was also limited overlap between tissues in a given disease and in the same tissue across the two diseases. These results indirectly imply that differences in transcription marks specific for HD and PD might be at least partially a consequence of methylation changes in the respective genes. Further analysis is in progress to evaluate candidate genes as epigenetic biomarkers for both neurodegenerative disorders.

19.5
Conclusions

Both a literature review and our analysis of gene candidates imply that epigenomic changes might potentially be used as biomarkers for neurodegenerative disorders. Candidate gene/mechanism approaches could be supplemented by global methylome and miRNA microarray analysis in the future. In this way epigenomic biomarkers have the potential to contribute, along with other types of biomarkers, to diagnostic and staging purposes and to be used as surrogate endpoints for new drug efficacy testing. Additionally, epigenomic biomarkers could contribute to our understanding of the complex interaction between genetic predisposition and environment in etiology and prognosis of neurodegenerative disorders and could be utilized as potential drug targets for this group of disorders.

References

1. Biomarkers Definitions Working Group (2001) Biomarkers and surrogate endpoints: preferred definitions and conceptual framework. *Clin. Pharmacol. Ther.*, **69**, 89–95.
2. Wang, Y., Liang, Y., and Lu, Q. (2008) MicroRNA epigenetic alterations: predicting biomarkers and therapeutic targets in human diseases. *Clin. Genet.*, **74**, 307–315.
3. Mehler, M.F. (2008) Epigenetic principles and mechanisms underlying nervous system functions in health and disease. *Prog. Neurobiol.*, **86**, 305–341.
4. Garber, K.B., Visootsak, J., and Warren, S.T. (2008) Fragile X syndrome. *Eur. J. Hum. Genet.*, **16**, 666–672.
5. Penagarikano, O., Mulle, J.G., and Warren, S.T. (2007) The pathophysiology of fragile X syndrome. *Annu. Rev. Genomics Hum. Genet.*, **8**, 109–129.
6. Willemsen, R., Mientjes, E., and Oostra, B.A. (2005) FXTAS: a progressive neurologic syndrome associated with Fragile X premutation. *Curr. Neurol. Neurosci. Rep.*, **5**, 405–410.
7. Santos-Rebouças, C.B., and Pimentel, M.M. (2007) Implication of abnormal epigenetic patterns for human diseases. *Eur. J. Hum. Genet.*, **15**, 10–17.
8. Mattson, M.P. (2004) Pathways towards and away from Alzheimer disease. *Nature*, **430**, 631–639.
9. Stack, E.C., Del Signore, S.J., Luthi-Carter, R., Soh, B.Y., Goldstein, D.R., Matson, S., Goodrich, S., Markey, A.L., Cormier, K., Hagerty, S.W., Smith, K., Ryu, H., and Ferrante, R.J. (2007) Modulation of nucleosome dynamics in Huntington's disease. *Hum. Mol. Genet.*, **16**, 1164–1175.
10. Ryu, H., Lee, J., Hagerty, S.W., Soh, B.Y., McAlpin, S.E., Cormier, K.A., Smith, K.M., and Ferrante, R.J. (2006) ESET/SETDB1 gene expression and histone H3 (K9) trimethylation in Huntington's disease. *Proc. Natl. Acad. Sci. U. S. A.*, **103**, 19176–19181.
11. Sugars, K.L., and Rubinsztein, D.C. (2003) Transcriptional abnormalities in Huntington disease. *Trend Genet.*, **19**, 233–238.
12. Anderson, A.N., Roncaroli, F., Hodges, A., Deprez, M., and Turkheimer, F.E. (2008) Chromosomal profiles of gene expression in Huntington's disease. *Brain*, **131**, 381–388.
13. Kontopoulos, E., Parvin, J.D., and Feany, M.B. (2006) Alpha-synuclein acts in the nucleus to inhibit histone acetylation and promote neurotoxicity. *Hum. Mol. Genet.*, **15**, 3012–3023.
14. Chen, K.L., Wang, S.S., Yang, Y.Y., Yuan, R.Y., Chen, R.M., and Hu, C.J. (2009) The epigenetic effects of amyloid-beta(1–40) on global DNA and neprilysin genes in murine cerebral endothelial cells. *Biochem. Biophys. Res. Commun.*, **2** (378), 57–61.
15. Steffan, J.S., Bodai, L., Pallos, J., Poelman, M., McCampbell, A., Apostol, B.L., Kazantsev, A., Schmidt, E., Zhu, Y.Z., Greenwald, M., Kurokawa, R., Housman, D.E., Jackson, G.R., Marsh, J.L., and Thompson, L.M. (2001) Histone deacetylase inhibitors arrest polyglutamine-dependent neurodegeneration in Drosophila. *Nature*, **413**, 739–743.
16. Ferrante, R.J., Kubilus, J.K., Lee, J., Ryu, H., Beesen, A., Zucker, B., Smith, K., Kowall, N.W., Ratan, R.R., Luthi-Carter, R., and Hersch, S.M. (2003) Histone deacetylase inhibition by sodium butyrate chemotherapy ameliorates the neurodegenerative phenotype in

Huntington's disease mice. *J. Neurosci.*, **23**, 9418–9427.

17 Gardian, G., Browne, S.E., Choi, D.K., Klivenyi, P., Gregorio, J., Kubilus, J.K., Ryu, H., Langley, B., Ratan, R.R., Ferrante, R.J., and Beal, M.F. (2005) Neuroprotective effects of phenylbutyrate in the N171-82Q transgenic mouse model of Huntington's disease. *J. Biol. Chem.*, **280**, 556–563.

18 Outeiro, T.F., Kontopoulos, E., Altmann, S.M., Kufareva, I., Strathearn, K.E., Amore, A.M., Volk, C.B., Maxwell, M.M., Rochet, J.C., McLean, P.J., Young, A.B., Abagyan, R., Feany, M.B., Hyman, B.T., and Kazantsev, A.G. (2007) Sirtuin 2 inhibitors rescue alpha-synuclein-mediated toxicity in models of Parkinson's disease. *Science*, **317**, 516–519.

19 Kim, D., Nguyen, M.D., Dobbin, M.M., Fischer, A., Sananbenesi, F., Rodgers, J.T., Delalle, I., Baur, J.A., Sui, G., Armour, S.M., Puigserver, P., Sinclair, D.A., and Tsai, L.H. (2007) SIRT1 deacetylase protects against neurodegeneration in models for Alzheimer's disease and amyotrophic lateral sclerosis. *EMBO J.*, **11** (26), 3169–3179.

20 Fischer, A., Sananbenesi, F., Wang, X., Dobbin, M., and Tsai, L.H. (2007) Recovery of learning and memory is associated with chromatin remodelling. *Nature*, **447**, 151–152.

21 Zoccolella, S., Martino, D., Defazio, G., Lamberti, P., and Livrea, P. (2006) Hyperhomocysteinemia in movement disorders: current evidence and hypotheses. *Curr. Vasc. Pharmacol.*, **4**, 237–243.

22 Bleich, S., Lenz, B., Ziegenbein, M., Beutler, S., Frieling, H., Kornhuber, J., and Bönsch, D. (2006) Epigenetic DNA hypermethylation of the HERP gene promoter induces down-regulation of its mRNA expression in patients with alcohol dependence. *Alcohol. Clin. Exp. Res.*, **30**, 587–591.

23 Hodges, A., Strand, A.D., Aragaki, A.K., Kuhn, A. *et al.* (2006) Regional and cellular gene expression changes in human Huntington's disease brain. *Hum. Mol. Genet.*, **15**, 965–977.

24 Moran, L.B., Duke, D.C., Deprez, M., Dexter, D.T. *et al.* (2006) Whole genome expression profiling of the medial and lateral substantia nigra in Parkinson's disease. *Neurogenetics*, **7**, 1–11.

25 Grünblatt, E., Zander, N., Bartl, J., Jie, L., Monoranu, C.M., Arzberger, T., Ravid, R., Roggendorf, W., Gerlach, M., and Riederer, P. (2007) Comparison analysis of gene expression patterns between sporadic Alzheimer's and Parkinson's disease. *J. Alzheimers Dis.*, **12**, 291–311.

26 Borovecki, F., Lovrecic, L., Zhou, J., Jeong, H. *et al.* (2005) Genome-wide expression profiling of human blood reveals biomarkers for Huntington's disease. *Proc. Natl. Acad. Sci. U. S. A.*, **102**, 11023–11028.

27 Scherzer, C.R., Eklund, A.C., Morse, L.J., Liao, Z. *et al.* (2007) Molecular markers of early Parkinson's disease based on gene expression in blood. *Proc. Natl. Acad. Sci. U. S. A.*, **104**, 955–960.

20
Epigenetic Mechanisms in Asthma

Rachel L. Miller and Julie Herbstman

Abstract

Recent advances have provided critical clues for how environmental exposures may interact with asthma genes without changing their coding sequences to alter their transcription and expression, and, hence, the asthma phenotype. Epigenetic mechanisms, including DNA methylation, histone modifications, and noncoding RNAs may modify the transcription of genes involved in the metabolism of environmental compounds, the ensuing proinflammatory response, or even the efficacy of pharmacological treatment. Evidence supporting this hypothesis includes both epidemiological studies that associate prenatal exposures with later disease, as well as molecular studies that suggest DNA methylation and histone modifications may be important in asthma immunopathogenesis. Another promising area of research involves the recognition that epigenetic patterns may occur differentially across specific asthma effector cell types. In this chapter, we will address the basis for considering epigenetic mechanisms in asthma, including both supportive epidemiological data and mechanistic-based work that have emerged over recent years. Future asthma research has the potential to link environmental and epidemiological findings to mechanistic asthma epigenetic pathways.

20.1
Introduction

Our appreciation of the role of the environment in the pathogenesis of asthma dates back to Galen (130–200 A.D.) who cautioned about the toxic effects of air pollution and advised his readers to "take care of cleanliness of the surrounding air which enters the body [1]." In the 12th century, Moses Maimonides appreciated the importance of environmental triggers such as a common cold, the rainy season and air pollution to the development of asthma [1]. With time and great scientific strides, a more modern 20th century paradigm recognized that the onset of asthma involves an interaction between genetic susceptibility and environmental exposures. Twin and family studies clearly identified the genetic contribution to the disease many years ago [2]. Recently, many studies have documented specific gene–environment interactions. These include studies that found associations between the presence of polymorphisms in the β_2-adrenergic receptor and smoking on asthma development [3], and in the CD14 gene and smoking on asthma severity [4]. Polymorphisms in the interleukin (IL)-1 antagonist gene as well as glutathione S-transferase M1 have each been associated with the development of childhood asthma when occurring in association with prenatal exposure to cigarette exposure [5, 6].

But perhaps it is time once again to revise our view about the interaction of genes and the environment in the development and severity of asthma. While the mechanisms underlying this interface are not entirely understood, recent advances have provided critical clues for how environmental exposures may interact with genes without changing their coding sequences to alter the transcription and expression of genes associated with the asthma phenotype. Epigenetic mechanisms, including DNA methylation, histone modifications, and noncoding RNAs may modify the transcription of genes involved in the metabolism of environmental compounds, the ensuing proinflammatory response, or even the efficacy of pharmacological treatment (Figure 20.1). In this chapter, we will address the basis for considering epigenetic mechanisms in asthma pathogenesis, including both supportive epidemiological data and mechanistic-based work that have emerged over recent years.

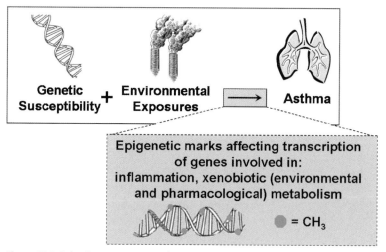

Figure 20.1 Role of epigenetics in asthma. Epigenetic mechanisms, including DNA methylation (shown here), may modify the transcription of genes involved in inflammation and the metabolism of xenobiotic compounds following environmental exposures.

20.2
Epigenetic Mechanisms

DNA methylation and histone modifications represent two distinct, but related, mechanisms of epigenetic modification. DNA methylation has been shown to occur in mammals at the cytosine residue of 5′-CpG-3′ dinucleotides in a reaction catalyzed by DNA (cytosine-5)-methyltransferases [7]. A methylation group from S-adenosyl-L-methionine (SAM) is covalently transferred to the fifth carbon of cytosines [8]. CpG methylation inhibits gene transcription by either blocking the ability of transcription factors to bind to the recognition sites on the CpG dinucleotides or by attracting methyl-binding proteins, obstructing or entirely silencing gene transcription [7, 9, 10]. CpG dinucleotides, under-represented in the genome, are present in a nonuniform pattern approximately once per 80 dinucleotides [7, 10]. A large proportion of the CpG dinucleotides are located in intergenic and intronic DNA regions and are constitutively methylated to prevent the transcription of these regions [11]. CpGs present at a higher frequency relative to the bulk of the genome are called CpG islands (CGIs) [12]. Lying in the promoter regions of transcribed genes, CpGIs are typically unmethylated [13].

Post-translational modifications of histones by means of acetylation, methylation, and phosphorylation are key elements in the chromatin packaging of DNA. The DNA is wrapped twice around an octomer core of histones (H2A, H2B, H3 and H4, two molecules each) [14]. As a result of this tight packaging, RNA polymerase II and transcription factors cannot reach their recognition sequences and turn on transcription. During acetylation, the DNA around the histone core unwinds, activators of transcription obtain access to DNA, and gene expression can then proceed [14, 15]. DNA methylation and histone modification mechanisms can operate independently or together to determine whether a gene or set of genes is transcribed or silenced [16]. Furthermore, microRNA may regulate asthma gene expression by inducing the degradation of target messenger RNA and blocking the translation of target proteins [17].

20.3
Fetal Basis of Adult Disease

Based on human data indicating that early or prenatal environmental exposures are associated with later disease, and data generated from animal models demonstrating that xenobiotic exposure during development alters epigenetic patterns, there is growing recognition that some adverse health effects associated with early exposures may be epigenetically regulated [18–21]. Early support for these claims has emerged from studies that found associations between lower birth weight and a greater risk for adult onset of cardiovascular disease [22], type 2 diabetes mellitus [23], osteoporosis [24], depression [25], and cancer [26]. In addition, prenatal exposures to allergens, antibiotics, and tobacco smoke have been implicated in the later life development of allergies, diabetes, as well as neurological and cardiovascular disorders [20, 27, 28]. Combined, these studies suggest that later childhood and adult disease may originate as a result of the fetal response to developmentally disruptive environmental stimuli.

Among the earliest and most seminal reports of epigenetic regulation *in vivo* were the publications of Cooney *et al.* and Waterland and Jirtle. Through their experiments with agouti mice, they demonstrated that supplementation of the maternal diet during pregnancy with methyl donors such as folic acid and vitamin B_{12} was associated with greater levels of CpG methylation of the region upstream of the agouti gene that regulates coat color. As a result, the coat color distribution of the offspring was affected [29, 30]. Exposure of mice to xenobiotic chemicals, endocrine disruptors, low dose radiation, and maternal stress have all been associated with epigenetic-induced phenotypic changes in the offspring (reviewed in [31, 32]).

Following this work, several groups have suggested that epigenetic mechanisms may play a role in the development of complex immune diseases such as asthma. Exposures associated with asthma occurring during susceptible prenatal and postnatal periods may coincide with time windows during which epigenetic modifications may be more likely to develop. As described in the next section, evidence supporting this hypothesis includes both epidemiological studies that associate prenatal exposures with later disease, as well as laboratory experiment studies that suggest DNA methylation and histone modifications may be important in asthma pathogenesis.

20.4
Fetal Basis of Asthma

Large prospective epidemiologic studies have provided the most convincing evidence that prenatal environmental exposures can influence the risk for subsequent asthma [33, 34]. For example, prenatal exposure to environmental tobacco smoke (ETS) has been associated with wheezy illnesses and asthma, reduced lung function, and respiratory infections in children [35, 36]. Moreover, prenatal air pollution exposure may augment the respiratory symptoms reported following postnatal exposure to ETS, as described by our group [37]. Possible transgeneration effects of ETS exposure have also been observed in an epidemiologic setting as well. In a study by Li *et al.*, researchers assessed the grandmaternal smoking histories of 338 children with asthma and 570 matched controls and found that children of mothers who had been exposed *in utero* to tobacco smoke were at an increased risk for developing asthma, independent of maternal smoking status. If both the mother and the child were exposed *in utero* to tobacco smoke, the child's risk of developing asthma was increased further [38]. A prenatal diet that is high fat also has been shown to predispose a child toward developing asthma. Most recently, this relationship was shown by Chatzi and colleagues who demonstrated that a Mediterranean diet in pregnancy is protective for childhood wheeze and atopy [39]. Other dietary influences, such as low maternal intake of foods containing vitamin

E, vitamin D and zinc [40, 41], as well as antibiotic use during pregnancy [42], have been associated with greater likelihoods that the children will develop asthma.

Other sets of studies linking prenatal exposure to asthma risk are those that show stronger associations of maternal, compared to paternal, inheritance of asthma-related phenotypes. For example, among children less than 5 years old, maternal, but not paternal, history of asthma placed children at increased risk for development of childhood asthma [43]. Also, maternal, but not paternal, immunoglobulin E (IgE) predicted elevated IgE levels in cord blood and at age 6 months, which are well-recognized biomarkers associated with allergy and asthma risk [44]. Finally, there are a number of maternally-inherited polymorphisms, including the β-chain of the high-affinity receptor for IgE and glutathione-S-transferase P1, that are more highly associated with childhood asthma markers (i.e., positive allergy skin prick tests, higher allergen-specific IgE levels [45], and reduced lung function in children [46], respectively.

20.5
Experimental Evidence

Animal models have confirmed that prenatal events and intrauterine exposures may influence the risk for airway inflammation in offspring. For example, offspring of female mice sensitized to ovalbumin exhibited greater airway hyperresponsiveness and allergen-specific IgE production that was blocked with antibodies against the proallergic cytokine IL-4 [47]. However, a similar phenotype was also found following prenatal exposure to the unrelated protein casein, suggesting these responses may not be allergen-dependent [47]. The same group also exposed female mice during pregnancy to residual oil fly ash (ROFA; extracted from a precipitator unit that removes particulate contaminates of an oil fired power plant). These offspring demonstrated greater levels of airway hyper-responsiveness and eosinophillic inflammation [48]. In another mouse model, prenatal exposure of female mice to lipopolysaccharide (i.e., endotoxin) suppressed the development of allergic immune responses and airway inflammation [49]. This group also showed similar responses in the offspring following a prenatal maternal diet supplemented with probiotic bacteria [50].

20.6
Epigenetic Mechanisms in Asthma

A growing body of literature implicates specific epigenetic mechanisms in asthma-related molecular pathways. These include several studies showing that DNA methylation can influence the regulation of T helper (Th) differentiation and cytokine production, and thus the polarization toward or away from a more allergic state. For example, Shin et al. showed that transcription factor STAT 4 expression in human T cells is regulated by DNA promoter methylation, resulting in impaired

Th1 responses [51]. In addition, hypomethylation at the proximal promoter and conserved intronic regulatory element of the proallergic interleukin (IL)-4 gene induced IL-4 production and Th2 polarization [52–55]. Hypermethylation of sites in the counter-regulatory interferon (IFN) γ promoter also upregulated Th2 differentiation. [55]. In contrast, hypermethylation of the DNaseI-hypersensitive region at the 3′ end of the IL-4 gene promoted Th1 differentiation [52]. Moreover, methylation of a CpG island that overlapped with a cyclic-AMP response element binding protein/activating transcription factor (CREB/ATF) site is inversely correlated with CREB binding and FoxP3 expression, which are important for regulatory T cell-mediated immune responses [56].

Histone acetylation and deacetylation also appear to be important epigenetic mechanisms involved in airway inflammation. In general, acetylation of histones is associated with gene induction, and deacetylation is associated with gene silencing [57]. In one study, stable patients with asthma possessed greater levels of histone acetyltransferase (HAT) and reduced levels of histone deacetylases (HDACs) in their bronchial biopsies. Levels normalized following treatment with inhaled corticosteroids [58]. Another group recently found that inhibition of endogenous HDAC and the resulting hyperacetylation using trichostatin A treatment of CD45RO+ T cells increased Th2-cytokine-associated (IL-13, IL-5), reduced Th1 cytokine-associated (IFNγ) recall responses, and increased expression of the transcription factor GATA-3 and spingosine kinase 1 [59]. Furthermore, under Th2 polarizing conditions, hypomethylation of several CpG sites of the proximal IL-13 promoter were identified; histone H4 acetylation levels were also higher. Hence, epigenetic regulation of the proallergic IL-13 gene seems to occur as well [60].

Furthermore, β2 adrenoreceptor agonists and glucocorticoids selectively inhibited tumor necrosis factor (TNF)-α induced histone H4 acetylation at the eotaxin promoter, and *in vivo* promoter binding of NF-kB, in human airway smooth muscle cells. This inhibition abrogated proallergic eotaxin transcription and gene expression. Therefore, inhibition of histone acetylation appears to be a mechanism by which two asthma medications influence inflammatory gene expression [15]. In subsequent work, IFNγ inhibited TNF-α-induced expression of proinflammatory genes (IL-6, IL-8, eotaxin) by modulating HDAC function and thereby NF-kB transactivation [61]

Finally, an intriguing paper by Cao and colleagues documented altered chromatin modification following exposure to the urban pollutant diesel particles. Exposure of human bronchial epithelial cell lines to diesel exhaust particles increased acetylation of histone H4 associated with the cyclooxygenase (COX)-2 promoter, leading to selective degradation of HDAC 1. Diesel exposure also induced recruitment of HAT p300 to the COX-2 promoter [62]. As a result, HDAC1 expression was downregulated at the post-transcriptional level. Cyclooxygenase is a key rate-limiting enzyme in the conversion of arachidonic acid to prostaglandins, and its expression is associated with airway inflammation [63]. Combined, these findings suggest that acetylation is an important regulator of the airway response to diesel exposure, linking a specific environmental exposure associated with asthma triggers with a specific epigenetic mechanism.

20.7
Cell-Specific Responses

Another promising area of research involves the recognition that epigenetic patterns may occur differentially across specific asthma effector cell types. A few examples from the literature have been published. In one, human alveolar macrophages exhibited reduced HDAC activity levels. Comparable effects were not observed when HDAC was measured in peripheral blood mononuclear cells derived from asthma patients [64]. The correlations between DNA methylation levels and the human high affinity receptor for leukotriene B4 receptor gene expression levels also varied according to cell type [65]. Most recently, this pattern has been observed in studies of the A disintegrin and metalloprotease (ADAM33) gene, associated with severe asthma [66]. The promoter CpG island (-362-+80) was hypermethylated in epithelial cells and hypomethylated in ADAM33-expressing fibroblasts. The methylation state in turn affected ADAM33 gene expression in a cell type-specific manner [67].

20.8
Conclusion

The etiology of asthma has long been recognized as extremely heterogeneous. The natural history of asthma, including the incidence and remission of symptoms, is also variable and often unpredictable. Yet, because the majority of asthma prevalence rates seem rooted in early and prenatal exposures, hypotheses implicating epigenetic mechanisms fit the clinical observations about the disease. Now, supportive experimental data are emerging as well. Nonetheless, many fundamental questions such as how genetic and epigenetic mechanisms interact to influence the risk of developing disease need to be answered. How do we know which individuals are susceptible to epigenetically-mediated mechanisms? Do we all potentially possess a diagnostic epigenome that will mark our risk? Perhaps individually-designed environmental interventions and asthma pharmacological therapies someday will take advantage of our epigenetic make-up. Future asthma research has the potential to link environmental and epidemiological findings to mechanistic asthma epigenetic pathways. As new research emerges, hopefully these questions will eventually be answered.

Acknowledgments

This work was supported by R21ES013063, R01ES013163, P50ES015905.

Abbreviations

ADAM33	A disintegrin and metalloprotease
CGI	CpG islands
COX	cyclooxygenase
CREB/ATF	cyclic-AMP response element binding protein/activating transcription factor
ETS	environmental tobacco smoke
HAT	histone acetyltransferase
HDAC	histone deacetylase
IFN	interferon
IgE	immunoglobulin E
IL	interleukin
MiRNA	microRNA
NF-kB	nuclear factor–kappa B
ROFA	residual oil fly ash
SAM	S-adenosyl-L-methionine
Th	T helper
TNF	tumor necrosis factor

References

1 Bernstein, L. (2003) Proceedings of the First Jack Pepys Occupational Asthma Symposium. American Thoracic Society Workshop. *Am. J. Respir. Crit. Care med.*, **167**, 450–471.

2 Los, H., Postmus, P.E., and Boomsma, D.I. (2001) Asthma genetics and intermediate phenotypes; a review from twin studies. *Twin. Res.*, **4**, 81–93.

3 Wang, Z., Chen, C., Niu, T., Wu, D., Yang, J., Wang, B., Fang, Z., Yandava, C., Drazen, J., Weiss, S. *et al.* (2001) Association of asthma with beta(2)-adrenergic receptor gene polymorphism and cigarette smoking. *Am. J. Respir. Crit. Care Med.*, **163**, 1404–1449.

4 Choudhry, S.P., Avila, P.C., Nazario, S., Ung, J., Kho, J., Rodriguez-Santana, R., Csal, J., Tsai, H.J., Torres, A., Ziv, E. *et al.* (2005) CD14 tobacco gene-environment interaction modifies asthma severity and immunoglobulin E levels in Latinos with asthma. *Am. J. Respir. Crit. Care Med.*, **172**, 173–182.

5 Ramadas, R., Sadeghnejad, A., Karmaus, W., Arshad, S., Matthews, S., Huebner, M., Kim, D., and Ewart, S. (2007) Interleukin-1R antagonist gene and pre-natal smoke exposure are associated with childhood asthma. *Eur. Respir. J.*, **29**, 502–508.

6 Gilliland, F.D., Li, Y.-F., Dubeau, L., Berhane, K., Avol, E., McConnell, R., Gauderman, W.J., and Peters, J.M. (2002) Effects of Glutathione S-Transferase M1, maternal smoking during pregnancy, and environmental tobacco smoke on asthma and wheezing in children. *Am. J. Respir. Crit. Care Med.*, **166**, 457–463.

7 Costello, J.F., and Plass, C. (2001) Methylation matters. *J. Med. Genet.*, **38**, 285–303.

8 Chiang, P.K., Gordon, R.K., Tal, J., Zeng, G.C., Doctor, B.P., Pardhasaradhi, K., and McCann, P.P. (1996) S-adenosyl-methionine and methylation. *FASEB. J.*, **10**, 471–480.

9 Momparler, R.L., and Bovenzi, V. (2000) DNA methylation and cancer. *J. Cell Physiol.*, **183**, 145–154.

10 Fazzari, M.J., and Greally, J.M. (2004) Epigenomics: Beyond CpG islands. *Nat. Rev. Genet.*, **5**, 446–455.

11 Wilson, A.S., Power, B.E., and Molloy, P.L. (2007) DNA hypomethylation and human diseases. *Biochim. Biophys. Acta.*, **1775**, 138–162.

12 Gardiner-Garden, M., and Frommer, M. (1987) CpG islands in vertebrate genomes. *J. Mol. Biol.*, **196**, 261–282.

13 Cooper, D.N., and Krawczak, M. (1989) Cytosine methylation and the fate of CpG dinucleotides in vertebrate genomes. *Hum. Genet.*, **83**, 181–188.

14 Grunstein, M. (1997) Future asthma research has the potential to link environmental and epidemiological findings to mechanistic asthma epigenetic pathways. *Nature*, **389**, 349–352.

15 Nie, M., Knox, A.J., and Pang, L. (2005) Beta2-adrenoceptor agonists, like glucocorticoids, repress eotaxin gene transcription by selective inhibition of histone H4 acetylation. *J. Immunol.*, **175**, 478–486.

16 Jones, P.L., and Wolffe, A.P. (1999) Relationships between chromatin organization and DNA methylation in determining gene expression. *Semin. Cancer Biol.*, **9**, 339–347.

17 Krek, A., Grün, D., Poy, M.N., Wolf, R., Rosenberg, L., Epstein, E.J., MacMenamin, P., Id, P., Gunsalus, K.C., Stoffel, M. et al. (2005) Combinatorial microRNA target predictions. *Nat. Genet.*, **37**, 495–500.

18 de Boo, H.A., and Harding, J.E. (2006) The developmental origins of adult disease (Barker) hypothesis. *Aust. N. Z. J. Obstet. Gynaecol.*, **46**, 4–14.

19 Lau, C., and Rogers, J.M. (2004) Embryonic and fetal programming of physiological disorders in adulthood. *Birth Defects Res. C. Embryo Today*, **72**, 300–312.

20 Hales, C., and Barker, D. (2001) The thrifty phenotype hypothesis. *Br. Med. Bull.*, **60**, 5–20.

21 Hales, C., Barker, D., Clark, P., Cox, L., Fall, C., Osmond, C., and Winter, P. (1991) Fetal and infant growth and impaired glucose tolerance at age 64. *Br. Med. J.*, **303**, 1019–1022.

22 Bateson, P., Barker, D., Clutton-Brock, T., Deb, D., D'Udine, B., Foley, R.A., Gluckman, P., Godfrey, K., Kirkwood, T., Lahr, M.M. et al. (2004) Developmental plasticity and human health. *Nature*, **430**, 419–421.

23 Ravelli, A.C., van der Meulen, J.H., Michels, R.P., Osmond, C., Barker, D.J., Hales, C.N., and Bleker, O.P. (1998) Glucose tolerance in adults after prenatal exposure to famine. *Lancet*, **351**, 173–177.

24 Dennison, E.M., Arden, N.K., Keen, R.W., Syddall, H., Day, I.N., Spector, T.D., and Cooper, C. (2001) Birthweight, vitamin D receptor genotype and the programming of osteoporosis. *Paediatr. Perinat. Epidemiol.*, **15**, 211–219.

25 Thompson, C., Syddall, H., Rodin, I., Osmond, C., and Barker, D.J. (2001) Birth weight and the risk of depressive disorder in late life. *Br. J. Psychiatry*, **179**, 450–455.

26 Ho, S.M., Tang, W.Y., Belmonte de Frausto, J., and Prins, G.S. (2006) Developmental exposure to estradiol and bisphenol A increases susceptibility to prostate carcinogenesis and epigenetically regulates phosphodiesterase type 4 variant 4. *Cancer Res.*, **66**, 5624–5632.

27 McKeever, T., Lewis, S.A., Smith, C., and Hubbard, R. (2002) The importance of prenatal exposures on the development of allergic disease: a birth cohort study using the west midlands general practice database. *Am. J. Respir. Crit. Care Med.*, **166**, 827–832.

28 Mone, S., Gillman, M.W., Miller, T.L., Herman, E.H., and Lipshultz, S.E. (2004) Effects of environmental exposures on the cardiovascular system: prenatal period through adolescence. *Pediatrics*, **113**, 1058–1069.

29 Cooney, C.A., Dave, A.A., and Wolff, G.L. (2002) Maternal methyl supplements in mice affect epigenetic variation and DNA methylation of offspring. *J. Nutr.*, **132**, S2393–S2400.

30 Waterland, R.A., and Jirtle, R.L. (2003) Transposable elements: targets for early nutritional effects on epigenetic gene regulation. *Mol. Cell Biol.*, **23**, 5293–5300.

31 Dolinoy, D., Weidman, J., and Jirtle, R. (2007) Epigenetic gene regulation: linking early developmental environment to adult disease. *Reprod. Toxicol.*, **23**, 297–307.

32 Tang, W., and Ho, S.M. (2007) Epigenetic reprogramming and imprinting in origins of disease. *Rev. Endocr. Metab. Disord.*, **8**, 173–182.

33. Magnusson, L., Olesen, A., Wennborg, H., and Olsen, J. (2005) Wheezing, asthma, hayfever, and atopic eczema in childhood following exposure to tobacco smoke in fetal life. *Clin. Exp. Allergy*, **35**, 1550–1556.

34. Alati, R.A.M.A., O'Callaghan, M., Najman, J.M., and Williams, G.M. (2006) In utero and postnatal maternal smoking and asthma in adolescence. *Epidemiology*, **17**, 138–144.

35. Lannero, E., Wickman, M., Pershagen, G., and Nordvall, L. (2006) Maternal smoking during pregnancy increases the risk of recurrent wheezing during the first years of life. *Respir. Res.*, **7**, 3.

36. Jedrychowski, W., and Flak, E. (1997) Maternal smoking during pregnancy and postnatal exposure to environmental tobacco smoke as predisposition factors to acute respiratory infections. *Environ. Health Perspect.*, **105**, 302–306.

37. Miller, R.L., Garfinkel, R., Horton, M., Camann, D., Perera, F., Whyatt, R.M., and Kinney, P.L. (2004) Polycyclic aromatic hydrocarbons, environmental tobacco smoke, and respiratory symptoms in an inner-city birth cohort. *Chest*, **126**, 1071–1078.

38. Li, Y.-F., Langholz, B., Salam, M.T., and Gilliland, F.D. (2005) Maternal and grandmaternal smoking patterns are associated with early childhood asthma. *Chest*, **127**, 1232–1241.

39. Chatzi, L., Torrent, M., Romieu, I., Garcia-Esteban, R., Ferrer, C., Vioque, J., Kogevinas, M., and Sunyer, J. (2008) Mediterranean diet in pregnancy is protective for wheeze and atopy in childhood. *Thorax*, **63**, 507–513.

40. Devereux, G., Turner, S.W., Craig, L.C., McNeill, G., Martindale, S., Harbour, P.J., Helms, P.J., and Seaton, A. (2006) Low maternal vitamin E intake during pregnancy is associated with asthma in 5-year-old children. *Am. J. Respir Crit. Care Med.*, **174**, 499–507.

41. Devereux, G., Litonjua, A., Turner, S.W., Craig, L.C., McNeill, G., Martindale, S., Helms, P.J., Seaton, A., and Weiss, S.T. (2007) Maternal vitamin d intake during pregnancy and early childhood wheezing. *Am. J. Clin. Nutr.*, **85**, 853–859.

42. Jedrychowski, W., Galas, A., Whyatt, R., and Perera, F. (2006) The prenatal use of antibiotics and the development of allergic disease in one year old infants. A preliminary study. *Int. J. Occup. Med. Environ. Health*, **19**, 70–76.

43. Litonjua, A., Carey, V.J., Burge, H.A., Weiss, S.T., and Gold, D.R. (1998) Parental history and the risk for childhood asthma. Does mother confer more risk than father? *Am. J. Respir. Crit. Care Med.*, **158**, 176–181.

44. Liu, C.A., Wang, C.L., Chuang, H., Ou, C.Y., Hsu, T.Y., and Land, K.D. (2003) Prenatal prediction of infant atopy by maternal but not paternal total IgE levels. *J. Allergy Clin. Immunol.*, **112**, 899–904.

45. Traherne, J.A., Hill, M.R., Hysi, P., D'Amato, M., Broxholme, J., Mott, R., Moffatt, M.F., and Cookson, W.O. (2003) Ld mapping of maternally and non-maternally derived alleles and atopy in Fc epsilon R1-beta. *Hum. Mol. Genet.*, **12**, 2577–2585.

46. Carroll, W.D., Lenney, W., Child, F., Strange, R.C., Jones, P.W., and Fryer, AA. (2005) Maternal Glutathione S-Transferase GSTP1 genotype is a specific predictor of phenotype in children with asthma. *Pediatr. Allergy Immunol.*, **16**, 32–39.

47. Hamada, K., Suzaki, Y., Goldman, A., Ning, Y.-Y., Goldsmith, C., Palecanda, A., Coull, B., Hubeau, C., and Kobzik, L. (2003) Allergen-independent maternal transmission of asthma susceptibility. *J. Immunol.*, **170**, 1683–1689.

48. Hamada, K., Suzaki, Y., Leme, A., Ito, T., Miyamoto, K., Kobzik, L., and Kimura, H. (2007) Exposure of pregnant mice to an air pollutant aerosol increases asthma susceptibility in offspring. *J. Toxicol. Environ. Health*, **70**, 688–695.

49. Blumer, N., Herz, U., Wegmann, M., and Renz, H. (2005) Prenatal lipopolysaccharide-exposure prevents allergic sensitization and airway inflammation, but not airway responsiveness in a murine model of experimental asthma. *Clin. Exp. Allergy*, **35**, 397–402.

50. Blumer, N., Sel, S., Virna, S., Patrascan, C., Zimmermann, S., Herz, U., Renz, H., and Garn, H. (2007) Perinatal maternal

application of lactobacillus rhamnosus GG suppresses allergic airway inflammation in mouse offspring. *Clin. Exp. Allergy*, **37**, 348–357.

51 Shin, H.J., Park, H.Y., Jeong, S.J., Park, H.W., Kim, Y.K., Cho, S.H., Kim, Y.Y., Cho, M.L., Kim, H.Y., Min, K.U. et al. (2005) Stat4 expression in human t cells is regulated by DNA methylation but not by promoter polymorphism. *J. Immunol.*, **175**, 7143–7150.

52 Lee, D.U., Agarwal, S., and Rao, A. (2002) Th2 lineage commitment and efficient IL-4 production involves extended demethylation of the IL-4 gene. *Immunity*, **16**, 649–660.

53 Agarwal, S., and Rao, A. (1998) Modulation of chromatin structure regulates cytokine gene expression during t cell differentiation. *Immunity*, **9**, 765–775.

54 Tykocinski, L.O., Hajkova, P., Chang, H.D., Stamm, T., Sozeri, O., Lohning, M., Hu-Li, J., Niesner, U., Kreher, S., Friedrich, B. et al. (2005) A critical control element for interleukin-4 memory expression in T helper lymphocytes. *J. Biol. Chem.*, **280**, 28177–28185.

55 Jones, B., and Chen, J. (2006) Inhibition of IFN-g transcription by site-specific methylation during T helper cell development. *EMBO J.*, **25**, 2443–2452.

56 Kim, H.P., and Leonar, W.J. (2007) CREB/ATF-dependent T cell receptor-induced FoxP3 gene expression: a role for DNA methylation. *J. Exp. Med.*, **204**, 1543–1551.

57 Bhavsar, P., Ahmad, T., and Adcock, I. (2008) The role of histone deacetylases in asthma and allergic diseases. *J. Allergy Clin. Immunol.*, **121**, 580–584.

58 Ito, K., Caramori, G., Lim, S., Oates, T., Chung, K., Barnes, P., and Adcock, I. (2002) Expression and activity of histone deacetylases in human asthmatic airway. *Am. J. Respir. Crit. Care Med.*, **166**, 392–396.

59 Su, R., Becker, A., Kozyrskyj, A., and Hayglass, K. (2008) Epigenetic regulation of established human type 1 versus type 2 cytokine responses. *J. Allergy Clin. Immunol.*, **121**, 57–63.

60 Webster, R.B., Rodriguez, Y., Klimecki, W., and Vercelli, D. (2007) The human IL-13 locus in neonatal cd4+ t cells is refractory to the acquisition of a repressive chromatin architecture. *J. Biol. Chem.*, **282**, 700–709.

61 Keslacy, S., Tliba, O., Baidouri, H., and Amrani, Y. (2007) Inhibition of tumor necrosis factor-alpha-inducible inflammatory genes by interferon-gamma is associated with altered nuclear factor-kappa B transactivation and enhanced histone deacetylase activity. *Mol. Pharmacol.*, **71**, 609–618.

62 Cao, D., Bromberg, P.A., and Samet, J.M. (2007) COX-2 expression induced by diesel particles involves chromatin modification and degradation of HDAC1. *Am. J. Respir. Cell Mol. Biol.*, **37**, 232–239.

63 Peebles, R.S. Jr., Hashimoto, K., Morrow, J.D., Dworski, R., Collins, R.D., Hashimoto, Y., Christman, J.W., Kang, K.H., Jarzecka, K., Furlong, J. et al. (2002) Selective cyclooxygenase-1 and -2 inhibitors each increase allergic inflammation and airway hyper-responsiveness in mice. *Am. J. Respir. Crit. Care Med.*, **165**, 1154–1160.

64 Cosío, B.G., Mann, B., Ito, K., Jazrawi, E., Barnes, P.J., Chung, K.F., and Adcock, I.M. (2004) Histone acetylase and deacetylase activity in alveolar macrophages and blood monocytes in asthma. *Am. J. Respir. Crit. Care Med.*, **170**, 141–147.

65 Kato, K., Yokomizo, T., Izumi, T., and Shimizu, T. (2000) Cell specific transcriptional regulation of human leukotriene B4 receptor gene. *J. Exp. Med.*, **192**, 421–431.

66 Foley, S., Mogas, A., Olivenstein, R., Fiset, P., Chakir, J., Bourbeau, J., Ernst, P., Lemière, C., Martin, J., and Hamid, Q. (2007) Increased expression of ADAM33 and ADAM8 with disease progression in asthma. *J. Allergy Clin. Immunol.*, **119**, 863–871.

67 Yang, Y., Haitchi, H., Cakebread, J., Sammut, D., Harvey, A., Powell, R., Holloway, J., Howarth, P., Holgate, S., and Davies, D. (2008) Epigenetic mechanisms silence A disintegrin and metalloprotease 33 expression in bronchial epithelial cells. *J. Allergy Clin. Immunol.*, **121**, 1393–1399.

Part VI
Ways to Translate the Concept

21
Public Health Genomics – Integrating Genomics and Epigenetics into National and European Health Strategies and Policies

Tobias Schulte in den Bäumen and Angela Brand

Abstract

In recent years epigenetic factors have been increasingly recognized as important determinants of health. The genomic research of the last decade has seen an important change of direction, from the high penetrance genes to the gene–environment interactions. So far Public Health and health policy makers are struggling to translate these emerging understandings of disease etiology into policies and practices. In this paper we argue that the different professions involved in this discourse are operating with different concepts. The analysis and recognition of small relative risks does not provide policy makers with the scientific base that they need to modify health policies. Relative risks may only be useful if researchers are able either to translate the relative risks into attributable risks or if they initially set up study designs which aim to identify attributable risks. In the language of law and policy makers these attributable risks are modifiable, therefore allowing collective and individual actions to be taken. Translating emerging scientific knowledge into policies is one of the main tasks of Public Health Genomics. Public Health Genomics is an emerging field of translational research which organizes the effective and responsible application of genome-based knowledge and technologies. In Europe the Public Health Genomics European Network (PHGEN), a DG SANCO-funded network of researchers, policy makers and their institutions is organizing this process. It has set up communications processes which ensure that epigenetic factors are increasingly recognized both by researchers and by policy makers.

21.1
Public Health and Genomics

The Human Genome Project, the development of new genetic tests, DNA chip technologies and related technologies offer new opportunities for the promotion of population health which will lead to fundamental challenges in the healthcare delivery systems. Medicine and Public Health get an increasing insight into the biological factors which drive disease mechanisms, in particular in the field of cancer [1]. The emerging genome-based knowledge calls for a paradigm shift in Public Health as we can see a clear need to adjust concepts of prevention and health service delivery. As a consequence we can describe a dichotomy: Genomics needs to understand how it can include Public Health aspects in its work program while Public Health needs to analyze how genomics changes the concepts of Public Health. The second approach is seen as the core task of Public Health Genomics. Still there is an interdependence between the two directions; for example, Public Health Genomics is also concerned about the organization of genetic services and the genetic health literacy of the population.

A comprehensive health care which regards, besides environmental, social and life style factors, genetic determinants will become essential as it creates new opportunities for individualized strategies in preventive medicine and early detection of illnesses. The integration of genome-based knowledge will change primary, secondary and tertiary prevention. Inter alia, disease prevention programs and clinical interventions will be specifically targeted at susceptible individuals and sub-entities of populations based on their genomic risk profile. So far health care systems, policy makers and industries are not prepared for the conceptual change and all stakeholders are struggling to transfer the emerging knowledge into clinical and technological applications. Public Health Genomics advocates the interdisciplinary discourse on and the understanding of genomics; it fosters progress in translational research and supports the introduction of new concepts of risk stratification and prevention.

21.2
The Bellagio Model of Public Health Genomics

Public Health Genomics (PHG) is an emerging multidisciplinary scientific approach which aims to integrate genome-based knowledge in a *responsible* and *effective* way into public health (Figure 21.1).

With the avalanche of emerging knowledge it is time to question whether we are offering the "right" health interventions (health promotion, prevention, therapy and rehabilitation), not only in the public health sector but also in the health care system as a whole for the benefit of population health. If we question the old concepts, we also have to ask whether our present and future public health strategies are evidence-based [2].

In the past 20 years, the advances in genetic and genomic research have revolutionized knowledge of the role of inheritance in health and disease. Nowadays, we know that our DNA determines not only the cause of single-gene disorders, but also predispositions to common diseases, responses to therapies as well as the onset and progression of diseases.

Whereas medicine is presently taking a breathtaking development from its morphological and phenotype orientation to a molecular and genotype orientation

The Enterprise

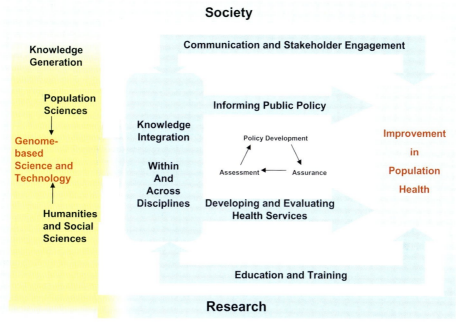

Figure 21.1 The Public Health Genomics Enterprise (Bellagio Model). The figure displays the core tasks of PHG between the "Knowledge Generation" and the overall goal "Improvement in Population Health". (source: Bellagio Statement. Genome-based Research and Population Health. Report of an expert workshop held at the Rockefeller Foundation Study and Conference Center, Bellagio, Italy, 14–20 April 2005).

promoting the importance of prognosis and prediction, public health practice has to date concerned itself with environmental determinants of health and disease and has paid scant attention to genomic variations between individuals, within populations and between populations [3, 4]. In particular, the advances brought about by the Human Genome Project, HapMap or Systems Biology are changing these perceptions [5, 6]. Many predict, that this knowledge will enable health promotion messages and disease prevention programs to be specifically directed toward families and individuals at risk, or to subgroups of the population, based on a genomic risk stratification – a paradigm shift in public health from subgroups to individual heath information management. This does not mean that we do not need public health as an interdisciplinary and inter-institutional task for translating basic research into policy and practice. The opposite is true, the management of individual health information and guidance for major individual and collective health risks needs to be organized as our health care system is not yet prepared.

The new technologies will allow researchers to examine genetic mutations at the level of the functional units of genes, and to better understand the significance of environmental factors such as noxious agents, nutrition and personal behavior for the cause of diseases such as cardiovascular diseases, psychiatric disorders or infectious diseases.

The new knowledge about pleiotropic effects of susceptibility genes in complex diseases being associated to more than one disease ("disease clusters"), about the role of individual genomic profiling, about the role of a genomic variant being a protective factor for some diseases and a risk factor for others, and about epigenomic effects will require a modification of both clinical and Public Health strategies. The post-genomic era also calls for a new knowledge base, inter alia for the integration of genome-based biobanks into public health surveillance systems as a tool for generating evidence on genome–environment interactions as well as for understanding diseases. The challenges we currently face can only be tackled by an interdisciplinary and systematic approach regarding the evaluation of the future impact of genome-based knowledge and technologies on health care systems [7]. Public Health follows the Trias which guides the assessment, policy development and assurance of Public Health actions (Figure 21.2).

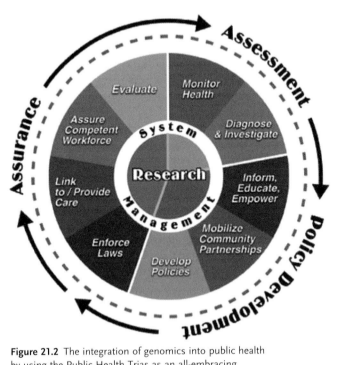

Figure 21.2 The integration of genomics into public health by using the Public Health Trias as an all-embracing methodology (adapted from IOM, 1998).

Obviously, the integration of genomics into public health research, policy and practice will be one of the most important future challenges for all health care systems. Clarifying the general conditions under which genomic knowledge can be put to best practice in the field of public health, and [8] particularly considering the economic, ethical, legal and social implications, is presently the most pressing task of the emerging field of public health genomics (PHG) [9], defined as the application of genetic and molecular science to the promotion of health and prevention of disease through the organized efforts of society [10]. Policymakers now have the opportunity to protect consumers, to monitor the implications of genomics for health, social, and environmental policy goals, and to assure that genomics advances will be used not only to treat medical conditions, but also to prevent disease and improve health. Policy must find an acceptable balance between providing strong protection for individuals' interests while enabling society to benefit from genomics [11].

The next decade will provide a window of opportunity to establish infrastructures, both in the health care sector and on a policy level across Europe and globally, that will enable the scientific advances to be effectively, efficiently and socially acceptably translated into evidence-based policies and interventions that improve population health.

The need for new products and processes is a major opportunity and a major risk at the same time. According to the Bellagio Model, Public Health Genomics wants to ensure the effective translation of genome-based knowledge. Health technology assessment (HTA) in combination with the concepts of health needs assessment (HNA) and health impact assessment (HIA) has the potential to assist Public Health Genomics fulfill these tasks. On the other hand, these methodological approaches themselves are challenged by genome-based knowledge and technologies and have to be developed further.

21.3
The Public Health Genomics European Network

Experts from the different fields and stakeholders have urged the EC to set up a European network on Public Health Genomics. In 2006 the EU-funded Public Health Genomics European Network (PHGEN) started its operations and has successfully implemented both working groups on a European level and *National Task Forces* on Public Health Genomics. The network will develop a comprehensive set of policies which will serve as a framework for the future implementation of Public Health Genomics [12]. Together with partner networks and projects, PHGEN will ensure the consistency and coherence of European policies.

Public Health Genomics has made substantial progress in recent years and is now seen as an integral part of Public Health. So far it has not fully reached its main goal, the transfer of genomic knowledge into all areas of Public Health. Still, the last two network meetings have proven the increasing acceptance of the challenge deriving from genomics. With the increasing genomic literacy of the population and the training of the workforces Public Health Genomics will continue to diffuse into all Public Health actions.

21.4
From Public Health Genomics to Public Health and Epigenetics/Epigenomics

In a recent paper Rothstein *et al.* called epigenetics the "Ghost in our genes", describing how genetic and environmental factors affect the well-being of individuals and their offspring [13]. Rothstein *et al.* focused on ethical and legal aspects of epigenetics and if we assess the Public Health dimension of epigenetics we move forward in a similar direction. Epigenetics was described by Conrad Waddington in 1942 as "the interaction of genes with the environment, which brings the phenotype into being" [14]. Today epigenetics is rather understood as the modification of the genome which does not require a modification of the DNA itself. In this chapter we will not analyze the biological factors which modify the expression of genes, rather we will try to transfer this knowledge to the Public Health domain. Epigenetics, from a "positive" perspective, allows the individual to react to the environmental factors, to develop the organism. If we look at the risk factors as well as protective factors which determine common complex diseases we see huge Public Health issues such as diet, lifestyle, toxic substances, stress and other exposures to environmental factors. So far the scientific community has been striving for an understanding of the interaction of the different determinants of health; as this knowledge becomes more systematic we see a need to assess potential consequences for Public Health and health policies.

Much ethical and legal reasoning will be needed to highlight all aspects of individual freedom and the responsibility for third persons and the offspring. Before we can determine the concordance between freedom and responsibility we should first question how we measure risks in epigenetics and whether these risk categories are able to help us when we shape policies which aim to reduce epigenetic risks. Setting up policies which refer to epigenetic knowledge may sound futuristic but it is not. The European Commission is following a "Health in all Policies" approach which is based in Articles 95 and 152 of the EC Treaty. As Health in all Policies is based on emerging genome-based knowledge it can serve as a viable tool for the transformation of knowledge into practice.

21.5
Health in All Policies – Translating Epigenetics/Epigenomics into Policies and Practice

Within the last decade the European Commission has paid increasing attention to both genomics and health regulations. The efforts of the European Commission, the WHO and the Finnish government have resulted in a new doctrine named "Health in all Policies" [15]. According to the "inventors" of this new approach "Health in all Policies" shall stimulate health regulations outside the traditional field of health law. The European Commission should not be perceived as a completely altruistic institution as it follows this path; due to Art. 152 of the EC Treaty the Commission has no competence in the field of health law and health systems. Using the "Health in all Policies" initiative and the increasing importance of cross-border health care this strict separation of competences seems to come to an end. At first it looks like a specifically European problem, but if we look at areas like tobacco policies, infectious diseases, obesity and chemicals we see a global expansion of health oriented policies in areas other than health law. Still, if we address regulatory bodies or politicians they would at least look surprised (if not deny) that these policies are part of the enterprise of Public Health Genomics. Legislatory projects like the European program on chemicals "REACH" address directly the interaction between the human genome and toxic substances. The same pattern applies when the City Council of South Los Angeles places a moratorium for new fast-food restaurants within the city limits [16]. The City Council argues that 30% of the male population is obese, more than double the average figures in California. Politicians and legal experts are often not aware when we speak about gene–environment interactions, and if we look at the way they perceive health risks, and the way scientists communicate health risks, we can understand better why Public Health Genomics has not yet reached the mainstream of the legal and political discourse.

21.6
Health in All Policies as a Guiding Concept for European Policies

The Health in all Policies (HiAP) concept was developed under the Finnish EU presidency in 2006. Following the Nordic experience with alcohol and tobacco policies the Finnish government supported a joint research program of the Finnish health authorities STAKES [17] and KTL [18], the WHO Europe and its Observatory and the EC. The concept has already been reflected in the Amsterdam recast of the EC Treaty; both Art. 95 and Art.152 of the EC Treaty demand the Commission to strive for a high level of health protection and to assess the health impact of EC policies. Still, the consequences of the new wording of the Treaty have been of minor importance. Following the Finnish presidency the European Commission set up a new health strategy in the year 2007 named "Together for Health" [19]. Health in all Policies plays a prominent role in the new strategy and many efforts have been made to translate the concept into practice. In particular Art. 95 para. 3 of the Treaty encourages the Commission to integrate the upcoming knowledge from genomics into health policies as it reads: " [The Commission].. will take as a base a high level of protection, taking account in particular of any new development based on scientific facts". The scientific evidence base is the key to the concept, but it must also be ensured that the evidence base can be translated into health policies. The norms must have a definable addressee or a target group, negative side effects must be excluded. With the ongoing individualization of genomics, law is challenged to find an answer to these technical difficulties in the regulation of genomics and health determinants. Law may limit itself as it focuses on the governance of processes rather than delivering content-oriented norms. Still, procedural laws are often seen as insufficient if we want to reduce health risks for major parts of the population. Therefore it needs to be analyzed how Public Health Genomics can filter the available evidence and how it can provide normative sciences with an aggregated knowledge which can be translated into coherent health policies.

21.7
Relative Risk and Risk Regulation – A Model for the Regulation of Epigenetic Risks?

Public Health Genomics is teaching us that we can only understand health problems if we look at both genetic and environmental (including lifestyle and social) factors systematically. If we transfer this idea to the area of epigenetics we need to discuss whether these categories are able to structure a regulatory discourse which aims to minimize epigenetic risk factors. In the field of genomics we may describe a ladder of risk, starting with high genetic risks in rare diseases and ending with small genetic contributions in many common complex diseases. Still, these risks are not so small if we look into subgroups or if we try to understand disease clusters ("diseasomes" or "disease nodes") [20]. The assessment of health risks comprises the study of inherited genetic variation, epigenetic changes, various fields of "-omics" and systems biology. But can we really expect that lawyers, ethicists and politicians will understand this?

Sceptical scientists tend to argue that genetics is useless if we look at common complex diseases and this is currently the message that politicians do understand. We know that we have to be physically active, that we should eat healthy food and that we must not smoke a single cigarette. We do not really need genetics to come to these conclusions. So, what is the additional benefit of Public Health Genomics? It seems to depend on the concept of risk! Colleagues from epidemiology have a clear focus on relative risks (RR) [21]. Individuals can have a high genomic predisposition or they may have a low genomic predisposition (high or low RR). From a legal point of view it looks like we do not have to care about both groups. If a person has a high relative risk (for a rare disease), the genetic predisposition reaches a level which is often technically equated with a disability in law. We aim to soften the condition, we offer help in the social security system and we protect the person from discrimination. If a person has a low relative risk law does not tend to accept the need for action as we are all affected by low relative risks and we would need an unpractical matrix of regulations to cover all these risks. In addition, legal instruments are not flexible enough to cover the interaction between different risks and the potentially protective side effects of certain genomic risks. Thus, we may assume that lawyers and politicians cannot work with the concept of individual relative risks as they fail to translate these into political actions.

21.8
Attributable Risks and Risk Regulation

To determine whether a certain disease is associated with a certain exposure we must determine, using data obtained in case-control or cohort studies, whether there is an excess risk of the disease in persons who have been exposed to a certain (epigenetic) agent. Whereas the concept of relative risks is primarily important as measuring the strength of the association between a risk factor and a disease and a concept of the medical setting, the attributable risks (AR) are relevant for decision making in Public Health [22]. The attributable risk is defined as the amount of proportion of disease incidence (or disease risk) that can be attributed to a specific risk factor. This "specific" risk factor does not necessarily need to consist of one factor, it can also be a group of risk factors which form a risk "bundle". Public Health faces a similar dilemma as lawyers and politicians if it is confronted with relative risks. Public Health aims to offer: the right interventions, at the right time, for the right population, in the right way. Therefore, the question of genetic susceptibility to environmental factors and the possible interaction of risks need to be addressed in terms of the magnitude of the "genomic predisposition" or "genomics risk fraction" in patients and the whole population. While the RR tells us something about the association of a risk factor and the disease, the AR helps Public Health to opt for the right interventions for the right target groups as well as to set priorities for action. The question is: how much of a disease can be attributed to a certain risk factor? How much of the disease incidence can we prevent if we are able to eliminate the exposure?

21.9
Translating Attributable Risks into Policies

After this excursion to the concept of risk in Public Health Genomics and the methods which are used to quantify risks, we should come back to the initial question: how are genomics and epigenetics affecting the "health in all policies" approach and how can law translate the emerging knowledge in these fields? We separated the relative risks from the attributable risks as the concept of the attributable risks empowers Public Health to choose the right intervention for the right target group. Thus, attributable risks are the bridge from Public Health to health policies. In the legal discourse we may probably not use the term attributable risk, instead we would rather differentiate between modifiable and non-modifiable risks. Health in all Policies aims to lower modifiable risks as it encourages policy makers to assess the health impact of policies in sectors other than the health sector itself. "Health is an outcome of a multitude of determinants, including those relating to individual, genetic and biological factors, and those relating to individual lifestyles, as well as those relating to the structures of society, policies and other social factors. The term is used much more in the context of addressing structural rather than individual, genetic or biological determinants of health, but public policies also influence or guide individual behavior and lifestyle choices" [23]. Still, it would be unacceptable to set up policies without a sound evidence base.

Within the scope of this chapter we can only briefly determine how law can participate and possibly support the translational research in the field of Public Health Genomics. As we are dealing with new and emerging technologies legal scholars should take a closer look at the political and legal instruments which govern the implementation of genome-based knowledge and technologies in Europe. The implementation of innovations in the health system and in markets follows different concepts, from market diffusion to state controlled approval systems. This issue has a significant impact on the Single Market in Europe; therefore we touch a field of major interest for the EC Commission and the Member States as the new Art. 152 of the Lisbon Treaty does not clarify the uncertainties of the prior version. As many Member States control the access of innovative technologies to the publicly funded part of the health care sector, these Member States control the implementation of genome-based technologies such as high-throughput, low-cost sequencing, which are being applied to increasingly large human genome and phenotypic data sets [24]. If we discuss the regulation of modifiable risks and the evidence base of health policies we should also note that the state control may turn out to be a starting point of a vicious circle. As policy makers and health systems do not apply emerging genome-based knowledge, they obstruct the generation of a solid evidence base for genome-based applications. Thus, genomics is used, like in the tobacco politics, but it is not part of the label. Researchers in the life sciences and politicians tend to forget that these processes are legally pre-structured and that adoption decisions may hamper both progress and markets. Still, these control mechanisms also enable Member States to consider the ethical and social consequences of innovations and their impact on the "ordre public".

As we have discussed earlier the borderlines between the traditional medical setting and the lifestyle decisions of individuals are changing as we analyze the interaction between modifiable and non-modifiable risks. To exploit the full potential of Public Health Genomics it might be necessary to set up a coherent system which addresses all types of risks and risk factors as well as protective factors. We see such developments already in areas like nutrigenomics or toxicogenomics and we may see broader applications in fields like cancer, type 2 diabetes, etc. The integration of genome-based knowledge, the increasing complexity and the diffusion of "health in all policies" will challenge the current governance of innovation. The methodology of "Health Technology Assessment (HTA)" has been developed to structure the processes which generate the evidence for evidence-based health policies [25]. This procedural equity could turn out to be an important step but it may also hamper genome-based innovations as HTA requires an evidence base that many new technologies are not even able to generate. The current legal discourse seems to be too premature to propose any solutions for the manifold problems and challenges in the area of health innovations and their regulation. With the new Art. 152 of the Lisbon Treaty this field will remain as problematic as it has been in the past 15 years.

21.10
Limits to the Concept of Health in All Policies in Genomics and Epigenetics

Directly linked to the issue of innovations and their implementation is a more conceptual and more complex issue. Legal decision-making does not generate evidence in the field of genomics; instead it depends on the existing evidence which is provided by researchers and Public Health experts. Yet, in the field of genomics we see a lot of preliminary evidence; sceptics may even say promises. Once the new evidence emerges it should be applied immediately for the benefit of affected or at risk individuals. For the legal discourse we see a dilemma arising as legal and political health care decision making has the potential to hinder or delay the implementation of upcoming genome-based knowledge. The core dilemma exists due to the uncertainty of genome-based innovations, the risk of adverse reactions, insufficient information and inappropriately educated professionals. The Health in All Policies approach can only reach its full potential if risks can be identified and erased. Technically, the regulation can hardly work with soft principles which need to be balanced. Instead the "Health in All Policies" approach requires clear norms and technical standards which can be applied throughout Europe. Thus, the advances in genomics may require preparatory and procedural legal governance which balances a priori the potential individual and societal legal positions which may be at odds with each other [11].

While ethical reasoning may work sufficiently with scenarios, many legal scholars will insist that decisions which affect fundamental rights of individuals need to be sufficiently evidence based and justified by reasons of general public interest. Thus, the complexity of the evidence and the probabilistic nature of much genome-based information is a regulatory problem of its own account. In the genetic setting ethical and legal scholars have discussed how law should handle the dichotomy of hopes and fears in genetics and predictive medicine. Within the traditional medical context, medical law has adapted the concept of "indication" as a scientific "benchmark". Within the context of genomics it needs to be discussed whether indication as a "scientifically good reason to act" is still the concept which we should or could apply. The normal legal reaction to this degree of uncertainty would be the application of the precautionary principle [26]. Depending on the concept of risk and the translation of evidence into health policies, the precautionary principle may have adverse consequences: either we put the burden of definite proof on genomics before we allow any application in practice or we advocate the application of genomics as any delay would put patients and risk groups at risk.

With the combination of genomics and Public Health we see new layers of the discourse arising which need to be further analyzed by all professions involved. Individuals are directly affected by Public Health decisions, in particular in areas like education of professionals or the allocation of funds. These Public Health policies are not yet evidence-based and concepts like HTA may not be sufficient to take account of the exploratory nature of genomics. Genomics is rapidly advancing and new knowledge is generated every week. The half life of genomic knowledge is a major burden for the translation of genome-based knowledge into health policies.

21.11
Conclusion

We see a need to foster the communication between discourse in law, ethics, politics and all other relevant stakeholders of the PHG enterprise. While the long term perspective is rather clear, less certainty exists in terms of the immediate and mediate time line. In the long run, Public Health Genomics, epigenetics and the normative sciences shall support individuals and society at large to understand diseases, to reduce health risks and to combat the major burden of disease in Europe. The Health in All Policies approach of the EC is one cornerstone in this strategy as it demonstrates that health goals may not be achieved if health policies do not leave the traditional domain of health law. Still, the task ahead is rather the integration and legal management of interfaces. Law can be one tool when stakeholders from all profession in this enterprise try to organize coherent and consistent policies in all areas of health care and prevention. As a first step, we need to explore these interfaces and offer procedural solutions for the management of genome-based knowledge and technologies and the societal trade-offs which seem to be inevitable as stakeholders pursue progress under the condition of scientific uncertainty. As a second step it will be necessary to improve the analysis and communication of health risks (including risks as positive and negative disease or health outcome modifiers). Health in All Policies can only reach its full potential if science is able to identify modifiable risks which are open for regulatory efforts.

Acknowledgment

This paper is a result of the work of the Public Health Genomics European Network (PHGEN), which is funded in the Public Health Program of the European Commission (Project Number 2005313). The paper reflects the views of the authors but not necessarily the view of the European Commission.

References

1 Hanahan, D., and Weinberg, R.A. (2000) The hallmarks of cancer. *Cell*, **100** (1), 57–70.

2 Collins, F.S., Patrinos, A., Jordan, E., Chakravarti, A., Gesteland, R., and Walters, L. (1998) New goals for the U.S. genome project: 1998–2003. *Science*, **282**, 682–689.

3 Khoury, M.J. (1996) From genes to public health: the applications of genetic technology in disease prevention. *Am. J. Public Health*, **86** (12), 1717–1722.

4 Burke, W. (2003) Genomics as a probe for disease biology. *N. Engl. J. Med.*, **349**, 969–974.
5 Khoury, M.J., Little, J., and Burke, W. (2004) *Human Genome Epidemiology. A Scientific Foundation for Using Genetic Information to Improve Health and Prevent Disease*, Oxford University Press, Oxford, New York, Tokyo.
6 Baird, P.A. (2000) Identification of genetic susceptibility to common diseases: the case for regulation. *Perspect. Biol. Med.*, **45** (4), 516–528.
7 Beskow, L.M., Khoury, M.J., Baker, T.G., and Thrasher, J.F. (2001) The integration of genomics into public health research, policy and practice in the United States. *Community Genet.*, **4**, 2–11.
8 Moldrup, C. (2002) Medical technology assessment of the ethical, social, and legal implications of pharmacogenomics. A research proposal for an internet citizen jury. *Int. J. Technol. Assess Health Care*, **18** (3), 728–732.
9 Public Health Genomics Foundation, www.phgfoundation.org.uk (accessed Sept. 4th 2009).
10 Brand, A. (2005) Public health and genetics – dangerous combination? View-point section. *Eur. J. Public Health*, **15** (2), 114–116.
11 Schulte in den Bäumen, T. (2006) Governance in genomics: a conceptual challenge for public health genomics law. *Ital. J. Publ. Health*, **4** (3), 46–52.
12 Public Health Genomics European Network (PHGEN), www.phgen.eu (accessed Sept. 4th 2009).
13 Rothstein, M.A., Cai, Y., and Marchant, G.E. (2008) The Ghost in our Genes: Legal and Ethical Implications of Epigenetics, online available at SSRN: http://ssrn.papers.com/abstract=1140443 (accessed 1 November 2008).
14 Wadington, C.H. (1942) The epigenotype. *Endeavour*, **1**, 18–20.
15 Puska, P. (2007) Health in all policies. *Eur. J. Public Health*, **17** (4), 328.
16 Hennessy-Fiske, M., and Zahniser, D. (2008) Council bans new fast-food outlets in South L.A., Los Angeles Times, online accessible at www.latimes.com (accessed Sept. 4th, 2009).
17 National Institute of Health and Welfare (THL), www.thl.fi (accessed Sept. 4th, 2009).
18 KTL is now part of THL, see reference 17.
19 European Commission, General Directorate Health and Consumers (DG SANCO) EU Health Strategy (2007) ec.europa.eu/health/ph_overview/strategy/health_strategy_en.htm (accessed Sept. 4th, 2009).
20 Barabasi, A.L. (2007) Network medicine – from obesity to the "diseasome". *N. Engl. J. Med.*, **357**, 1866–1868.
21 Yng, Q., and Khoury, M.J. (1997) Evolving methods in genetic epidemiology. III. Gene-environment interactions in epidemiologic research. *Epidemiol. Rev.*, **19** (1), 33–43.
22 Northridge, M.E. (1995) Public health methods – attributable risk as a link between causality and public health action. *Am. J. Public Health*, **85** (9), 1202–1204.
23 Sihto, M., Ollila, E., and Koivusalo, M. (2006) Principles and challenges of health in all policies, in *Health in All Policies* (eds T. Stahl, M. Wismar *et al.*), Finnish Ministry of Social Affairs and Health, Helsinki, p. 7.
24 Lunshof, J.E., Chadwick, R., Vorhaus, D.B., and Church, G.M. (2008) From genetic privacy to open consent. *Nat. Rev. Genet.*, **9** (5), 406–411.
25 Oliver, A., Mossialos, E., and Robinson, R. (2004) Health technology assessment and its influence on health care priority setting. *Int. J. Technol. Assess Health Care*, **20** (1), 1–10.
26 Raffensberger, C., and Tickner, J. (eds) (1999) Protecting Public Health and the Environment: Implementing the Precautionary Principle, Island Press, Washington DC.

22
Taking a First Step: Epigenetic Health and Responsibility
Astrid H. Gesche

Abstract

Confronted with epigenetic challenges, individuals are likely to opt for a number of possible responses: they could ignore potential risk, avoid detrimental triggers, maximize risk reduction measures, seek out pharmacological interventions, or take the opposite approach and pursue strategies for epigenetic enhancement. After introducing these responses, the chapter reflects on the interplay between response and responsibility. We continue by recommending the establishment of an integrated public health policy framework that is capable of protecting the public interest as well as guiding a process of transformational change toward assigning greater responsibility to individuals for their and their family's short and long-term epigenetic health outcomes.

22.1
Introduction

The methods by which phenotypes are created, sustained or modulated by hereditary and epigenetic mechanisms have been of intense interests to scientists for some time. In contrast, until recently, public knowledge of epigenetics has been scant. In the public domain, there have been few discussions regarding the impact and consequences of epigenetics. However, Brand et al. report that currently significant inroads are being made with initiatives taken by the Public Health Genomics Foundation (PHG Foundation, UK) in Cambridge, the Public Health Genomics European Network (PHGEN) stationed in Bielefeld (Germany), and the US National Office of Public Health Genomics in Atlanta [1]. Their efforts and those of others will gradually see a shift toward greater awareness. Such a shift entails that scientists, health care providers and governments not only educate citizens about yet another facet of genetics, but, perhaps more importantly, requires them to motivate their citizens to engage with the new epigenetic knowledge in ways that avoid or reduce possible detrimental outcomes for themselves and their offspring and to accept a share of that responsibility. How a given society and the individuals embedded within react to epigenetic findings will be influenced by different world views, personal and cultural values, economic resources, government structures, and socio-economic realities. They in turn will shape expectations and responses. Here we can only introduce one such set of preconditions.

22.2
Responding to Epigenetic Challenges

As public knowledge of epigenetics increases together with a growing awareness that at least some adverse epigenetic modifications might be preventable or reversible, different reactions from individuals can be expected. By and large, individuals are likely to respond to epigenetic challenges in two ways: they can either remain passive or become active. Remaining passive and ignoring potential risks could lead to potential ill health. Choosing a more proactive path instead might result in more satisfactory health outcomes.

Different proactive options are possible. A first option is avoidance of detrimental triggers. If certain environmental influences, such as excessive stress, are known to cause unfavorable changes to the genome/epigenome interface, a simple strategy would be to deliberately avoid them. A second option is to take actions that minimize potential or known risks. As evidence is growing that nutritional factors can modulate epigenetic outcomes (reviewed by Jirtle and Skinner [2]), affected individuals may seek to actively down-modulate or reverse known detrimental influence(s). Eating healthy, highly specific, even targeted foods is emerging as a key interventionist strategy for trying to modulate the genome/epigenome interaction [1]. Achieving and maintaining a healthy body weight before and during pregnancy is another. For example, maternal undernutrition or malnutrition during critical periods of fetal development has been associated with obesity [3, 4] and subsequent metabolic disturbances in the offspring, such as an increased risk of cardiovascular disease and type 2 diabetes. In contrast, maternal overnutrition, obesity or abnormal weight gain during pregnancy have been implicated in both fetal hyperinsulinemia and larger than normal neonates (LGA), potentially leading to obesity or the metabolic syndrome in the offspring later in life [5].

A third option with which to counteract negative epigenetic influences is using pharmacological substances (reviewed by Szyf [6]). Not every individual can be motivated to lead a stress-reduced lifestyle and to consume healthy foods. Not every individual has the capacity to be an active partner in such health care measures and disease prevention and therapy strategies. In these cases, pursuing pharmacological possibilities may shape up as an alternative option with which to modify deleterious genes or adverse environmental influences. Although "epigenetic therapy" approaches using pharmacological agents are still in their infancy, they may one day offer new opportunities for directing epigenetic outcomes.

A fourth option pertains to taking advantage of epigenetic processes. Instead of seeking countermeasures to minimize potentially harmful outcomes, some individuals could pursue the opposite. Their goal could be to take advantage of the epigenetic potential for human enhancement of physical and mental traits, cosmetic attributes, adaptive fitness, or others. For example, epigenetic enhancement might improve an organism's adaptive fitness and chance of survival, perhaps by strengthening its capacity to adjust more readily to a particular internal or external environment. Bjorklund [7], reporting on research on cognition that used chimpanzees as animal model, notes that parental adaptive behavior, precipitated by long-term changes in the environment, can lead to epigenetic modifications in the mother that are passed on to her offspring and provide them with a significant boost to their adaptive fitness [2]. Another scenario which could be exploited for epigenetic enhancement pertains to how an animal responds to stress. Research in the rat model has demonstrated that close mother–infant interactions, especially during the immediate postpartum period, produce offspring that are less anxious in novel environments and have a weaker stress response compared with offspring of mothers whose postpartum maternal care was low [8].

22.3
Responsibility and Public Health Care Policy

In its simplest form, responsibility can be understood as being accountable for one's actions and decisions. However, responsibility can also be equated with free will, a sense of duty, or an awareness of obligation. Acting informed and responsibly presupposes that affected individuals as well as associated health care professionals have access to information based on scientific evidence, to risk and impact assessments, to economic forward estimates, and so forth, so that they can make informed decisions and exercise choice. It also presupposes that they have the capacity to act, that they are able to influence epigenetic regulatory mechanisms and, through them, their health and the health of their offspring.

The responses and examples above indicate that considering epigenetic factors in public health strategy and management is likely to become an important focus of not only preventative medicine, but also for ethical deliberation and social and fiscal policy in the future. Will failing to act on epigenetic data be viewed in the future as failing to prevent likely harm? Will it be treated as negligence? If governments attempt to prevent epigenetic harm from occurring by exerting financial or other pressure on the individual to, for example, conform to known nutritional standards, would this erode the principle of individual freedom? Will it lead to discrimination and stigmatization, should the individual fail to act in the best interest of his or her own or offspring's genome or, if asked to pay, be unable or unwilling to make a monetary contribution? Rothstein alerts us to further ethical issues and mentions environmental justice, equity, privacy and confidentiality [9].

As individuals ponder epigenetic implications and respond or do not respond to options, and as health care costs continue to soar, a near future seems likely in which civil society, industry, regulatory bodies and government link together and demand greater personal accountability for health outcomes and request individuals to share some of the responsibility for actions taken. Indeed, given the current climate of fiscal constraint, there might be an expectation that at risk individuals, who have the capacity to deal with epigenetic risk, but ignore their responsibility, could be found negligible and asked to substantially contribute to those health costs that reasonably can be assumed to have arisen from their negligence.

How desirable is such a future? Is it not true that states of health and disease are multi-factorial and epigenetic factors are just one set of determinants (even if they are strong) among many others? Is it thus preferable to follow the path of solidarity and relational care and co-responsibility, not least because epigenetic modulation reaches back into the generational past (over which we have no influence) and stretches far into an unknowable future? If we ignore this reality and ask for co-contributions and a sharing of responsibility, do we further close the door to an "open future" for many individuals and their families whose biological, social, and economic realities limit their options?

When social and ethical factors of concern to the public collide with government's concerns about financial expenditure and an already heavily strained public health budget, the likelihood of discontent on both sides can be considerable. It is at this junction that ethical principles could guide decisions, especially the principles of transparency, equity, distributive justice and procedural fairness. Public health officials usually invite the community to participate in the process of formulating public health policy. One of the benefits of community participation in public health policy formulation is that it promotes a sense of ownership of decisions taken and might lead participants to a greater understanding of why sharing responsibility might constitute the best option. The financial pressure on governments to rein in their spending on public health might, however, complicate deliberations and could be exacerbated by public argument that meaningful participation and sustainable transformation can only develop within an environment where options are grounded in real capacity and are based on genuine choices, which might sometimes be difficult or impossible to attain without financial support.

As with so many complex issues emerging in science and technology, it is desirable to have an integrative approach to strategy and policy development that is based on scientific evidence and characterized by an overarching, ethically sound, co-operative policy framework. Such framework should be underpinned by continuous feedback loops that combine emerging knowledge and expertise with stakeholder input and expectations for risk analysis processes that include ongoing monitoring, assessment and evaluation processes, and should have the necessary quality control mechanisms in place that foster ethically fair, equitable and sustainable outcomes.

22.4
Conclusion

A web of complex and finely tuned interactions of hereditary and epigenetic mechanisms plays a pivotal role in gene expression. Unfavorable environments can negatively influence epigenetic regulation, leading to a higher risk of developing chronic diseases. In addition, epigenetic changes can be passed on to successive generations. There is a growing awareness that each individual may be able to partially influence epigenetic events and minimize particular negative epigenetic outcomes. Public health needs to respond to these emerging challenges by taking a first step and building an integrated, ethically sound policy framework from which to deal with emerging social, ethical and legal issues in the rapidly expanding area of epigenetics. A health policy framework is desirable that not only protects the public interest, but is also capable of guiding a process of transformational change that has at its center a commitment to share responsibilities and obligations sustainably and fairly.

References

1 Brand, A., Brand, H., and Schulte in den Bäumen, T. (2008). The impact of genetics and genomics on public health. *Eur. J. Hum. Genet.*, **16**, 5–13.

2 Jirtle, R.L., and Skinner, M.K. (2007) Environmental epigenomics and disease susceptibility. *Nat. Rev. Genet.*, **8**, 253–262.

3 Anguita, R.M., Sigulern, D.M., and Sawaya, A.L. (1993) Intrauterine Food restriction is associated with obesity in young rats. *J. Nutr.*, **123**, 1421–1428.

4 Gallou-Kabani, C., and Junien, C. (2005) Nutritional epigenomics of metabolic syndrome: new perspectives against the epidemic. *Diabetes*, **54**, 1899–1906.

5 King, J.C. (2006) Maternal obesity, metabolism, and pregnancy outcomes. *Annu. Rev. Nutr.*, **26**, 271–291.

6 Szyf, M. (2009) Epigenetics, DNA methylation, and chromatin modifying drugs. *Ann. Rev. Pharmacol. Toxicol.*, **49**, 243–263.

7 Bjorklund, D.F. (2006) Mother knows best: epigenetic inheritance, maternal effects, and the evolution of human intelligence. *Dev. Rev.*, **26** (2), 213–242.

8 Meaney, M.J. (2001) Maternal care, gene expression, and the transmission of individual differences in stress reactivity across generations. *Ann. Rev. Neurosci.*, **24**, 1161–1192.

9 Rothstein, M.A., Cai, Y., and Marchant, G.E. (2009) Ethical implications of epigenetics research. *Nat. Rev. Genet.*, **10** (4), 224, http://www.nature.com/nrg/journal/v10/n4/full/nrg2562.html (accessed 21 May 2009).

Index

a

A1298C polymorphism   41–47
abdominal obesity   211
acceptable daily intake   92, 93
acetaldehyde   111
acrylamide   110
adaptive fitness   283
adaptive immune responses   167, 171
adipocyte hyperplasia   160
adipocytokines   211
adiponectin   211, 216
adipose tissue   211
adolescence, health determinants   160, 161
β2 adrenoreceptor agonists   258
adult diseases *see* human diseases
adults, health determinants   160, 161
aflatoxin B1   92–94, 105, 109, 113, 117
aging
– anti-aging modalities   215–217
– cancer, epigenetic dysregulation   209–218
– cancer predisposition model   217
– cancer-prone metabolic phenotype   210–212
– diseases associated   158
– DNA hypermethylation during   183, 184
– epigenetic silencing via DNA methylation   212, 213
– extrinsic   27
– health determinants   160, 161
– inflammatory control of epigenetic regulators   214, 215
– intrinsic   27
– malignant cell development   180
– metabolism-associated genes   181–183
– nutrition   26–28
– *see also* life span; neurodegenerative diseases
agouti viable yellow murine model   136

air pollution   22, 253, 256, 258
alcohol
– aging   27
– cancer deaths   106
– carcinogens   111
– folate metabolism   43
– MTHFR variants   46
– NAFLD   56, 57
allergens/allergies   132, 255, 257
ALSPAC data set   68, 73, 75
aluminium   231
Alzheimer's disease   225–228, 245, 246
– diet   237, 238
– environmental factors   231, 232
– epigenetic dysregulation   247, 248
– epigenetics   234–236
– gene expression   249
– susceptibility genes   229, 230
amyloid beta (Aβ) peptides   226, 228, 234, 236
amyloid precursor protein gene (APP)   226–228, 234–236
amyotrophic lateral sclerosis   225–228
– diet   238
– environmental factors   231, 232
– sporadic   236, 237
– susceptibility genes   229, 230
ancestors' nutrition   71–74, 81, 82, 136
antiapoptosis   215
antibiotics   257
anticancer therapy   141, 142, 145–152
antioxidants   112, 113, 118, 151, 235
AP-1 transcriptional complex   148, 149
Apolipoprotein E (APOE)   229, 231
aristolochic acid   109
arsenic   5, 238
asthma
– epigenetic mechanisms   253–260
– fetal basis   256, 257

Epigenetics and Human Health
Edited by Alexander G. Haslberger, Co-edited by Sabine Gressler
Copyright © 2010 WILEY-VCH Verlag GmbH & Co. KGaA, Weinheim
ISBN: 978-3-527-32427-9

ATRA 144–149
attachment theory 79
attributable risks 267, 275, 276
5-aza-2´-deoxycytidine 195, 202, 203
5-azacytidine 202

b
babies *see* newborn
Bellagio Model of Public Health Genomics 268–271
benzo[a]pyren 113
biobanking and biomolecular resources 55
biobanks
– background 51, 52
– disease- and population-oriented 5, 52, 54, 55
– formats and purpose 52, 53
– genome-based 270
– key components and applications 53
– need for networks 53–55
– role in gene–environment interactions 51–60
– *see also* genetic epidemiology
bioflavonoids 142, 149, 150
bioinformatics 127, 128, 132, 214, 249
biological plausibility 77
biomarkers
– chemoprevention 204
– definition 53, 246
– identification 58
– neurodegenerative disorders 245–250
– *see also* epigenetic marks
birth weight 75, 159, 161, 175, 255
bisphenol A 21, 237
bladder cancer 187
blood cell transcriptome 130, 131
blood samples, DNA methylation 200
brain cancer 187
brain injuries 232
breast cancer
– aging 209
– diet 145
– enzyme polymorphisms 114–117
– estrogen receptor gene 148, 212, 213
– genes methylated 187
– MTHFR 44, 45
butyrate 25

c
C677T polymorphism 37, 41–47
CAG repeat length 228, 229
calcitriol 145
caloric restriction 26, 28, 209, 210, 215–217

cancer
– aging, epigenetic dysregulation 209–218
– aging predisposition model 217
– anticancer therapy 141, 142, 145–152
– caloric restriction 28
– causes of death 105, 106, 152, 210, 217
– deaths 158
– epigenetic progenitor model 200, 201
– genetic/epigenetic mechanisms 184
– genetic factors 106, 184
– HDAC role 189, 190
– histone modifications 19, 20
– hypermethylated gene promoters 187, 188
– hypo and hypermethylation *see* hypermethylation; hypomethylation
– low birth weight 255
– metabolic phenotype, cancer-prone 210–212
– *see also* carcinogenesis; malignancy; promoter hypermethylation
Cancer and Leukemia Group B 202
cancer cells
– DNA methylation 185, 196, 197
– genetics and epigenetics 180
– nuclear structure 179
cancer genome 195
cancer risk 209
– enzyme polymorphisms 114–117
– from genotoxic dietary carcinogens 111, 112
– MTHFR gene polymorphisms 37, 41–47
candidate genes *see* gene candidates
carbon tetrachloride 92
carcinogenesis
– hepatocarcinogenesis 90–99
– multistage concept 89, 90, 91–95
– *see also* cancer
carcinogenic potency 107
carcinogens
– dietary protection 112–114
– DNA methylation 89, 96, 97
– in foods 105–118
– in plant-derived foods 109–111
– *see also* genotoxic carcinogens; non-genotoxic carcinogens
cardiovascular disease
– caloric restriction 28
– deaths 72, 158
– epigenetic inheritance 72, 78, 81, 82
– low birth weight 255
– obesity 8

– overweight in childhood   160
– see also coronary heart disease
β-carotene   145
case-control study design   38
cases   52
catechins   142, 149, 150
catechols   142, 149–151
causative genes in neurodegenerative diseases   226–231
cell-specific responses in asthma   259
cellular aging   27
chemoprevention   97, 112, 141, 142, 145–152
– biomarkers   204
childhood
– epigenetic inheritance and life span   75
– growth velocity   69, 70
– nutritional and environmental effects   20–22
– overweight   160, 161
– proband   68
– see also newborn; slow growth period (SGP)
cholesterol   128, 129
chromatin
– DNA methylation interaction   188–190, 196, 197
– histone modifications   18, 258
– packaging   14, 15, 255
chronic noncommunicable diseases   158–162
cirrhosis   56, 57
coffee   133, 134, 138
coffee polyphenols   142, 149, 150
cohort studies   38
colon cancer
– aging   209
– C677T polymorphism   45, 47
– DNA methyltransferases   197
– genes methylated   187
– MGMT gene   212
common disease genetic and epigenetic hypothesis   233
conjugated linoleic acids   25
control mechanisms
– epigenetic   169, 170
– evolutionary need   167, 168
– RNA silencing   168, 169
controls   52
coronary heart disease
– coffee consumption   133, 134
– deaths   78
– etiology   64, 65, 68
– obesity   161

– prenatal determinants   159
– see also cardiovascular disease
cortisol/cortisone   177
CpG islands   143, 255
– during aging   212
– in cancer cells   185
– methylation   3, 15–17
– neurodegenerative disorders   249
cutaneous T-cell lymphoma   195, 203
cycasin   110
cyclooxygenase-2 promoter   258
cytochrome P450   92, 109
– CYP1A1   114, 115
– CYP1A2   133, 134
cytokine transforming growth factor β1   95
cytosine methylation pathway   23, 24
cytotoxic stress   201, 202

d

dairy products, transcriptome   131
Data Schema and Harmonization Platform for Epidemiological Research (DataSHaPER)   54
de novo methylation   16, 143
deaths
– cancer   105, 106, 152, 158, 210, 217
– cardiovascular and diabetes-related   72, 158
– from coronary heart disease   78
– leading causes   68, 158
– lung cancer   198
defense mechanisms   167–171
depression   255
developmental origins of adult disease   65, 79
diabetes
– deaths   72, 158
– epigenetic inheritance   63, 72, 77, 81, 82
– epigenetic programming   28
– inflammatory control   214
– low birth weight   255
– NAFLD   56, 57
– obesity   8, 161
– pregnancy   175
– prenatal determinants   159
Dicer   168–170
diet
– age-related diseases   158, 213, 216
– cancer deaths   105, 106, 152
– DNA methylation   8, 22–26, 213
– epigenetic effects   8, 22–28, 282
– epigenetics in neurodegenerative diseases   237, 238

– genotoxic carcinogens 111, 112
– pregnancy 24, 159, 256, 257, 282
– protective effects towards carcinogens 112–114
– *see also* caloric restriction; foods; nutrition
dietary natural compounds 141, 142, 144–152
disease clusters 270, 274
disease-oriented biobanks 5, 52, 55
diseases *see* human diseases; neurodegenerative diseases
DNA-adduct measurements 107
DNA methylation 15–18, 142–144, 151, 152, 255
– age-related epigenetic silencing via 212, 213
– asthma 254, 257–259
– carcinogens 89, 96, 97
– chromatin interaction 188–190, 196, 197
– clinical applications 203, 204
– CpG islands 3, 15–17
– diet 8, 22–26, 213
– epigenetic modification during lifetime 183, 184
– gene silencing 17, 19
– methods for detecting 197, 198
– monitoring in blood samples 200
– neurodegenerative diseases 236
– normal and cancer cells 185, 196, 197
– pathway 17
– regulation by natural compounds 141, 142, 144–151
– therapeutic applications of inhibitors 202, 203
– *see also* hypermethylation; hypomethylation; promoter hypermethylation
DNA methyltransferases (DNMTs) 16, 17, 135, 141, 143, 144
– activities of 181, 183
– colon cancer 197
– DNMT1 activity regulation 147–151
DNA-reactive carcinogens 112–114
DNA-repair genes 114, 116
DNA sequencing 129, 130, 133, 174
dUMP/dTMP ratio 42

e

ecology 11, 12
EGCG 150, 151
endocrine disruptors 98, 233
endogenous retroviruses 168
environmental epigenomics 238, 239
environmental health 11

environmental influences
– asthma 253, 254, 256
– complex diseases 272
– early life conditions 20–22
– ecology 11, 12
– epigenetics, and human diseases 233, 234
– fetal development 174–176, 282
– health 158
– neurodegenerative diseases 231–233, 239
– pollution and toxins 22
– *see also* alcohol; diet; foods; gene–environment interactions (GEI); nutrition; smoking
enzyme polymorphisms and genotoxic carcinogen metabolism 114–117
epidemiology *see* biobanks; genetic epidemiology
epigenetic dysregulation
– aging and cancer 209–218
– human diseases 233
– neurodegenerative diseases 236, 247, 248
epigenetic health 281–285
epigenetic inheritance 77–79, 82
– future directions 80, 81
– methodology 65–70
– patterns 70–77
– transgenerational models 28, 29
epigenetic marks 13, 21, 180, 238
– in neurodegenerative disorders 245–250
– *see also* biomarkers
epigenetic mechanisms 15–20
– asthma 253–260
– cancer 184
– control mechanisms 169, 170
– neurodegenerative diseases 230
– *see also* DNA sequencing; histone modifications; microRNA (miRNA)
epigenetic misprogramming 28
epigenetic modifications 8, 15, 21, 174, 212
– DNA methylation 183, 184
– neurodegenerative diseases 233, 238
– nutrition 76, 77
epigenetic network, metabolism role 181, 182
epigenetic plasticity 200, 201
epigenetic progenitor model, cancer 200, 201
epigenetic reprogramming cycle 23
epigenetic risk factors 274
epigenetic "signatures" 209, 210, 213
epigenetic therapy approach 282

epigenetics
– definition 4
– environment and human diseases 233, 234
– evolutionary aspects 28, 29
– food metabolism 7, 8
– genetics in cancer cells 180
– neurodegenerative diseases 234–237, 245–248
– new paradigm 3–5
– nutri-epigenetics 135–137
– public health policy 272, 273, 277, 278
epigenome 14, 15, 135, 233
epigenomics 246
– environmental 238, 239
– public health policy 272, 273
estrogen receptor (ER) gene 148, 210, 212, 213
ethics 272, 277, 283, 284
– nutrition research 137, 138
ethnicity
– gene–environment analysis 56, 57
– MTHFR gene polymorphisms 43
– study design 38
– subgroup analysis 40
– urine metabolite profile 133
European Community
– biobank collaboration 55
– Public Health Genomics European Network (PHGEN) 267, 271, 272, 281
– see also "Health in all Policies"
evidence base 268, 271, 273, 276, 277
evolution theory, and self-organization 5
evolutionary aspects
– control mechanisms 167, 168
– epigenetics 28, 29
extrinsic aging 27

f

familial forms, neurodegenerative diseases 226–228
famine 21
feedforward control loop 63, 72
fetal basis
– adult disease 255, 256, 282
– asthma 256, 267
fetal development 159, 160
– environmental influences 174–176, 282
fetal programming 64
– epigenetic inheritance 78, 79
Finnish government 273
flavones 137, 146
flavonoids 25, 26
folate 23, 24, 118, 237–239
– MTHFR role in metabolism 37, 41–43, 46, 47, 158
folate-methionine cycle 135
folic acid 175, 237
food additives 109–111
food availability 69
– adult diseases 63, 81, 82
– life span 71–74
– scarcity 26, 28, 63
– during the slow growth period 76, 77
– see also caloric restriction
food metabolism, and epigenetics 7, 8
foods
– carcinogens 105–118
– diversity 157, 161, 162
– functional 118, 137
– genotoxic carcinogens 106–111
forest plot 44, 45
fragile X syndrome 246, 247
fragile X tremor and ataxia syndrome 246, 247
free radical theory of development 235
fumonisin B1 109
functional foods 118, 137
fundamental rights 277
fungi, RNA silencing 170
fungicides 20, 177, 233

g

gas chromatography-mass spectrometry 132
gastrointestinal cancers
– C677T polymorphism 44–47
– genes methylated 187
gene candidates 57, 229, 231
– epigenetic biomarkers 248, 249
gene–environment interactions (GEI) 4, 43
– asthma 254
– NAFLD development 57, 58
– role of biobanks 51–60
gene expression 7, 65, 66
– diet 141, 142
– hereditary and epigenetic interactions 13–29
– neurodegenerative disorders 248, 249
gene polymorphisms see MTHFR gene polymorphisms; single nucleotide polymorphisms (SNPs); tandem repeat polymorphisms
gene silencing 184, 188
– age-related via DNA methylation 212, 213

- DNA methylation 17, 19
- see also RNA silencing
General Systems Theory 29
generations
- epigenetic inheritance studies 66–68
- intergenerational effects 161, 162
- proband 66–68, 75, 82
- see also transgenerational effects; transgenerational responses
genetic and epigenetic plasticity 200, 201
genetic association studies
- MTHFR gene polymorphisms 44–47
- population-based 37–41
genetic epidemiology
- definition and goals 37, 38
- Human Genome Epidemiology Network 40, 41
- Mendelian randomization approach 41
- meta-analysis 39, 40
- study designs 38
- see also biobanks
genetic modifiers 51, 57
genetic selection 76
genetics
- asthma 254
- cancer cells 180
- cancer incidence 106, 184
genistein 144–147, 149, 237
genome see cancer genome; epigenome; human genome
genome-wide association studies (GWAS) 3, 4, 14
genomic biomarkers 246
genomic imprinting 63, 72, 77
genomics 128
- nutrition research 129, 130
- population genomics data 54
- public health genomics 267, 278
genotoxic carcinogens 89–92, 99
- dietary 111, 112
- epigenetic effects 96, 97
- foods 106–111
- non-parenchymal liver cells 97, 98
- polymorphism affecting metabolism 114–117
genotypes
- combinations, and cancer risk 114, 116, 117
- thrifty 28
glucocorticoids 176, 177, 258
glutathione 216, 235
glycidamide 110

growth velocity, childhood 69, 70
gynaecology 173–177

h

haplotype blocks 133
HapMap 13, 14, 269
harmonized standard operating procedures 54
harvests, quality 67, 69, 71
HDACs see histone deacetylases (HDACs)
health
- environmental 11
- epigenetic 281–285
- inheritance 268
- nutrition 7–9
- population health improvements 269
- prerequisites 157
- transgenerational effects 67
health care 268
health determinants, during life 147–162
Health Genome Concept 118
health impact assessment 271
"Health in all Policies" 272–274
- attributable risks 275, 276
- limits in genomics and epigenetics 277, 278
health needs assessment 271
health promotion 157, 162
health technology assessment (HTA) 271, 276, 277
hematopoietic malignancies 186, 187
hepatocarcinogenesis 90–99
hepatocellular carcinomas 56, 57, 109
hereditary dispositions 13, 14
herpes viruses 171
heterochromatin 196
heterocyclic aromatic amines (HAAs) 105, 108–110, 113, 117
high birth weight 159
high pressure liquid chromatography 132
histone acetylation 19, 181, 258
histone code 19
histone deacetylases (HDACs) 17, 19, 188–190
- asthma 258, 259
- cancer 189, 190
- dietary compounds, effect 25, 152
- inhibitors 248
histone methyltransferases (HATs) 17, 19
histone modifications 18–20, 24, 25, 255
- asthma 258
- cancer 19, 20
HIV, defense mechanisms acting 167, 170
homocysteine 23, 24, 42, 43, 237, 238, 248

human diseases
– developmental origins   65, 79
– epigenetics, environment   233, 234
– fetal basis   255, 256, 282
– public health issues   272
– susceptibility genes   270
– transgenerational explanations   63, 64, 81, 82
– see also neurodegenerative diseases
human epigenome see epigenome
human genome   38, 39, 129
– biobanks   270
– whole-genome association studies   231
Human Genome Epidemiology Network   40, 41
Human Genome Organization   127, 129
Human Genome Project   13, 127, 129, 267, 269
human hepatocellular carcinoma   96
human hepatoma cell line HCC-1.2   95
Human Metabolome Database   132
human microbiome   130
Human Relevance Framework Concept   94
Huntington's disease   226, 228, 229, 245, 246
– epigenetic dysregulation   247, 248
– epigenetics   235
– gene expression   249
hyperlipidemia   128, 129
hypermethylation
– aging   183, 184
– aging and cancer   210
– cancer   4, 8, 9, 142–144, 146, 151
– cancer cells   185
– prostate cancer   213
– see also DNA methylation; promoter hypermethylation
hypertension   68, 161, 175
hypomethylation and cancer   4, 8, 9, 42, 142

i
ideal study design   66, 67
identical (monozygotic) twins   78, 136, 233
immune competent cells   168, 171
immune responses
– adaptive   167, 171
– T cell   257, 258
immune system   171
– epigenetic mechanisms   256
– nutrition   26
immunoglobulin E   257
improvement in population health   269
in silico experiments   128

in vitro fertilization   176
"indication"   277
individual freedom   272
infancy see childhood; newborn
inflammation
– aging and cancer   209–211, 213
– diet   26
– hepatocarcinogenesis   97
– proinflammatory response   254
inflammatory control, age-related epigenetic regulators   214, 215
influenza A virus   170
inheritance
– health and disease   268
– hereditary dispositions   13, 14
– neurodegenerative disorders   246, 247
– see also epigenetic inheritance
"initiated" cells   89–93, 95, 97, 98
insecticides   232
insulin-like growth factor   182
insulin resistance
– aging and cancer   209, 211, 214
– pregnancy   175
interferon system   170
intergenerational effects   161, 162
interleukin 8   26
international collaboration in biobanks   54, 55
International HapMap Project   13, 14, 269
intestinal microflora   130, 159
intrauterine growth retardation   159, 161
intrauterine stress   177
intrinsic aging   27
Inuit populations   5
isochor maps   15
isogenic line   66

k
knowledge generation   269, 270
Kupffer cells   98, 99

l
latent early-life association regulation model   235
lead exposure   231, 236, 239
legislation   272–274, 277, 278
– attributable risks   275, 276
leptin   211
leukemias
– acute myeloid   186, 202
– lymphoid   186
life span
– epigenetic inheritance and childhood circumstances   75

– epigenetic modification by DNA methylation 183, 184
– food scarcity 26, 28, 63
– health determinants 157–162
– interactions throughout 161
– paternal ancestors' nutrition 71–74, 81, 82
– see also adolescence; aging; childhood
d-limonene 110
lineage priming 171
lipokines 209, 211
lipoproteins 128, 129
liquid chromatography-mass spectrometry 132
Lisbon Treaty 276
longevity see aging; life span
low birth weight 75, 159, 161, 175, 255
lung cancer 22
– diagnostic tools 195
– genes methylated 187
– individualization of therapy 203, 204
– paradigm 198–200
– therapeutic inhibitors 202
lymphoid leukemia 186
lymphoma 186, 195

m

malignancy
– aging 180
– genetic and epigenetic factors 27
– hematopoietic 186, 187
– see also cancer; carcinogenesis
mammals, RNA silencing 170
manganese 231
margin of exposure concept 92
mass spectrometry 132
MDS see myelodysplastic syndrome
meat, cooking methods 108
Mendelian randomization approach 41
Mendelian traits 226, 228
mercury 231
meta-analyses
– definition 39
– genetic association studies 37–41
– MTHFR gene polymorphisms 44–47
metabolic programming 79
metabolic syndrome 209
– epigenetic programming 28
– non-alcoholic fatty liver disease 56, 58
metabolism
– aging 210–212
– folate metabolism 37, 41–43, 46, 47, 158
– food metabolism 7, 8
– genotoxic carcinogens 114–117
metabolome analysis 59

metabolomics 128
– nutrition research 132, 133
metallothionein proteins 237
metals, neurodegenerative diseases 231, 232, 236, 239
methionine 175, 181, 237
– folate-methionine cycle 135
– s-adenosylmethionine 97, 125, 143, 237
methionine restriction 210, 215–217
methyl-CpG-binding domains 17
methylation see DNA methylation; hypermethylation; hypomethylation; promoter hypermethylation
methylation-specific PCR 197, 198
methylome 236
MGMT gene 212
microarray analysis 59, 130
microenvironment, hepatic 97, 98
microRNA (miRNA) 20, 168–170, 255
mitochondrial efficiency 216
mitochondrial superoxide dismutase (MnSOD) 24
mobile elements 168
mode of action concept 94, 95
modifiable risks 275, 276
molecular nutrition research 127–129
mortality see deaths
motor neuron disease see amyotrophic lateral sclerosis
MTHFR gene polymorphisms 37, 41–47, 158, 213
multifactorial disorders 226
multiple inheritance systems 80
multiple myeloma 186
mycotoxins 109
myelodysplastic syndrome (MDS) 181–183, 186, 195, 202
myeloid leukemias 186, 202
myeloma, multiple 186
myricetin 150, 151

n

NAFLD see non-alcoholic fatty liver disease (NAFLD)
national task forces 271
natural compounds and DNA methylation 141, 42, 144–152
neoplasia, development 90, 91
nephrones 175, 176
neurodegenerative diseases 225, 226
– caloric restriction and 28
– causitive and susceptibility genes 226–231
– environmental factors 231–233, 239
– epigenetic markers 245–250

– epigenetic role of diet   237, 238
– epigenetics   234–237, 247, 248
neuromelanin   226
neurotoxins   232, 238
newborn
– birth weight   75, 159, 161, 175, 255
– overfeeding   175
– postnatal development   159, 160
– see also fetal development
NF-KB see nuclear factor kappa B (NF-KB)
nitrosamines   105, 107, 108, 113, 117
no-threshold concept   89, 90, 92, 98
non-alcoholic fatty liver disease (NAFLD)   51, 55–59
– gene–environment risk factors   57
non-communicable diseases, chronic   158–162
non-genotoxic carcinogens   89–93, 99
– epigenetic effects   96, 97
– non-parenchymal liver cells   97, 98
non-modifiable risks   275, 276
non-parenchymal liver cells   97, 98
non-small cell lung cancers (NSCLC)   198–200, 202
Nrf2 transcriptional factor   112
nuclear factor kappa B (NF-KB)   209–211, 214–217
nuclear magnetic resonance   132
nuclear structure in a cancer cell   179
nutri-epigenetics   135–137
nutrigenetics   133, 134, 158, 237
nutrigenomics   127, 129–133
nutrition
– aging   26–28
– ancestors'   71–74, 81, 82, 136
– early life conditions   20–22
– epigenetic modification of   76, 77
– health   7–9
– immune system   26
– infancy and childhood   160
– life-stage   159
– during the slow growth period   63, 67, 72, 76, 77, 81, 82, 136
– undernutrition   26, 28, 63
– see also caloric restriction; diet; food availability; foods
nutrition research
– ethics and socio-economics   137, 138
– genomics   129, 130
– metabolomics   132, 133
– molecular   127–129
– proteomics   131, 132
– transcriptomics   130, 131
nutritional systems biology   137

o

obesity   8
– childhood/adolescence   160, 161
– deaths from cancer   217
– epigenetic programming   28
– inflammatory control   214
– intergenerational effects   161
– maternal diet   282
– NAFLD   56, 57
– visceral/abdominal   211
– see also overweight
ochratoxin A   109, 113
older people see aging
oncogenes   184
one-carbon metabolic pathway   135
ornithine transcarbamylase gene   236
osteoporosis   28, 175, 255
ovarian cancer   187
overfeeding, premature babies   175
Överkalix cohorts of 1890, 1905 and 1920   67, 68, 73, 75
overweight   8
– cancer risk   111, 112
– childhood/adolescence   160, 161
– deaths from cancer   217
– see also obesity
oxidative stress   214–216, 234, 235

p

$p16$ gene   199
$p21^{WAF1/CIP1}$ 147, 148
p53 protein   94, 109
2D-PAGE   132
PAHs see polycyclic aromatic hydrocarbons (PAHs)
Parkinson's disease   225–228, 245, 246
– environmental factors   231, 232
– epigenetic dysregulation   247, 248
– gene expression   249
– susceptibility genes   229, 230
paternal ancestors' nutrition   71–74, 81, 82
paternal smoking   75
patulin   113
peroxisome proliferation   94, 97
peroxisome proliferator-activated receptors   21, 211
persistent organic pollutants   5
pesticides   231, 232
pharmacogenomics   127, 128
phase I enzymes   114
phase II enzymes   112, 114
phenotype, cancer prone, aging   210–212
phenotypic data   54

phthalates 94
phytoestrogens 144–149
plant-derived foods
– additives and carcinogens 109–111
– protective effects 112, 113
plants, RNA silencing 169, 170
pollution
– air pollution 22, 253, 256, 258
– persistent organic pollutants 5
polycyclic aromatic hydrocarbons (PAHs) 105, 107, 113, 117
– enzyme polymorphisms 114
polymorphisms *see* MTHFR gene polymorphisms; single nucleotide polymorphisms (SNPs); tandem repeat polymorphisms
polyphenols 141, 142, 145, 146, 151
– catechol-containing 149–151
pooled analysis 39, 46
population-based biobanks 5, 52, 54, 55
population-based genetic association studies 37–41
population genomics data 54
population stratification 38, 40
postnatal development 159, 160
precautionary principle 277
pregnancy
– ALSPAC data set 68
– epigenetic changes 175, 176
– maternal diet 24, 159, 256, 257, 282
– paternal smoking 75
premature babies *see* newborn
prenatal development *see* fetal development
preneoplastic (initiated) cells 89–93, 95, 97, 98
presenilins 227, 228, 234
presymptomatic phase 245
prevention programs
– individualized strategies 29, 268, 269
– risk factors 64, 158
proactive approach 282
proband generations 66–68, 75, 82
proinflammatory response 254
proliferating cell nuclear antigen 147
promoter hypermethylation 22, 143, 144, 146, 151
– cancers 187, 188, 212
– hematopoietic malignancies 186, 187
prostate cancer
– aging 209
– genes methylated 187, 213
protective factors
– complex diseases 272
– dietary 112–114

proteomics 128, 246
– nutrition research 131, 132
public health challenges
– neurodegenerative diseases 245
– responding 282, 283
public health expenditure 284
Public Health Genomics 267–278
– Bellagio Model 268–271
– definition 271
– epigenetics/epigenomics 272
– risk 274–276
Public Health Genomics European Network (PHGEN) 267, 271, 272, 281
Public Health Genomics Foundation 281
public health policy 272–278
– epigenetic health and responsibility 281–285
Public Health Trias 270
Public Population Project in Genomics 54
publication bias 39, 40
pyrethroid insecticides 232

r
RASSF1A gene 199
rat liver models 91–97
Rb/E2F pathway 147, 148
reactive oxygen species 91, 92, 112, 113, 214–216, 234, 235
real-time quantitative PCR 198
regulations *see* legislation
relative risks 267, 274
renal cancer 187
reproductive medicine 173–177
reprogramming cycle, epigenetic 23
response 281
responsibility 281–285
resveratrol 144–149, 216
retinoic acid 144, 145
retroviruses, endogenous 168
risk assessment
– lung cancer 204
– toxicological 90, 91
risk factors
– attributable and relative 267, 274–276
– complex diseases 272, 274
– coronary heart disease 64, 65
– epigenetic 274
– hepatocarcinogenesis 92
– life cycle 157–159, 161
– NAFLD development 56, 57
– neurodegenerative diseases 230–232
– public health policy 275
– responding 282
– *see also* cancer risk

risk regulation 274, 275
RNA silencing 167
– control 168, 169
– fungi 170
– mammals 170
– plants 169, 170
– *see also* gene silencing

s

s-adenosylmethionine (SAM) 97, 125, 143, 237
saccharin 110
safety evaluation
– genotoxic chemicals 97, 98
– tumor promotion 89, 90, 93, 94
salicylates 214
secretases 226–228, 234, 235
self-organization, and evolution theory 5
senile plaques 226, 228, 247
sex chromosomes, and epigenetic inheritance 73–75, 77
short-chain fatty acids 25
short interfering RNAs 169, 170
single nucleotide polymorphisms (SNPs) 13, 14, 39
– cancer risk 114
– nutrigenetics 133
– in whole-genome association studies 231
sirtuins 26, 28, 209, 216, 217
slow growth period (SGP) 70, 81, 82
– male 73
– nutrition 63, 67, 72, 76, 77, 136
small cell lung cancers 199
smoking 22
– asthma 254, 256
– folate metabolism 43
– intergenerational effects 161
– lung cancer 198, 199
– nitrosamines 108
– paternal and pregnancy outcome 75
SNPs *see* single nucleotide polymorphisms
soccer players 232
social context, transgenerational responses 70, 71
socio-economics, nutrition research 137, 138
spices 110, 111
sporadic amyotrophic lateral sclerosis 236, 237
sporadic forms, neurodegenerative diseases 226, 229
stakeholders 138, 268, 278
steatohepatitis 56–58
steatosis 56–58
stress response 283

stroke 28
susceptibility, diseases, epigenetics, and the environment 233, 234
susceptibility genes 51, 57
– neurodegenerative diseases 226–231
– public health 270
systems biology 128, 137, 269

t

T cell immune responses 257, 258
tamoxifen 96
tandem repeat polymorphisms 134
tea catechins 142, 149, 150
6-thioguanine 201
thrifty genotype 28
"Together for Health" 273
transcription, DNA methylation and chromatin 196, 197
transcriptome 130, 131
transcriptomics 128, 246
– neurodegeneration 249
– nutrition research 130, 131
transgenerational effects
– adult diseases 63, 64
– epigenetic inheritance 28, 29
– on human health 67
transgenerational responses
– ancestors' nutrition 71, 81, 136
– asthma 256, 257
– nutrition during slow growth period 72, 76, 81, 82, 136
– paternal ancestors' nutrition 71–74, 81, 82
– social context 70, 71
transitions, concept 11, 12
translational research 267, 276
trichostatin A 203
trichotecenes 109
tumor initiation 91, 92
– cellular and molecular mechanisms 93–95
tumor progression 92, 93
tumor promotion 89, 90, 92, 93
– cellular and molecular mechanisms 93–95
– *see also* non-genotoxic carcinogens
tumor suppressor genes 181, 184, 188
– lung cancer 199

u

undernutrition and life span 26, 28, 63
urine metabolite profile 133
US National Office of Public Health Genomics 281
uterine cancer 187

v

valproic acid 203
vegans 24
vegetarians 24
vinclozolin 20, 233
viral infections 167–171
visceral/abdominal obesity 211
vitamin B$_{12}$ 23, 24, 175, 237–239
vitamin D$_3$ 144–149
vitamins 141, 144–149, 151
vorinostat 195, 203

w

Waddington 3, 4
whole-genome association studies 231